T0344448

AI and IoT for Proactive Disaster Management

Mariyam Ouaissa
Chouaib Doukkali University, Morocco

Mariya Ouaissa
Cadi Ayyad University, Morocco

Zakaria Boulouard
Hassan II University, Casablanca, Morocco

Celestine Iwendi
University of Bolton, UK

Moez Krichen
Al-Baha University, Saudi Arabia

A volume in the Advances in
Computational Intelligence and
Robotics (ACIR) Book Series

Published in the United States of America by
   IGI Global
   Engineering Science Reference (an imprint of IGI Global)
   701 E. Chocolate Avenue
   Hershey PA, USA 17033
   Tel: 717-533-8845
   Fax: 717-533-8661
   E-mail: cust@igi-global.com
   Web site: http://www.igi-global.com

Library of Congress Cataloging-in-Publication Data

CIP DATA PROCESSING

AI and IoT for Proactive Disaster Management
  Mariyam Ouaissa, Mariya Ouaissa, Zakaria Boulouard, Celestine Iwendi, Moez Krichen
    2024 Engineering Science Reference

ISBN: 9798369338964(hc) I ISBN: 9798369348789(sc) I eISBN: 9798369338971

This book is published in the IGI Global book series Advances in Computational Intelligence and Robotics (ACIR) (ISSN: 2327-0411; eISSN: 2327-042X)

British Cataloguing in Publication Data
A Cataloguing in Publication record for this book is available from the British Library.

All work contributed to this book is new, previously-unpublished material.
The views expressed in this book are those of the authors, but not necessarily of the publisher.

For electronic access to this publication, please contact: eresources@igi-global.com.

Advances in Computational Intelligence and Robotics (ACIR) Book Series

ISSN:2327-0411
EISSN:2327-042X

Editor-in-Chief: Ivan Giannoccaro, University of Salento, Italy

MISSION

While intelligence is traditionally a term applied to humans and human cognition, technology has progressed in such a way to allow for the development of intelligent systems able to simulate many human traits. With this new era of simulated and artificial intelligence, much research is needed in order to continue to advance the field and also to evaluate the ethical and societal concerns of the existence of artificial life and machine learning.

The **Advances in Computational Intelligence and Robotics (ACIR) Book Series** encourages scholarly discourse on all topics pertaining to evolutionary computing, artificial life, computational intelligence, machine learning, and robotics. ACIR presents the latest research being conducted on diverse topics in intelligence technologies with the goal of advancing knowledge and applications in this rapidly evolving field.

COVERAGE

- Fuzzy Systems
- Neural Networks
- Machine Learning
- Robotics
- Brain Simulation
- Artificial Life
- Artificial Intelligence
- Agent technologies
- Synthetic Emotions
- Algorithmic Learning

IGI Global is currently accepting manuscripts for publication within this series. To submit a proposal for a volume in this series, please contact our Acquisition Editors at Acquisitions@igi-global.com or visit: http://www.igi-global.com/publish/.

Titles in this Series

For a list of additional titles in this series, please visit:
http://www.igi-global.com/book-series/advances-computational-intelligence-robotics/73674

Shaping the Future of Automation With Cloud-Enhanced Robotics
Rathishchandra Ramachandra Gatti (Sahyadri College of Engineering and Management, India) and Chandra Singh (Sahyadri College of Engineering and Management, India)
Engineering Science Reference • © 2024 • 431pp • H/C (ISBN: 9798369319147) • US $345.00

Bio-inspired Swarm Robotics and Control Algorithms, Mechanisms, and Strategies
Parijat Bhowmick (Indian Institute of Technology, Guwahati, India) Sima Das (Bengal College of Engineering and Technology, India) and Farshad Arvin (Durham University, UK)
Engineering Science Reference • © 2024 • 261pp • H/C (ISBN: 9798369312773) • US $315.00

Comparative Analysis of Digital Consciousness and Human Consciousness Bridging the Divide in AI Discourse
Remya Lathabhavan (Indian Institute of Management, Bodh Gaya, India) and Nidhi Mishra (Indian Institute of Management, Bodh Gaya, India)
Engineering Science Reference • © 2024 • 355pp • H/C (ISBN: 9798369320150) • US $315.00

Machine Learning Techniques and Industry Applications
Pramod Kumar Srivastava (Rajkiya Engineering College, Azamgarh, India) and Ashok Kumar Yadav (Rajkiya Engineering College, Azamgarh, India)
Engineering Science Reference • © 2024 • 307pp • H/C (ISBN: 9798369352717) • US $365.00

Intelligent Decision Making Through Bio-Inspired Optimization
Ramkumar Jaganathan (Sri Krishna Arts and Science College, India) Shilpa Mehta (Auckland University of Technology, New Zealand) and Ram Krishan (Mata Sundri University Girls College, Mansa, India)

701 East Chocolate Avenue, Hershey, PA 17033, USA
Tel: 717-533-8845 x100 • Fax: 717-533-8661
E-Mail: cust@igi-global.com • www.igi-global.com

Table of Contents

Chapter 10
        Naser Hussein, University of Tunis El Manar, Tunisia
        Hella Kaffel Ben Ayed, University of Tunis El Manar, Tunisia

Detailed Table of Contents

Chapter 1

> *Mariyam Ouaissa, Laboratory of Information Technologies, Chouaib
> Doukkali University, El Jadida, Morocco*
> *Mariya Ouaissa, Computer Systems Engineering Laboratory, Cadi
> Ayyad University, Marrakech, Morocco*
> *Sarah El Himer, Sidi Mohammed Ben Abdellah University, Fez,
> Morocco*
> *Zakaria Boulouard, LIM, Hassan II University, Casablanca, Morocco*

Natural and man-made disasters have become more frequent than ever and the need for disaster preparedness has increased over the years. New technologies, such as artificial intelligence (AI) and the internet of things (IoT), have the potential to revolutionize the field of disaster risk prevention, reduction, and response. This chapter explores the potential of artificial intelligence and IoT technologies for disaster relief operations. Potential applications of these technologies in disaster response are discussed; they include the use of IoT to detect early warning signs of natural disasters, alerting populations of risks and facilitate communication between emergency responders, as well as use of AI-based systems to provide data in real time to rescue workers. The authors discuss future directions and research directions for AI and IoT-based disaster management, as well as present case studies in the context of wildfire and flooding. Finally, we conclude by highlighting the need to advance AI and IoT-based technologies for their use in disaster management.

Chapter 2

M.A. Jabbar, Vardhaman College of Engineering, India
Ruqqaiya Begum, Vardhaman College of Engineering, India
K. Tejasvi, Vardhaman College of Engineering, India

A disaster is an unexpected event that disrupts a society's functioning while also harming the human environment and causing financial and material losses. It can be caused by either natural or human factors. In today's society, disasters are seen as the product of good planning, which leads to hazards and vulnerabilities. The term "disaster management" refers to the planning and management of disasters. Artificial intelligence is the ability of computers to perform the tasks that are usually done by humans. Artificial intelligence has been used in many industrial applications and also used in everyday interactions with technology. Artificial intelligence is being used in sustainable development, and disaster risk management. Machine learning and artificial intelligence models can be used in both ways in disaster management, i.e., pre-in disaster management and post-in disaster management.

Chapter 3

Muhammad Usman Tariq, Abu Dhabi University, UAE & University of Glasgow, UK

This chapter outlines the complete way to deal with flood estimating and alleviation through the mix of man-made brainpower (artificial intelligence) and the web of things (IoT). Floods are among the most pulverizing catastrophic events universally, requiring progressed techniques for expectation and reaction. Computer-based intelligence and IoT innovations offer promising arrangements by empowering precise estimating, constant observation, and compelling relief measures. This part investigates the groundwork of computer-based intelligence and IoT in calamity the board, featuring their assembly for upgraded information examination and navigation. It digs into artificial intelligence models, such as profound learning and AI for flood expectation, alongside the organization of IoT sensors and gadgets for continuous observing in flood-inclined regions.

Predicting forest fire occurrences can bolster early detection capabilities and improve early warning systems and responses. Currently, forest and grassland fire prevention and suppression efforts in China face significant hurdles due to the complex interplay of natural and societal factors. While existing models for predicting forest fire occurrences typically consider factors like vegetation, topography, weather conditions, and human activities, the moisture content of forest fuels is a critical aspect closely linked to fire occurrences. Additionally, it introduces forest fuel-related factors, including vegetation canopy water content and evapotranspiration from the top of the vegetation canopy, to construct a comprehensive database for predicting forest fire occurrences. Furthermore, the study develops a forest fire occurrence prediction model using machine learning techniques such as the random forest model (RF), gradient boosting decision tree model (GBDT), and adaptive augmentation model (AdaBoost).

Chapter 5

The main goal is to appropriately utilize advanced algorithms to analyze environmental data, improve early disease detection and intervention tactics, and reduce the harmful effects of forest fires on human beings. Analyze the challenges faced by traditional methods in addressing the constantly evolving nature of wildfires and the need for more adaptable and proactive approaches, and highlight the advantages of AI. Discusses the main constituents incorporated into the AI model, comprising meteorological data, satellite imagery, and historical fire records. It analyzes the selection of AI algorithms specifically tailored for forest fire prevention, considering parameters. Analyze the challenges faced during the creation and implementation of AI models for forest fire prevention and viability of integrating artificial intelligence models into existing fire management infrastructure and emergency response systems. It showcases the current research, progress, and use of AI-driven solutions to address the challenges posed by wildfires and provides a concise overview of the chapter's findings.

Chapter 6
Explicit Monitoring and Prediction of Hailstorms With XGBoost Classifier
for Sustainability 107

Peryala Abhinaya, *Stanley College of Engineering and Technology for Women, India*

C. Kishor Kumar Reddy, *Stanley College of Engineering and Technology for Women, India*

Abhishek Ranjan, *Faculty of Engineering and Technology, Botho University, Botswana*

Ozen Ozer, *Department of Mathematics, Faculty of Science and Arts, Turkey*

Hailstorms are extremely dangerous for both people and property, hence precise forecasting techniques are required. To increase hailstorm forecast accuracy, this study suggests utilizing the XGBoost algorithm. The gradient boosting technique XGBoost is well-known for its effectiveness at managing intricate datasets and nonlinear relationships. The suggested approach improves prediction abilities by incorporating many meteorological factors and historical hailstorm data. The model outperforms conventional approaches through thorough evaluation utilizing cross-validation techniques. XGBoost, or extreme gradient boosting, is an excellent technique for hailstorm prediction because of its scalability, robustness, and proficiency with complicated datasets. By using the XGBoost algorithm, there is a chance to increase the accuracy of hailstorm predictions and decrease the socio-economic effects of these occurrences. To increase forecasting accuracy and mitigation tactics, this work demonstrates advances in hailstorm prediction using numerical weather models and machine learning approaches.

Chapter 7
Reviewing the Potential of GeoAI for Post-Earthquake Land Prospection:
Lessons From the El Haouz Earthquake 133

Ouchlif Ayoub, *Hassan II Agronomic and Veterinary Institute, Morocco*

Hicham Hajji, *Hassan II Agronomic and Veterinary Institute, Morocco*

Hamid Khalifi, *Faculty of Sciences, Mohammed V University in Rabat, Morocco*

Kenza Aitelkadi, *Hassan II Agronomic and Veterinary Institute, Morocco*

The El Haouz 2023 earthquake caused extensive damage and displaced many people. A strategic roadmap is necessary for successful adoption of advanced technologies in the post-earthquake context. Land prospection is a critical step in reconstruction, but it can encounter obstacles such as limited land availability, geological and environmental risks, and social and economic constraints. This study explores the potential of geospatial artificial intelligence (GeoAI) in modernizing land prospection in post-earthquake contexts. A systematic review of literature identified previous GeoAI applications in post-earthquake interventions. The findings indicate that GeoAI can significantly contribute to land selection by providing accurate and real-time information on terrain features, hazards, and limitations. The study proposes a methodological framework for integrating GeoAI into post-earthquake land selection challenges based on lessons learned from prior experiences. The implications of this study are significant for decision-makers, practitioners, and researchers in the field.

    Murtala Ismail Adakawa, Bayero University, Kano, Nigeria
    N. S. Harinarayana, University of Mysore, India

This research investigates scholarly communication theories used during the COVID-19 pandemic as a strategy for building effective disaster-resilience infrastructure. The study employed scientometric and content analysis to understand the behavior of data in this regard. For scientometric analysis part, using Scopus from 16-23rd January 2024 employing search strategy "COVID-19 and theories" OR "Community resilience" OR "Disaster-resilience Infrastructure," it yielded about 5,266,065 documents and reduced to 10,053 through data pruning. The findings showed that 2023 has been the year with the highest number of publications 4369(43.45%) followed by 2021 accounting for 2277(22.65%). For content analysis, types of theories, constructs of importance, methodologies, etc. employed by researchers were studied. The study concludes that, even though theories are deterministically used to direct policy formulations and implementation in indeterministic disaster conditions, they quickly provide a means of understanding and enumerating possible variables for tackling such hazards.

    Ravinder Singh, Indian Army, India
    Geetha Manoharan, SR University, India
    Samrath Singh, Vellore Institute of Technology, Chennai, India

The emergence of artificial intelligence (AI) and unmanned aerial vehicles (UAVs), commonly known as drones, has revolutionized disaster management practices, particularly in the context of natural disasters. In recent years, there has been a notable surge in the integration of innovative smart connected devices and platforms, including drones and UAVs, into the extensive network of the internet of things (IoT). The integration of AI and drones into the IoT network presents numerous challenges, opportunities, and implications for leveraging them in disaster management. This study offers an in-depth overview of the several applications of AI and drones in various stages of disaster management. In this review, the authors have explored the evolution of disaster management paradigms, the applications of AI and drones in disaster management, and the challenges and disruptive technologies shaping this field. The study reveals that the application of AI and drones has great potential in disaster management and can enhance the resilience of the community.

Chapter 10

    Naser Hussein, University of Tunis El Manar, Tunisia
    Hella Kaffel Ben Ayed, University of Tunis El Manar, Tunisia

The portability and automation of the internet of drones (IOD) have drawn increasing attention in recent years, and it is being used in various fields (such as military, rescue and entertainment, and disaster management). Unmanned aerial vehicles (UAVs) have recently established their capacity to provide cost-effective and credible solutions for various real-world scenarios. since security and privacy are among the main concerns for the IOD, in this chapter, the authors perform a complete analysis of security issues and solutions for IOD security, analyzing IOD-related security in disaster requirements and identifying the newest improvement in IOD security research. This analysis explores many essential security technologies emphasizing authentication mechanisms and blockchain-powered schemes. Based on a rigorous review, the authors discuss the issues faced by current approaches in disaster management and offer future IOD security research areas.

Preface

In our contemporary era, the looming threats of natural calamities like hurricanes, earthquakes, and floods are unavoidable. While the idea of utilizing technology for disaster preparedness might seem distant, the truth is that advancements in artificial intelligence (AI) and the internet of things (IoT) are steadily bringing it closer to fruition. By harnessing these technologies to pinpoint risk zones, monitor evolving situations, and deliver instantaneous data, we can elevate the effectiveness of disaster response. The tools and sensors powered by AI and IoT could enable us to react quicker to calamities and predict them with unparalleled precision, potentially reducing the impact and facilitating rapid recovery.

The fusion of AI and IoT is dramatically reshaping our approach to crisis management. By tapping into live data streams, predictive analytics, and smart systems, we can bolster early warning mechanisms, fine-tune our responses to disasters, and fortify our communities against these threats. Nevertheless, it's imperative to tread carefully, considering ethical implications and ensuring the safety and welfare of individuals and communities during this transition.

The purpose of this book is to shed light on the most groundbreaking research avenues, focusing on the potential of AI and IoT to address and surmount the hurdles of disaster management. It will curate a series of original insights about AI frameworks, IoT infrastructures, the synergy of cutting-edge technologies, and their pivotal roles in averting natural catastrophes.

This book will help the UG/PG Students, Academicians, Research Scholars, Industry Professionals, Technology Enthusiasts gain useful insights from the domain of AI, IoT, and Natural Disaster.

ORGANIZATION OF THE BOOK

Chapter 1: AI and IoT Integration for Natural Disaster Management: A Comprehensive Review and Future Directions

This chapter delves into the potential of artificial intelligence (AI) and the Internet of Things (IoT) to revolutionize disaster relief operations. It explores various applications of these technologies in disaster response, such as early warning systems, risk detection, and real-time data provision to emergency responders. Case studies focusing on wildfire and flooding illustrate the practical implementation of AI and IoT in disaster management. Moreover, the chapter discusses future research directions and emphasizes the need to advance these technologies for effective disaster preparedness and response.

Chapter 2: Demystifying the Applications of Artificial Intelligence in Disaster Management: A Review

Disasters, whether natural or human-induced, disrupt societies and pose significant challenges to disaster management efforts. This chapter provides an overview of artificial intelligence (AI) applications in disaster risk reduction and management. It explores how AI, particularly machine learning models, can contribute to both pre-disaster planning and post-disaster response. The chapter highlights the importance of incorporating AI into disaster management strategies to enhance preparedness and mitigate the impact of disasters.

Chapter 3: AI and IoT in Flood Forecasting and Mitigation: A Comprehensive Approach

Floods rank among the most devastating natural disasters globally, demanding advanced techniques for prediction and response. This chapter examines the integration of artificial intelligence (AI) and the Internet of Things (IoT) in flood forecasting and mitigation efforts. It explores how AI models, such as deep learning, coupled with IoT sensors, enable accurate forecasting and real-time monitoring of flood-prone areas. Case studies illustrate the practical application of AI and IoT in enhancing flood management strategies.

Chapter 4: An Innovative Investigation on predicting Forest Fire Using Machine Learning Approach

Forest fires pose significant challenges to early detection and effective response efforts. This chapter introduces a machine learning-based approach for predicting forest fire occurrences. It discusses the incorporation of various factors, including vegetation, weather conditions, and human activities, into predictive models. The chapter demonstrates the effectiveness of machine learning techniques in improving forest fire prediction accuracy and underscores their potential in enhancing forest fire management strategies.

Chapter 5: Artificial Intelligence Models to Prevent Forest Fires

Addressing the complex and evolving nature of wildfires requires proactive and adaptable approaches. This chapter explores the application of artificial intelligence (AI) algorithms to analyze environmental data for early detection and intervention in forest fires. It discusses the integration of meteorological data, satellite imagery, and historical fire records into AI models tailored for forest fire prevention. The chapter examines the challenges and opportunities associated with implementing AI-driven solutions in forest fire management.

Chapter 6: Explicit Monitoring and Prediction of Hailstorms with XGBoost Classifier for Sustainability

Hailstorms pose significant risks to both lives and property, necessitating accurate forecasting techniques. This chapter proposes the use of the XGBoost algorithm for predicting hailstorm occurrences. It highlights the effectiveness of gradient boosting techniques in managing complex datasets and nonlinear relationships. The chapter demonstrates how the XGBoost algorithm can improve prediction accuracy by incorporating meteorological factors and historical data, thereby enhancing resilience to hailstorms.

Chapter 7: Reviewing the potential of GeoAI for post-earthquake land prospection: Lessons from the El Haouz earthquake

Post-earthquake reconstruction efforts rely on efficient land prospection methods to address various challenges. This chapter explores the potential of geospatial artificial intelligence (GeoAI) in modernizing land prospection in post-earthquake contexts. It reviews previous GeoAI applications and proposes a methodological framework for integrating GeoAI into land selection processes. Case studies and lessons learned from prior experiences inform the development of strategies for leveraging GeoAI in post-earthquake reconstruction efforts.

Chapter 8: Scholarly Communication Theories for Building Effective Disaster-Resilient Infrastructure: A Study

Effective disaster-resilient infrastructure requires informed decision-making based on scholarly communication theories. This chapter investigates the application of scholarly communication theories in building disaster-resilient infrastructure, with a focus on the COVID-19 pandemic. It employs scientometric and content analysis to understand the behavior of data in this context. By analyzing publication trends and theoretical frameworks, the chapter offers insights into leveraging scholarly communication theories for disaster resilience.

Chapter 9: Strategic Deployment of AI and Drones Enhancing Disaster Management in Natural Disasters

The integration of artificial intelligence (AI) and drones holds great potential for enhancing disaster management practices, particularly in natural disasters. This chapter provides an overview of AI and drone applications at various stages of disaster management, from preparedness to response and recovery. It explores the challenges and opportunities associated with deploying AI and drones in disaster management and highlights their potential to improve community resilience.

Chapter 10: Blockchain technology with the Internet of Drones (IOD) to address privacy and security issues in disaster management

The portability and automation of the Internet of Drones (IOD) offer promising solutions for disaster management, but security and privacy concerns remain significant barriers. This chapter examines the use of blockchain technology to address security issues in IOD-based disaster management systems. It analyzes security challenges and proposes blockchain-powered solutions to enhance data integrity and privacy protection. The chapter discusses future research directions for improving IOD security in disaster management applications.

IN CONCLUSION

As we conclude this edited reference book on *AI and IoT for Proactive Disaster Management*, it is evident that the integration of artificial intelligence (AI) and the Internet of Things (IoT) holds immense potential in revolutionizing disaster management practices. Through the exploration of various chapters, we have delved into the applications of AI and IoT across different domains of disaster preparedness, response, and mitigation.

From forecasting natural disasters like floods and forest fires to enhancing urban air quality monitoring, the contributions presented in this book showcase the versatility and effectiveness of AI and IoT technologies. We have seen how machine learning algorithms, deep learning models, and sensor networks can be harnessed to improve early warning systems, optimize resource allocation, and facilitate real-time decision-making during crises.

Moreover, the discussions on the ethical considerations, challenges, and future research directions underscore the importance of responsible innovation in leveraging AI and IoT for disaster management. It is imperative for stakeholders, including researchers, policymakers, and practitioners, to collaborate and address these challenges while maximizing the benefits of these technologies for the greater good.

As editors, our aim in compiling this book was to provide a comprehensive overview of the latest research and developments in the field, catering to a diverse audience ranging from students and academics to industry professionals and technology enthusiasts. We hope that the insights presented herein will inspire further exploration and innovation in leveraging AI and IoT for proactive disaster management, ultimately contributing to safer and more resilient communities worldwide.

Mariyam Ouaissa
Chouaib Doukkali University, Morocco

Mariya Ouaissa
Cadi Ayyad University, Morocco

Zakaria Boulouard
Hassan II University, Casablanca, Morocco

Celestine Iwendi
University of Bolton, UK

Moez Krichen
Al-Baha University, Saudi Arabia

Chapter 1
AI and IoT Integration for Natural Disaster Management:
A Comprehensive Review and Future Directions

Mariyam Ouaissa

iD https://orcid.org/0000-0002-3993-8405
Laboratory of Information Technologies, Chouaib Doukkali University, El Jadida, Morocco

Mariya Ouaissa

iD https://orcid.org/0000-0002-0088-3742
Computer Systems Engineering Laboratory, Cadi Ayyad University, Marrakech, Morocco

Sarah El Himer
Sidi Mohammed Ben Abdellah University, Fez, Morocco

Zakaria Boulouard

iD https://orcid.org/0000-0002-4891-3760
LIM, Hassan II University, Casablanca, Morocco

ABSTRACT

Natural and man-made disasters have become more frequent than ever and the need for disaster preparedness has increased over the years. New technologies, such as artificial intelligence (AI) and the internet of things (IoT), have the potential to revolutionize the field of disaster risk prevention, reduction, and response. This chapter explores the potential of artificial intelligence and IoT technologies for disaster relief operations. Potential applications of these technologies in disaster response are discussed; they include the use of IoT to detect early warning signs of natural disasters, alerting populations of risks and facilitate communication between emergency responders, as well as use of AI-based systems to provide data in real time to rescue workers. The authors discuss future directions and research directions for AI and IoT-based disaster management, as well as present case studies in the context of wildfire and flooding. Finally, we conclude by highlighting the need to advance AI and IoT-based technologies for their use in disaster management.

DOI: 10.4018/979-8-3693-3896-4.ch001

INTRODUCTION

In a world where natural and human disasters seem to be increasing in frequency and intensity, the urgency of finding innovative solutions to better anticipate, understand and respond to them is more pressing than ever (Ray et al., 2017).

The urgent need to develop disaster prediction and management strategies has become increasingly important due to the increasing frequency and severity of these events, largely attributed to the effects of climate change. Artificial Intelligence (AI) offers an approach to effectively address this pressing problem (Fan et al., 2021).

Through its advanced analysis of big data, AI can identify precursor patterns of natural disasters. By combining historical data with real-time information from sensors and satellites, it is able to predict with unprecedented accuracy the location and potential severity of catastrophic events (Shah et al., 2019). The Internet of Things (IoT) offers solutions to detect the warning signs of natural disasters, to alert populations in the event of risks and to facilitate communication between emergency services. Artificial Intelligence (AI) can quickly analyze large data sets to identify areas requiring priority intervention, optimize rescue routes, and efficiently manage rescue logistics (Renugadevi, R., & Medida, L. H., 2024). It plays a key role in minimizing the impacts of disasters and improving response times (Bail et al., 2021).

This study examines in detail how artificial intelligence techniques are used in flood forecasting and management. It explores the use of data sources such as satellite imagery and Internet of Things sensors to improve the accuracy of AI-based predictions. However, like any advancement using artificial intelligence in this field, it presents its own unique challenges. The aim of this research is to study the issues that need attention in the field of AI-powered flood management (Kamruzzama, et al, 2017). These challenges include aspects such as ensuring data integrity across geographic regions, efficiently managing the computational demands of deep learning models, and addressing issues related to equitable distribution of AI-powered flood control technologies. By examining the methodologies used in this area as well as the challenges associated with them, this chapter aims to provide an understanding of how artificial intelligence is profoundly influencing the future direction of flood forecasting and management.

This chapter presents a comprehensive review of the integration of Artificial Intelligence and Internet of Things technologies in natural disaster management. It explores the current landscape of AI and IoT applications, their effectiveness, limitations, and future directions for enhancing disaster preparedness, early warning systems, response strategies, and community resilience.

The structure of this chapter is organized as follows: The subsequent section provides an overview of key concepts of this study. The third section presents a description of AI Techniques for disaster prediction in IoT applications. Section 4 describes the techniques of management disaster. In Section 5, we discuss the role IoT in disaster management. We present some challenges of AI and IoT in natural disaster management and we discuss two case studies in section 6 and 7 respectively. Finally, we present the future directions of this study. Conclusions are drawn in section 8.

LITERATURE REVIEW

In this section, we will provide an overview of key concepts related to disaster management, AI and IoT that are relevant to our study.

Disaster Management

A disaster is an event that occurs suddenly and causes widespread destruction of lives, property and ecosystems. Affected populations can suffer terrible losses from both natural disasters such as hurricanes, earthquakes and floods and man-made disasters such as terrorist attacks and industrial accidents (Sharma, K., & Anand, D., 2023). Preparedness, emergency response and post-disaster reconstruction are all part of the broader scope of disaster management. Government agencies, Non-Governmental Organizations (NGOs) and community groups are just some of the many actors who must coordinate and collaborate for effective disaster management (Adnan et al., 2015).

Artificial Intelligence

The field of AI focuses on programming computers to imitate human intelligence in areas such as learning and problem solving (Boulouard et al., 2022). Machine learning is a branch of AI that uses statistical methods to teach computers new skills without being explicitly taught. By using artificial neural networks with many layers, deep learning is able to learn hierarchical representations of data (Kuglitsch et al., 2022). AI can be used in many different industries, such as medicine, banking and transportation. The process of building and deploying ML models usually involves various steps, including collecting relevant data, preprocessing and formatting data, selecting and transforming features, choosing an appropriate ML algorithm, training and evaluation of the model, and its deployment for predictions or decisions on new data. These steps provide a general framework for using ML in various applications (Linardos et al., 2022).

Internet of Things

The IoT refers to the network of physical objects such as devices, instruments, vehicles, buildings, and other elements (Ouaissa et al., 2018). These objects are integrated with electronics, circuits, software, sensors, and network connectivity, enabling them to collect and exchange data (Ghasemi et al., 2020). This interconnectedness allows objects to be remotely sensed and controlled over existing network infrastructure, creating opportunities for a more direct integration of the physical world into computing systems. The result is improved efficiency and accuracy in various sectors. Preliminary IoT applications have already been developed in sectors such as healthcare, transportation, industry, and agriculture. Although IoT technologies are still in their early stages, significant progress has been made in integrating objects with sensors on the Internet (Ali et al., 2019).

AI TECHNIQUES FOR DISASTER PREDICTION

The rapid advancement of artificial intelligence has ushered in an era of flood prediction. AI uses methodologies and data resources to achieve efficient computing capacity as well as adaptability and learning capabilities. By integrating data streams such as satellite observations from space and data from sensors on the ground, AI offers an approach to understanding and predicting flood events. In this section, we will explore AI techniques used for disaster prediction and their respective contributions to the development of forecasting models (Pang, G., 2022).

Data Acquisition

In today's era, where we have an abundance of large-scale data and the creation of disaster prediction models relies heavily on the integration of data sources, obtaining relevant and timely data is crucial to the development of prediction models (Sinha et al., 2019).

- **Satellite images:** Satellites provide observations obtained from space. By examining captured images and temporal patterns, artificial intelligence algorithms have the ability to identify ordinary behaviors as well as deviations, making it easier to anticipate susceptible periods and areas. Additionally, the use of satellite data provides comprehensive global coverage, enabling effective disaster risk monitoring even in the most remote and inaccessible regions.

- **IoT sensors:** The advent of the Internet of Things has significantly increased the level of detail of data that can be used for disaster prediction. Strategically positioned sensors in disaster-prone areas provide real-time data collection regarding various factors.
- **Social media and platform data:** The widespread prevalence of social media platforms and mobile phones has proven to be a valuable and unexpected resource for the purpose of predicting disasters. Analyzing testimonials, photos and videos posted on social media platforms such as Twitter or Facebook can provide real-time information.

Machine Learning Models

The wide range of data sources accessible for disaster prediction makes it an area particularly suited to the use of machine learning algorithms. Various techniques can be customized to fit different data types or specialized prediction needs (Chai et al., 2023).

- **Linear Regression:** Linear regression is a fundamental model used in predictive analytics, especially in situations where variables have obvious linear associations. By analyzing the relationship between certain factors such as precipitation levels and their impact on different outcomes. However, the simplicity of this approach may not account for complex interconnections.
- **Decision trees:** These models show remarkable performance in scenarios where decisions must be made based on certain conditions. Researchers derive anticipated predictions for likely disaster scenarios by analyzing previous disaster occurrences in relation to current environmental variables. This thinking helps authorities determine which areas are at higher risk and effectively allocate appropriate resources.
- **Support Vector Machines (SVM):** Support vector machines work by classifying data into distinct categories and are used in pattern recognition problems. Regions can be classified according to their susceptibility to disasters. The classification process helps improve policy formulation by directing attention and devoting resources to sectors that present high levels of risk. Using artificial intelligence to effectively predict and mitigate disasters.
- **Neural Networks:** Neural networks have the ability to efficiently analyze large amounts of data and identify complex patterns similar to the computational capabilities of the human brain. They offer significant advantages in regions characterized by complex environmental factors because they provide global predictions that take into account multiple contributing variables (Rao et al., 2024).

Deep Learning Approaches

Deep learning is a special branch of machine learning, it uses multi-layer neural networks to analyze and interpret complex and multi-dimensional data sets. This particular characteristic makes it very suitable for disaster prediction (Sun, W. et al., 2020).

- **Convolutional Neural Networks (CNN):** CNN have mainly been used in the field of image processing. They have also proven useful in predicting disasters. More specifically, researchers used CNNs to analyze satellite and aerial photos to identify areas at higher risk of disaster in environments where traditional data may be insufficient.
- **Recurrent Neural Networks (RNN) and Long Short Term Memory (LSTM) Networks:** RNN and LSTM are types of networks that excel in understanding the importance of order in prediction tasks. These models are ideal for analyzing time series data. By incorporating trends, RNNs and LSTMs can effectively predict scenarios while maintaining contextual relevance.
- **Hybrid models:** Hybrid models represent a combination of learning techniques integrated together. Hybrid models provide in-depth disaster predictions by combining data from sources such as satellite images, weather forecasts and historical records. This integrated approach ensures that all relevant factors are taken into account, resulting in accurate predictions.

ARTIFICIAL INTELLIGENCE IN DISASTER MANAGEMENT

Accurate prediction of disasters is of considerable importance and is equally crucial for the effective management of these phenomena. Artificial intelligence is not limited to predictive capabilities for future events, it also has the power to create effective strategies, improve response mechanisms and assist recovery operations after disasters. This section examines the multifaceted role of artificial intelligence in the field of disaster management, with a focus on the ways in which AI helps to effectively address the many challenges posed by disasters. The application of artificial intelligence in global disaster management is of considerable importance and is continuously advancing. It encompasses several aspects, including optimizing resource allocation in emergencies and scenario analysis to improve future preparedness (Abid et al., 2021).

The use of digital technologies and artificial intelligence to anticipate and manage the risks of natural or human disasters revolves around several strategic axes.

- **Real-Time Data Collection and Analysis**

The rapid computing capabilities of AI enable real-time collection and analysis of data from Internet of Things sensors, satellite images and weather information. The continuous process of simultaneously receiving and analyzing data provides authorities with real-time updates, allowing them to predict the onset and progression of disasters (Sun, R. et al., 2020).

- **Monitoring and Sensors:** Deploy sensor networks to monitor in real time critical environmental indicators, namely temperature, humidity and tectonic movements, that can signal the impending risk of disaster.
- **Satellite Imagery and Drones:** Use satellite imagery and drones to obtain detailed views of areas at risk, enabling accurate assessment of potential or ongoing damage and mapping of vulnerable areas.
  - **Predictive Modeling and Simulation**

A significant advantage of artificial intelligence is its ability to efficiently perform simulations in a limited time frame. Using data-driven models, these systems can generate a range of disaster scenarios (Tan el al., 2021).

- **Climate and Geological Models:** Develop and refine predictive models to anticipate events such as hurricanes, floods, earthquakes, and volcanic eruptions.
- **Scenario Simulation:** Simulate various disaster scenarios to plan emergency responses and test the resilience of critical infrastructure.
  - **Artificial Intelligence and Machine Learning**
- **Forecasting:** Use AI to analyze huge volumes of data and predict catastrophic events with improved accuracy and anticipation.
- **Social Media Analysis:** Monitor social media and online communications to detect early signals of disasters or to assess the mood and needs of the population in real time.
  - **Early Warning and Communication Systems**

The integration of AI into warning systems has transformed the prediction and understanding of disaster events. Using data and algorithms, AI enables a proactive approach to disaster management (Zafar et al., 2019).

- **Dissemination of Alerts:** Establish early warning systems that use mobile and internet technologies to quickly inform the population of imminent risks.

- **Information Platforms:** Create applications and digital platforms to provide vital information, preparedness advice, and crisis updates.
- **Community-based alerts:** Often, generic notifications are ineffective in eliciting a meaningful response from individuals. With this in mind, AI-powered alert systems have been developed to personalize messages based on the unique characteristics of community sites. This approach ensures that alerts are tailored to the expected path of an event, making notifications more relevant. Therefore, people are motivated to act in response.
  - **AI-Assisted Emergency Planning and Response**
- **Relief Coordination:** Use AI to optimize logistics and coordination of relief efforts, including resource allocation and volunteer management.
- **Rescue Robots and Drones:** Deploy robots and drones to carry out search and rescue in dangerous areas or areas inaccessible to humans.
  - **Training and Awareness**

Use digital platforms to train citizens and emergency management professionals on best practices in disaster preparedness and response.

Integrating these technologies not only improves the accuracy of forecasts and the speed of responses in the event of a disaster, but also fosters a culture of preparedness and resilience within communities and organizations.

INTERNET OF THINGS FOR DISASTER MANAGEMENT

The Internet of Things is a network of things, physical devices equipped with electronic sensors, software and other equipment, connected to each other over the Internet and which exchange information with other devices and systems. Advances in cloud computing, broadband wireless networks, the sensors themselves, and data analysis have given rise to powerful, integrated, real-time IoT systems. Today, IoT applications are used across all sectors, be it healthcare, education, transportation, agriculture, industry, etc. When it comes to disaster management, IoT can be used to monitor sudden natural disasters, such as earthquakes and landslides, to issue emergency alerts and transmit data to control centers and management of emergency situations in near real time, thereby increasing capacity for disaster prevention and mitigation. As part of the 3rd Generation Partnership Project (3GPP), they have already implemented a set of narrowband IoT technologies based on LTE; namely narrow-band IoT (NB-IoT) technologies and advanced machine-type communications, which expand the range of LTE technologies, enabling wider use of more energy-efficient IoT services (Adeel et al., 2019).

In recent years, the use of the IoT has grown in the field of environmental protection. Applications for monitoring extreme weather conditions, forest fires, water security and the protection of endangered species have emerged. IoT has become an ally for environmental protection and contributes to the fight against climate change. IoT sensors are deployed in risky areas to monitor environmental conditions in real time. For example, IoT weather sensors collect data on air pressure, temperature, humidity and precipitation. This data feeds into disaster prediction models (Sharma, M., & Kaur, J., 2019). Data collected by IoT sensors also powers emergency communications systems. These systems provide rapid alerts to residents, businesses and response agencies in the event of an impending disaster. In the event of a disaster, real-time data from IoT sensors helps rescue and relief teams coordinate their efforts. Accurate information about conditions on the ground is essential for effective response (Park et al., 2018).

CHALLENGES OF AI AND IOT IN DISASTER MANAGEMENT

Despite the many benefits of AI and IoT-based technologies in disaster management, there are also several challenges that need to be addressed. Some of these challenges include:

Technical limitations

Emerging technologies, such as AI and IoT, may not yet be fully developed or tested for use in crisis management. Their usefulness may be limited by technical factors such as battery life and communication issues.

Privacy Concerns

There are privacy concerns related to the use of AI-based technology in disaster management, particularly regarding the collection and use of personal data. Data should only be used for appropriate purposes, and protecting the privacy of affected populations should be a top priority.

Regulatory challenges

Regulations, such as permitting and compliance with airspace rules, apply to the use of drones in disaster management. These rules can be difficult to decipher and may differ from jurisdiction to jurisdiction.

Ethical considerations

Ethical concerns, such as the possibility of bias and discrimination in decision-making, are raised by the use of AI-based technology in disaster management. To prevent further inequality, it is crucial that these technologies are used openly and responsibly.

CASE STUDIES

Natural disasters, such as hurricanes, floods, wildfires and earthquakes, can cause considerable loss of life and property. Accurately forecasting these events is crucial to minimizing their impacts. Data analytics through the Internet of Things has transformed the way we predict and manage natural disasters. IoT sensors enable real-time monitoring, accurate forecasting, and effective coordination of disaster responses. With these technologies, cities can better protect their citizens and property.

Case Study One: IoT Sensors for Flood Monitoring:

IoT sensors can detect sudden increases in water levels, which may indicate a risk of flooding. The data collected is analyzed to predict the scale and extent of impending flooding, allowing authorities to take preventive measures, such as evacuating areas at risk (Boulouard et al., 2023).

In this context, IoT plays a vital role in water leak detection, preventive maintenance, real-time measurement and temperature management. By optimizing each step of the chain, it allows more intelligent management of water resources. Connected water meters reduce the need for labor, facilitate frequent sampling and sophisticated on-site testing, while integrating response efforts with detection and monitoring systems. Just as in the case of fires, malfunctions can be reported through alerts, thus providing better control of water use, both for individuals and professionals (Boulouard et al., 2022).

In addition, sensors are also used to ensure the health and safety of water in wastewater treatment plants. Additionally, they help farmers by indicating which lands should be irrigated, thereby avoiding overwatering and saving valuable time.

Thanks to IoT, many advances have been made in environmental protection. Connected devices provide improved monitoring, early detection of environmental problems and more efficient management of natural resources. The continued expansion of IoT will have a significant impact on our planet, helping to mitigate the consequences of global warming and promote a more sustainable future (Goyal et al., 2021).

LPWAN networks play a particularly crucial role in environmental monitoring applications. For example, they allow water levels to be measured using water meters, helping to prevent potential flooding. Their ability to send messages effectively makes them extremely useful in these contexts (Pham et al., 2021).

Case Study Two: Forest Fires

In wildfire-prone regions, IoT sensors monitor drought, wind speed, and the presence of abnormal heat. When these factors reach critical thresholds, alerts are triggered to warn of fire risks. This allows fire departments to prepare to fight potential fires (Dhall et al., 2020).

Climate change is having a significant impact on weather patterns and the natural balance of our planet. Rising temperatures are fueling wildfires, disrupting rainfall patterns and increasing the frequency and intensity of extreme weather events such as storms and floods. Faced with these challenges, IoT is a key instrument for improving the monitoring, analysis and detection of environmental changes.

Thanks to the connectivity offered by cellular IoT, it is possible to deploy a wide range of sensors to monitor and analyze the environment. This allows for a better understanding of soils, trees and weather conditions. In-depth analysis of this data facilitates environmental monitoring and contributes to informed decision-making.

Cellular IoT technologies provide extensive coverage even in the most remote and densely populated areas, such as forests. The forestry sector is undergoing transformation, notably thanks to the growing adoption of connected tools. Cellular IoT sensors make it possible to automate data collection, which until now was a manual and time-consuming task, while still being manageable remotely. The integration of these sensors makes it possible to make safer and more efficient decisions in terms of forest management, within shorter time frames, and even to anticipate threats such as fires.

Traditional methods of monitoring fires, such as satellite images and ground cameras, have proven insufficient to cope with the rapid spread of wildfires, which is being accelerated by drought. IoT sensors offer a more effective solution by detecting fire outbreaks in real time and alerting the relevant authorities.

LTE-M and NB-IoT are LPWA (Low Power Wide Area) technologies that are characterized by their long range and low power consumption. IoT devices using these networks benefit from better penetration capability and can communicate with other devices, even in hard-to-reach areas. Thanks to these energy-efficient networks, some sensors can remain in place for almost 10 years without requiring maintenance, thanks to their long battery life.

FUTURE DIRECTIONS

Future research and development avenues for AI and IoT in disaster management will be discussed.

Integration of AI and IoT-based Technologies

Combining AI with IoT-based technology is one of the most promising avenues for the future in disaster management. The objects collect a wealth of data, including photos and sensor readings, which can be processed by AI-based systems and transmitted in real time to disaster response teams. Objects can also be used to gather information in dangerous or inaccessible environments. When these technologies are combined, they can make rescue operations faster and more precise.

Multi-Agent Systems

The creation of multi-agent systems is another potential avenue for disaster management. Autonomous agents in multi-agent systems can coordinate their efforts to accomplish a task. Multi-agent systems can be used to better coordinate response efforts and allocate resources in the Disaster Management Context. Drones and ground robots, for example, can cooperate in searching for survivors and assessing damage.

Explanatory AI

The field of explanatory AI aims to create intelligent machines capable of justifying their choices and actions. Explanatory AI can help establish credibility between human responders and automated systems during emergency situations. An AI-based system that reports the current location of survivors, for example, should be able to justify its reasoning.

Privacy and Ethical Considerations

It is crucial to think about privacy and ethics as technologies like AI and IoT become more common in disaster management. There are privacy concerns associated with using the objects for surveillance, for example. Additionally, discrimination and bias must be explicitly eliminated from AI-based systems.

CONCLUSION

In this research, we explored how AI and drone-based technologies can be used in emergency response. Future avenues and research directions for AI and drone-based disaster management were also discussed, such as the integration of AI and drone-based technologies, the development of multi-agent systems and the importance of explanatory AI and ethical considerations. The potential benefits of AI- and drone-based technologies in disaster management are considerable, despite the many challenges ahead. These technologies have the potential to save lives and reduce the impact of disasters by increasing the efficiency and accuracy of response activities. In conclusion, we believe that the continued development and integration of AI and drone-based technologies in disaster management will be crucial to reduce the impact of disasters on communities around the world.

REFERENCES

Abid, S. K., Sulaiman, N., Chan, S. W., Nazir, U., Abid, M., Han, H., Ariza-Montes, A., & Vega-Muñoz, A. (2021). Toward an integrated disaster management approach: How artificial intelligence can boost disaster management. *Sustainability (Basel)*, *13*(22), 12560. doi:10.3390/su132212560

Adeel, A., Gogate, M., Farooq, S., Ieracitano, C., Dashtipour, K., Larijani, H., & Hussain, A. (2019). A survey on the role of wireless sensor networks and IoT in disaster management. *Geological disaster monitoring based on sensor networks*, 57-66.

Adnan, A., Ramli, M. Z., & Abd Razak, S. K. M. (2015, November). Disaster management and mitigation for earthquakes: are we ready. In *9th Asia Pacific structural engineering and construction conference (APSEC2015)* (pp. 34-44). IEEE.

Ali, K., Nguyen, H. X., Shah, P., Vien, Q. T., & Ever, E. (2019). Internet of things (IoT) considerations, requirements, and architectures for disaster management system. *Performability in internet of things*, 111-125.

Bail, R. D. F., Kovaleski, J. L., da Silva, V. L., Pagani, R. N., & Chiroli, D. M. D. G. (2021). Internet of things in disaster management: Technologies and uses. *Environmental Hazards*, *20*(5), 493–513. doi:10.1080/17477891.2020.1867493

Boulouard, Z., Ouaissa, M., Ouaissa, M., Krichen, M., Almutiq, M., & Algarni, M. (2023, December). Streamlining River Flood Prevention with an Integrated AIoT Framework. In *2023 20th ACS/IEEE International Conference on Computer Systems and Applications (AICCSA)* (pp. 1-6). IEEE. 10.1109/AICCSA59173.2023.10479264

Boulouard, Z., Ouaissa, M., Ouaissa, M., Krichen, M., Almutiq, M., & Gasmi, K. (2022). Detecting hateful and offensive speech in arabic social media using transfer learning. *Applied Sciences (Basel, Switzerland)*, *12*(24), 12823. doi:10.3390/app122412823

Boulouard, Z., Ouaissa, M., Ouaissa, M., Siddiqui, F., Almutiq, M., & Krichen, M. (2022). An integrated artificial intelligence of things environment for river flood prevention. *Sensors (Basel)*, *22*(23), 9485. doi:10.3390/s22239485 PMID:36502187

Chai, J., & Wu, H. Z. (2023). Prevention/mitigation of natural disasters in urban areas. *Smart Construction and Sustainable Cities*, *1*(1), 4. doi:10.1007/s44268-023-00002-6

Dhall, A., Dhasade, A., & Nalwade, A., VK, M. R., & Kulkarni, V. (2020). A survey on systematic approaches in managing forest fires. *Applied Geography (Sevenoaks, England)*, *121*, 102266. doi:10.1016/j.apgeog.2020.102266

Fan, C., Zhang, C., Yahja, A., & Mostafavi, A. (2021). Disaster City Digital Twin: A vision for integrating artificial and human intelligence for disaster management. *International Journal of Information Management*, *56*, 102049. doi:10.1016/j.ijinfomgt.2019.102049

Ghasemi, P., & Karimian, N. (2020, April). A qualitative study of various aspects of the application of IoT in disaster management. In *2020 6th International Conference on Web Research (ICWR)* (pp. 77-83). IEEE. 10.1109/ICWR49608.2020.9122323

Goyal, H. R., Ghanshala, K. K., & Sharma, S. (2021). Post flood management system based on smart IoT devices using AI approach. *Materials Today: Proceedings*, *46*, 10411–10417. doi:10.1016/j.matpr.2020.12.947

Kamruzzaman, M. D., Sarkar, N. I., Gutierrez, J., & Ray, S. K. (2017, January). A study of IoT-based post-disaster management. In *2017 international conference on information networking (ICOIN)* (pp. 406-410). IEEE.

Kuglitsch, M. M., Pelivan, I., Ceola, S., Menon, M., & Xoplaki, E. (2022). Facilitating adoption of AI in natural disaster management through collaboration. *Nature Communications*, *13*(1), 1579. doi:10.1038/s41467-022-29285-6 PMID:35332147

Linardos, V., Drakaki, M., Tzionas, P., & Karnavas, Y. L. (2022). Machine learning in disaster management: recent developments in methods and applications. *Machine Learning and Knowledge Extraction, 4*(2).

Ouaissa, M., Rhattoy, A., & Chana, I. (2018, October). New security level of authentication and key agreement protocol for the IoT on LTE mobile networks. In *2018 6th International Conference on Wireless Networks and Mobile Communications (WINCOM)* (pp. 1-6). IEEE. 10.1109/WINCOM.2018.8629767

Pang, G. (2022). Artificial intelligence for natural disaster management. *IEEE Intelligent Systems, 37*(6), 3–6. doi:10.1109/MIS.2022.3220061

Park, S., Park, S. H., Park, L. W., Park, S., Lee, S., Lee, T., Lee, S., Jang, H., Kim, S., Chang, H., & Park, S. (2018). Design and implementation of a smart IoT based building and town disaster management system in smart city infrastructure. *Applied Sciences (Basel, Switzerland), 8*(11), 2239. doi:10.3390/app8112239

Pham, B. T., Luu, C., Van Phong, T., Nguyen, H. D., Van Le, H., Tran, T. Q., & Prakash, I. (2021). Flood risk assessment using hybrid artificial intelligence models integrated with multi-criteria decision analysis in Quang Nam Province, Vietnam. *Journal of Hydrology (Amsterdam), 592*, 125815. doi:10.1016/j.jhydrol.2020.125815

Rao, T. V. N., Jakkam, P., & Medipally, S. (2024). Future Trends and Innovations in Natural Disaster Detection Using AI and ML. In *Predicting Natural Disasters With AI and Machine Learning* (pp. 110–134). IGI Global.

Ray, P. P., Mukherjee, M., & Shu, L. (2017). Internet of things for disaster management: State-of-the-art and prospects. *IEEE Access : Practical Innovations, Open Solutions, 5*, 18818–18835. doi:10.1109/ACCESS.2017.2752174

Renugadevi, R., & Medida, L. H. (2024). Artificial Intelligence and IoT-Based Disaster Management System. In Predicting Natural Disasters With AI and Machine Learning (pp. 135-146). IGI Global. doi:10.4018/979-8-3693-2280-2.ch006

Shah, S. A., Seker, D. Z., Hameed, S., & Draheim, D. (2019). The rising role of big data analytics and IoT in disaster management: Recent advances, taxonomy and prospects. *IEEE Access : Practical Innovations, Open Solutions, 7*, 54595–54614. doi:10.1109/ACCESS.2019.2913340

Sharma, K., & Anand, D. (2023). AI and IoT in Supply Chain Management and Disaster Management. In Artificial Intelligence in Cyber-Physical Systems (pp. 275-289). CRC Press. doi:10.1201/9781003248750-16

Sharma, M., & Kaur, J. (2019). Disaster management using internet of things. In *Handbook of research on big data and the IoT* (pp. 211–222). IGI Global. doi:10.4018/978-1-5225-7432-3.ch012

Sinha, A., Kumar, P., Rana, N. P., Islam, R., & Dwivedi, Y. K. (2019). Impact of internet of things (IoT) in disaster management: A task-technology fit perspective. *Annals of Operations Research*, *283*(1-2), 759–794. doi:10.1007/s10479-017-2658-1

Sun, R., Gao, G., Gong, Z., & Wu, J. (2020). A review of risk analysis methods for natural disasters. *Natural Hazards*, *100*(2), 571–593. doi:10.1007/s11069-019-03826-7

Sun, W., Bocchini, P., & Davison, B. D. (2020). Applications of artificial intelligence for disaster management. *Natural Hazards*, *103*(3), 2631–2689. doi:10.1007/s11069-020-04124-3

Tan, L., Guo, J., Mohanarajah, S., & Zhou, K. (2021). Can we detect trends in natural disaster management with artificial intelligence? A review of modeling practices. *Natural Hazards*, *107*(3), 2389–2417. doi:10.1007/s11069-020-04429-3

Zafar, U., Shah, M. A., Wahid, A., Akhunzada, A., & Arif, S. (2019). Exploring IoT applications for disaster management: identifying key factors and proposing future directions. *Recent trends and advances in wireless and IoT-enabled networks*, 291-309.

Chapter 2
Demystifying the Applications of Artificial Intelligence in Disaster Management:
A Review

M.A. Jabbar
Vardhaman College of Engineering, India

Ruqqaiya Begum
Vardhaman College of Engineering, India

K. Tejasvi
Vardhaman College of Engineering, India

ABSTRACT

A disaster is an unexpected event that disrupts a society's functioning while also harming the human environment and causing financial and material losses. It can be caused by either natural or human factors. In today's society, disasters are seen as the product of good planning, which leads to hazards and vulnerabilities. The term "disaster management" refers to the planning and management of disasters. Artificial intelligence is the ability of computers to perform the tasks that are usually done by humans. Artificial intelligence has been used in many industrial applications and also used in everyday interactions with technology. Artificial intelligence is being used in sustainable development, and disaster risk management. Machine learning and artificial intelligence models can be used in both ways in disaster management, i.e., pre-in disaster management and post-in disaster management.

DOI: 10.4018/979-8-3693-3896-4.ch002

INTRODUCTION

A disaster is an unexpected event that Disrupts a society's functioning and additionally purpose Human surroundings, Economic and cloth loss. It may result from nature or via way of means of human causes. There are forms of Disasters. Natural and Man-made screw-ups. Natural Disaster: These take place via way of means of herbal Pollution and end in asset damage, harm effects, lack of livelihood, Environmental harm, etc. Those screw-ups are like Earthquakes, Floods, Cyclones, and Droughts. Various phenomena likely earthquakes, avalanches, volcanic eruptions, inundation, hurricanes, tornadoes, blizzards, tsunamis, cyclones, wildfires, and pandemics are all herbarium dangers that destroy lots of humans and shatter billions of greenbacks of abode and property every year. However, the fast growth of the Earth's commonality and its extended application often in dangerous environments has escalated the frequency and severity of screw-ups. With the tropical weather and volatile landforms, conjugated with deforestation, and unplanned increment proliferation, non-engineered buildings mate the catastrophe-susceptive provinces major prone (Sun et al., 2020).

Manmade Disasters: These sorts are Disasters result from Human Faults like Stamping, oil spills, Burning, Transport incidents, etc (Tan et al., 2021). Other forms of prompted screw-ups consist of the greater cosmic eventualities of catastrophic weather changes, nuclear war, and bioterrorism. In one opinion, all mistakes can be considered man-made because of the inability to take appropriate emergency controls (Nunavath and Goodwin, 2018). Domestically, famine can be caused by droughts, floods, fires, or epidemics, but food is plentiful around the world, and persistent regional shortages are government mismanagement, fierce conflicts, or necessity. Often the cause is a financial institution that does not distribute food. Earthquakes are especially dangerous due to man-made houses and dams. Protecting against tsunamis and landslides caused by earthquakes is essentially a decision of the situation.

Disaster risk mitigation refers to the actions adapted to address the root causes that make individuals vulnerable to disasters (Arslan et al., 2017). Preparation: Measures have been taken to prepare the mobilization of personnel, funds, equipment, and materials in a safe environment for corrective action. Disaster risk mitigation is the process of developing capacity before a disaster occurs to mitigate the impact. Countermeasures include the availability of food stocks, emergency stockpiles, seed stocks, fitness facilities, alert systems, backup infrastructure, repair handbooks, and project cabinets (Zhu et al., 2019).

Causes of disaster management: Below are some of these underlying causes of Disaster Management:

Table 1. Phases and approaches of disaster management

Phases of DM	Approaches for DM
Early warning and event detection	Temporary summarization, Earthquake monitoring, Crowdsourcing.
Post-disaster coordination and response	Crowd machine learning, data-driven application, SVM, Naïve Bayes Decision-tree, Unsupervised learning, Automated Machine Learning, CNN.

Table 2. Causes of disaster management

S. No	Causes	Description
1	Poverty	Almost all disaster studies show that the richest people survive disasters, either affected or recover quickly.
2	Population growth	More and more persons vying for a finite amount of resources leading to conflict and conflict can lead to crisis-related migration
3	Rapid urbanization	Competition for various resources is an inevitable consequence of rapid urbanization, leading to man-made disasters.
4	Cultural practices	Societies are always in a state of change and transformation. they are often extremely disruptive and unequal, leaving holes in the social and technological mechanisms of coping.
5	Environmental degradation	Deforestation results in fast rain which leads to floods.
6	Lagging of knowledge	Disasters occur, As the people who are at risk from them simply do not know how to avoid danger or take protective measures.

APPLICATIONS OF ICT WITHIN DISASTER MANAGEMENT

ICT performs a vital position in a variety of factors in disaster gamble management. At existing so is an increasing consciousness about the worth of ICT because of catastrophe gamble management. Disaster hazard management through ICT ambitions in imitation of minimizing the damage brought on by using herbal then artificial dangers such as earthquakes, floods, droughts stability cyclones epidemics, wars, and ethno- spiritual then political conflicts. The use concerning ICT may facilitate the administration of mess ups through presenting data on disaster prevention, promptly catastrophe prediction, communicating and durability disseminating catastrophe information after residents, or making sure a rapid verbal exchange rules before, through, and, below the disaster in imitation of both regime and non-government organizations because remedy materials (Nunavath and Goodwin, 2019). Information and Communication Technologies (ICTs) execute back according to aid the work about catastrophe jeopardy administration (DRM) within times of crisis, namely well as much within times over planning and within instances of reconstruction (Mohan and Mittal, 2013).

New ICTs: Computers, satellites, wireless communications (like cellular phones), Internet, e-bid then multimedia normally peruse within the New ICT category (Van Hentenryck, P., 2013) (Arinta, R.R. and WR, E.A, 2019). The use of smartphones and landlines because of disaster gambling is a famous device in the present age. The cellular network has been improved considerably worldwide. Terrestrial constant services, ball communications which includes moon cellular phones, as much nicely so cellular and wi-fi capabilities redact possible frame yet facts change within distinct remedy teams, because of put one's cards on the table then harmony concerning remedy activities.

The use of the Internet for catastrophe jeopardy management is an excellent supply because of records sharing, as like lengthy so the network is up. Weather forecasts, online newspapers then scenario reports are among the issues relevant after disaster preparedness or response. These are all reachable online. Some groups additionally use the Internet because of online databases, sharing records with their global counterparts and then sponsors. Use E-mail because disaster hazard management is old because Information sharing; is generally, at the corporation's managerial level. It varies in availability and use. Head places of work tend to bear access in conformity with e-mail then usage to that amount as much a natural means concerning communication.

The use of the managing facts rule because of catastrophe hazard management (MIS) is aged with the aid of some Firms. For example in Jamaica the Office on Disaster Preparedness then Emergency Management is raising its existing provision to function as an interactive data dividing dictation because of the resolution companies in disaster operations. This system, condition-based online, keeps logged beside a variety of attractions yet executes furnish statistics according to donors around the world or staff/volunteers in the field.

MACHINE LEARNING CONCEPTS

Machine learning (ML) has emerged as one of the most widely used computer technologies in daily life and in various fields (Munawar et am., 2019). ML is an artificial intelligence (AI) application that makes use of its functional algorithms. Several decisions need to be made, like bandwidth selection, power control, data rate selection, and user assignment to a base station. We can make use of these algorithms to solve these particular problems to reduce human intervention in hazardous and random environments.

The following are the advantages of Machine Learning in disaster management:

- ML algorithms can process large amounts of data and identify trends. The use of ML, like traffic forecasting, video monitoring, and online customer support, is becoming common.
- Rule-based technologies help to identify bogus information. The use of this helps to reduce the need for human intervention and making decisions. And also prevent rumors, usually during man-made disasters.
- As the amount of data increases Performance of ML algorithms is improving widely. For example, in the case of an earthquake prediction, the larger the volume of data, the higher the predictive power of the algorithm.
- ML algorithms can process multidimensional data and identify dissidents in the data set. In extremely dangerous situations, dissident analysis is an important technique. Machine learning algorithms are used for disaster and pandemic management. These algorithms are:

Supervised Learning

These algorithms report the training data fed to the systems and provide a set of Relatable outputs. We expect that the machine should learn the pattern of data and predict the output value for new data entries (Blaikie et al., 2003). Learning involves two main methodologies:

Unsupervised Learning

Unlike supervised learning algorithms, data or information transferred to systems during unsupervised learning will not be labeled. The user does not have any idea about the nature of the data.

MACHINE LEARNING APPLICATIONS IN DISASTER MANAGEMENT

Machine learning models may be used in both ways in disaster management, ie; pre-disaster management, and post-disaster management in disaster management prediction can be done with the help of IoT devices, i.e.; Machines connected to other things to gather data over various networks without human intervention, Sensors that measure properties such as temperature, CO_2 content, greenhouse gases, etc. Machine learning algorithms process the data acquired from IoT devices and ensure accuracy.

Table 3. Methodologies used in disaster management

S.NO	Methodology	Description
1.	Nearest Neighbors (KNN)	It takes assumptions that similar objects occur next to each other and evaluate their k nearest neighbors for similitude.
	Support Vector Machine (SVM)	It works by identifying a hyperplane that classifies the data points.
	Naive Bayes	These classifiers are a set of classification algorithms depending on Bayesian theory. It is known as naïve, as all the classified traits should not be dependent on each other.
2.	Logistic regression	With the use of a logistic function whose output values fall between 0 and 1, logistic regression generates a curve that forms the letter (Strzelecka et al., 2020) (Soofi and Awan, 2017).
	Decision trees	In this, the characteristics are termed by internal nodes, decision rules by branches, and results by leaf nodes (Ren and Gao, 2011).
	Bayesian back regression	This method is a probabilistic method that uses the estimation of statistical models. It is useful when there is a bad, inadequate, or poorly distributed data source (Jia et al., 2016).
	Deep Neural Networks	The DNN consists of an input layer, hidden layers, and an output layer. Its main advantage is it gradually reads features and modifies their results depending on a certain basis.
	CNN	Classifies images in pixels that are captured by drones from affected and non-affected areas.

Table 4. Machine learning methods and datasets used for disaster management

Reference	AI/ML Methods used	Data Set used	Remarks
1	PCOD-means algorithm	8 datasets from 26 various disaster events. (CrisisL exT26 public)	Precision, Recall, AUC
2	Data-driven approach	Datasets from 17 crisis events. (CrisisL exT26 public)	Cohen Kappa measure
5	Naïve Bayesian classification	Twitter dataset, 200K tweets, Joplin2011 tornado (CrisisL exT26 public)	Accuracy
6	Supervised classification	52 million Tweets from 19 Crisis events (CrisisL exT26 public)	Precision, Recall
8	CNN	13680 26 various disaster events. (CrisisL exT26 public)	F-1 Score

Table 5. Unsupervised learning methods

S. No	methodology		Description
1.	Clustering	K-means	In K clustering defines the number of clusters (k) that are to be initialized. The centers are randomly opted for N points by comparing the data set. (Sadhukhan et al., 2018) used K means to detect flood and damaged areas by aggregating data sets. The authors of (Assery et al., 2019) used K means to group the behavior of crowds. The Authors of (Bucholz M, 2019) used it to classify crowds into different classes. The authors of (Tang et al., 2005) used it for the prediction of the spread of cholera.
		Fuzzy C-means	K Medoids also starts with random assignment of centers and random initialization of data points to those clusters. Fuzzy C means the algorithm uses a special parameter αjk and a fuzzy parameter F. αjk is the degree of association of data point with clusters, and F is greater than 1.
2.	Reinforcement learning	Q learning	Decision-making is done consecutively. The output of this is based on the present input to give the best solution and the next input is determined by the previous output.

Table 6. Machine learning algorithms used in disaster management

Machine Learning Algorithms	Purpose
ANN	predicting flood risk in a water catchment area.
logistic regression	predicting flood risk in a water catchment area.
SVM	Classifies images in pixels that are captured by drones from affected and non-affected areas.
k-Means	Classifies images in pixels that are captured by drones from affected and non-affected areas.
CNN	The network classifies the data for the occurrence of floods as positive and negative. And also Detects Fire, Floods, and landslides.

CHALLENGES IN APPLYING MACHINE LEARNING IN DISASTER MANAGEMENT

Table 7. Challenges in adopting ML in disaster management

S. No	Challenges	Solutions
1	Social media data. Like (Twitter, Facebook, Instagram, WhatsApp)	As datasets aren't without difficulty shared from exclusive platforms, it's miles viable to get information from diverse assets like Facebook groups, which can be neighborhood to a catastrophe occasion place and Facebook pages of Emergency groups or government as there aren't any regulations to get entry to from those spaces.
2	Information from exclusive reasserts.	ought to have a devoted Public Information Officer (PIO) to make certain that catastrophe sufferers are receiving the assistance they wish making sure speedy transport of statistics and the use of an automatic method. Should take feedback from contemporary automatic processes and requests from the catastrophe sufferers and different coordinators approximately the transport of assistance.
3	Data integration: The integration of various datasets collected from social media, sensors, satellites, and authorized data offers greater possibilities for emergency help.	The HFN generation vicinity constructs strong networks to provide higher communications inside the effects of any catastrophe.
4	Recovery: Based on information from social media, that spreads with increased exposure; sharing messages several times could ensure the spread of information.	The neighborhood governments and catastrophe Management groups, neighborhood instructional researchers, and media ought to be worried about carrying the generation into catastrophe aid in growing countries

RESEARCH DIRECTION

1. To save time while deploying several models in real-world settings, the preprocessing time must be decreased. Preprocessing time, for instance, accounts for 60% to 80% of the whole time.
2. Successful models in actual catastrophe scenarios must be able to adjust to changes in the real environment. When the area's population density rises, there is no such mechanism for this.
3. ML models that employ high-resolution photos have the potential to improve accuracy. Cloudy and low-clarity photos might cause the model to underfit (Chamola et al., 2020).

CONCLUSION

Natural disasters and pandemics have become far more harmful and common in recent years. A rapid increase in the frequency of disasters and pandemics has put a tightness on emergency services, and this is where machine learning algorithms come into play to help them work more efficiently and make the greatest use of the resources which is available. Hence In this Chapter, we provide a concise, demystifying review of the applications of artificial intelligence and machine learning in disaster management. The use of artificial intelligence in disaster management will minimize the loss of human life and rescue operation time by using robotics, drones, sensors, etc.

REFERENCES

Arslan, M., Roxin, A. M., Cruz, C., & Ginhac, D. (2017). *A review on applications of big data for disaster management.* IEEE. doi:10.1109/SITIS.2017.67

Assery, N., Xiaohong, Y., Almalki, S., Kaushik, R., & Xiuli, Q. (2019). *Comparing learning-based methods for identifying disaster-related tweets.* doi:10.1109/ICMLA.2019.00295

Blaikie, P. (2003). *At Risk – Natural hazards, people's vulnerability, and disasters.* Research Gate.

Buchholz, M. (2019). Deep reinforcement learning. introduction. deep q network (dqn) algorithm. Research Gate.

Chamola, V., Hassija, V., Gupta, S., Goyal, A., Guizani, M., & Sikdar, B. (2020). Disaster and pandemic management using machine learning: A survey. *IEEE Internet of Things Journal, 8*(21), 16047–16071. doi:10.1109/JIOT.2020.3044966 PMID:35782181

Munawar, H. S., Hammad, A., Ullah, F., & Ali, T. H. (2019). *After the flood: A novel application of image processing and machine learning for post-flood disaster management.*

Nunavath, V. & Goodwin, M. (2019). The Use of Artificial Intelligence in Disaster Management Systematic Literature Review. (pp. 1–8). IEEE.

Nunavath, V. (2018). *The role of artificial intelligence in social media big data analytics for disaster management-initial results of a systematic literature review.* (pp. 1–4). IEEE.

Strzelecka, A., Kurdyś-Kujawska, A., & Zawadzka, D. (2020). Application of logistic regression models to assess household financial decisions regarding debt. *Procedia Computer Science, 176*, 3418–3427. doi:10.1016/j.procs.2020.09.055

Sun, W., Bocchini, P., & Davison, B. D. (2020). Applications of artificial intelligence for disaster management. *Natural Hazards, 103*(3), 1–59. doi:10.1007/s11069-020-04124-3

Tan, L., Guo, J., Mohanarajah, S., & Zhou. (2021). Can we detect trends in natural disaster management with artificial intelligence? *A review of modeling practices, 107*(3), 2389-2417

Zhu, X., Zhang, G. & Sun, B. (2019). *A comprehensive literature review of the demand forecasting methods of emergency resources from the perspective of artificial intelligence.* Research Gate.

Chapter 3
AI and IoT in Flood Forecasting and Mitigation:
A Comprehensive Approach

Muhammad Usman Tariq

(iD) https://orcid.org/0000-0002-7605-3040
Abu Dhabi University, UAE & University of Glasgow, UK

ABSTRACT

This chapter outlines the complete way to deal with flood estimating and alleviation through the mix of man-made brainpower (artificial intelligence) and the web of things (IoT). Floods are among the most pulverizing catastrophic events universally, requiring progressed techniques for expectation and reaction. Computer-based intelligence and IoT innovations offer promising arrangements by empowering precise estimating, constant observation, and compelling relief measures. This part investigates the groundwork of computer-based intelligence and IoT in calamity the board, featuring their assembly for upgraded information examination and navigation. It digs into artificial intelligence models, such as profound learning and AI for flood expectation, alongside the organization of IoT sensors and gadgets for continuous observing in flood-inclined regions.

INTRODUCTION TO FLOOD MANAGEMENT

Prologue to Flood The board Difficulties Floods presents critical difficulties to networks worldwide, affecting huge numbers of lives, causing hato frameworks to frameworks and environments. As environmental change speeds up, the recurrence and force of flooding occasions are projected to increase, compounding existing

DOI: 10.4018/979-8-3693-3896-4.ch003

weaknesses and requiring creative ways to deal with Flooding on the board (IPCC, 2021). This segment outlines the worldwide difficulties related to flooding the board, featuring the complicated interchange of natural, social, and financial variables.

Floods are regular peculiarities described by the flood of water onto ordinarily dry land. They can result from different causes, including heavy precipitation, storm floods, snowmelt, and the disappointment of frameworks like dams and levees (UNDRR, 2019). Flood risks manifest in various structures, from streak floods that happen quickly and without advance notice to riverine floods that foster bit by bit after some time (Milly et al., 2002). Beachfront regions are especially powerless against storm floods and ocean-level ascent, while inland districts face chances related to waterway flooding and metropolitan immersion (Hallegatte et al., 2013).

Impacts of Flooding

The effects of Flooding are complex, influencing networks, economies, and environments in significant ways. Floodwaters can make broad harm homes, organizations, and basic frameworks like streets, scaffolds, and utilities (Shaper et al., 2008). Interruption with transportation organizations and fundamental administrations can block crisis reaction endeavours and impede admittance to clinical consideration, food, and clean water (Parker et al., 2011). In addition, floods can bring about a death toll, removal of populaces, and long-term mental injury (Councilman et al., 2008). Financially, floods can decimate neighbourhood economies by upsetting horticultural creation, disturbing stock chains, and expanding protection costs (Hochrainer-Stigler et al., 2019). Ecological effects incorporate soil disintegration, natural surroundings obliteration, and tainting of water sources, with suggestions for biodiversity and environment administrations (Winsemius et al., 2015).

Weakness and Strength

The effects of Flooding are not equitably disseminated, with underestimated networks frequently enduring the worst part of the harm (Füssel & Klein, 2006). Financial factors like destitution, deficient framework, and restricted admittance to assets worsen Weakness to floods (Adger et al., 2007). Moreover, urbanization and land-use changes can build the openness of populaces and resources for flood gambles (Ward et al., 2018). Building flexibility to floods requires a multi-layered approach that tends to physical, social, and financial weaknesses (Birkmann et al., 2013). This includes working on early advance notice frameworks, upgrading foundation flexibility, fortifying social well-being nets, and elevating biological system-based ways to deal with flooding the executives (Aerts et al., 2018).

Environmental Change and Future Dangers

Environmental change is supposed to enhance the recurrence and seriousness of flooding occasions, presenting new difficulties for flood executives (Hirabayashi et al., 2013). Increasing temperatures heighten precipitation designs, prompting more continuous and extraordinary precipitation occasions (Knutson et al., 2013). Ocean-level ascent is compounding waterfront flooding gambles, while changes in snowmelt designs adjust the timing and greatness of stream floods (Hirabayashi et al., 2013). The intensifying impacts of environmental change, quick urbanization, and populace development highlight the criticalness of executing versatile systems to relieve flood gambles (Dilley et al., 2005). In rundown, flood the executives is a complicated and diverse test that requires composed activity at neighbourhood, public, and worldwide levels. Tending to the effects of Flooding requires inventive methodologies that coordinate logical information, innovative headways, and local area commitment. The accompanying segments of this section will investigate the job of Man-made reasoning (simulated intelligence) and the Web of Things (IoT) in improving flood gauging and relief methodologies, offering experiences into how these advancements can add to building stronger and versatile networks despite developing flood chances.

BACKGROUD

Floods are among the most destructive natural disasters that cause significant damage to infrastructure, property, and human life. Traditional flood prediction methods rely on historical data and deterministic models, which may not be accurate or timely enough for effective disaster mitigation. However, the integration of artificial intelligence (AI) and the Internet of Things (IoT) has revolutionized the flood forecasting system by enabling real-time data collection, analysis, and predictive modeling. This chapter explores the background, development, and application of AI and IoT in flood forecasting systems. It discusses the evolution of these technologies, their key components, challenges, and future prospects. The chapter provides a comprehensive review of the existing literature and case studies to provide insight into the transformative potential of AI and IoT to improve flood forecasting accuracy, early warning systems, and disaster preparedness. AI techniques have been used in flood forecasting since the late 20th century. Researchers have studied neural networks, genetic algorithms, and fuzzy logic for hydrological modeling. Previous efforts have focused on improving precipitation and streamflow modeling and streamflow forecasting using computer intelligence techniques. For instance, neural network models can capture non-linear relationships between meteorological variables and hydrological responses, improving the accuracy of flood forecasting (Smith et al., 1996).

As computing capabilities advanced and artificial intelligence algorithms developed, researchers began integrating machine learning techniques such as support vector machines (SVM), random forests and deep learning into flood forecasting systems. In particular, deep learning has shown significant success in extracting complex patterns from large-scale environmental data, enabling more accurate and reliable flood forecasts (Noh et al., 2019). The use of Convolutional Neural Networks (CNN) and Recurrent Neural Networks (RNN) facilitated the analysis of spatio-temporal patterns of rainfall, river flow and other relevant variables, which improved the time and reliability of flood forecasting (Li et al., 2020).

At the same time, the proliferation of IoT technologies has ushered in a new era of real-time data collection and monitoring for flood forecasting applications. IoT devices such as sensors, gauges and drones enable continuous measurement of various environmental parameters such as rainfall intensity, water level, soil moisture and river flow velocity. These sensor networks form the backbone of IoT-based flood monitoring systems and provide high-resolution spatial and temporal data for accurate modeling and forecasting. The integration of IoT sensors with cloud computing platforms and wireless communication networks has enabled seamless transmission of data from remote monitoring stations to centralized servers for analysis and decision making. In addition, the development of low-power and cost-effective sensor technologies has democratized access to real-time environmental data, enabling local communities and authorities to take proactive measures to respond to imminent floods (Gao et al., 2018).

The deployment of IoT-enabled flood warning systems in flood-prone areas has shown significantly improved response times and disaster preparedness, reducing the impact of floods on vulnerable populations (Albuquerque et al., 2021). The convergence of AI and IoT in flood forecasting: The convergence of AI and IoT has tremendous potential to enhance and improve flood forecasting and disaster management. By integrating AI-based forecasting models with IoT-enabled sensor networks, stakeholders can harness the power of real-time data analytics to produce accurate and actionable flood forecasts. The synergy between AI and the Internet of Things enables adaptive and dynamic modeling methods that can continuously learn from new inputs and update predictive models to respond to changing environmental conditions.

One of the main advantages of combining AI and the Internet of Things for flood forecasting is the ability to incorporate different data sources and heterogeneous data types into predictive models. Traditional hydrological models often struggle to accommodate diverse data streams from different sources, leading to uncertainty and inaccuracies in forecasts. However, AI algorithms such as ensemble learning and hybrid models can effectively integrate data from IoT sensors, satellite images, social media and citizen observations, resulting in more comprehensive and reliable flood forecasts (Wang et al., 2020).

In addition, IoT provides a sensor-based real-time feedback loop to continuously validate and improve the model, improving the accuracy and sustainability of AI-based flood forecasts over time. For example, sensor data on rainfall intensity and river levels can be fed into AI models in real time, enabling on-the-fly calibration and adjustment to optimize forecast performance (Pandey et al., 2019). In addition, artificial intelligence techniques such as machine learning and pattern recognition can analyze historical flood data to identify recurring patterns and trends, enabling predictive risk assessment and adaptive decision making.

Despite the significant success of AI and IoT technologies in flood forecasting, there are still several challenges and research directions that require further investigation. One of the most important challenges is the integration of heterogeneous data sources and the interoperability of IoT devices from different manufacturers. To ensure seamless data exchange and interoperability between different sensor networks and platforms, standardization efforts and data harmonization protocols are needed (Crawford et al., 2022). Additionally, data protection, security, and ethical issues must be considered to increase the trust and adoption of AI-based flood forecasting systems among stakeholders and communities. The reliability and robustness of AI models under extreme weather conditions and unexpected scenarios are still areas of active research. Improving the robustness of AI-based flood forecasting systems against uncertainty and non-stationarity in climate models requires the development of adaptive learning algorithms and ensemble modeling methods (Liu et al., 2021).

Furthermore, efforts to enhance the interpretability and transparency of AI models are critical to increasing citizen trust and facilitating informed decision-making by policymakers and emergency responders. Moving forward, ongoing research focuses on the scalability, performance, and potential of new technologies such as edge computing, distributed ledger technology (DLT), and quantum computing in AI-based flood forecasting systems. To solve the computational complexity, edge AI platforms built into IoT devices enable real-time data processing and analysis at the edge of the network, reducing latency and bandwidth requirements when transferring data to centralized servers. Similarly, the application of DLT solutions such as blockchain can improve data provenance, integrity, and decentralized governance in IoT-based flood monitoring networks, ensuring data reliability and trustworthiness (Zhou et al., 2022).

The integration of AI and IoT technologies represents a paradigm shift in flood forecasting and disaster management, offering unprecedented opportunities to improve forecast accuracy, early warning systems, and community resilience. By leveraging the synergy between AI-based predictive modeling and IoT-based sensor networks, stakeholders can mitigate the impact of floods and improve preparedness and response strategies. However, addressing the technical, ethical, and management issues is critical to harnessing the full potential of AI and the Internet of Things to secure lives and livelihoods against the increasing risk of flooding in an era of climate change and urbanization. Continued interdisciplinary research and collaboration are essential to advance AI-based flood forecasting and ensure the fair and sustainable use of these technologies for the benefit of society.

Overview of the Global Impact of Flood Disasters

Flood debacles affect social orders, economies, and the climate worldwide. Understanding the scale and extent of these effects is significant for creating compelling flood-the-board systems and alleviating the related dangers. This segment gives a nitty gritty assessment of the worldwide effect of flood debacles, drawing on experimental proof and insightful exploration to explain the multi-layered nature of these occasions.

Financial Effects

Flood calamities have huge financial results, influencing networks both straightforwardly and by implication. Direct effects incorporate harm to the foundation, loss of vocations, and removal of populaces (UNDRR, 2019). Floodwaters can immerse homes, organizations, and basic frameworks like streets, scaffolds, and utilities, prompting exorbitant fixes and disturbances to fundamental administrations (Shaper et al., 2008). Uprooting populaces because of Flooding can strain social and emotionally supportive networks and worsen existing weaknesses, especially among underestimated networks (Council member et al., 2008). Aberrant financial effects of floods stretch out past quick actual harm, influencing more extensive monetary exercises and occupations. Upset transportation organizations and supply chains can block monetary recuperation endeavours and impede admittance to business sectors and fundamental administrations (Parker et al., 2011). Farming creation might be antagonistically impacted by Flooding, prompting crop misfortunes, food deficiencies, and expanded costs (Hochrainer-Stigler et al., 2019). Also, floods can intensify neediness and disparity by driving currently weak populaces further into difficulty (Füssel & Klein, 2006).

Ecological Effects

Flood catastrophes have critical ecological results, influencing biological systems, biodiversity, and normal assets. Floodwaters can cause soil disintegration, sedimentation, and annihilation of natural surroundings, changing the scene and disturbing biological cycles (Winsemius et al., 2015). Tainting water sources with poisons, flotsam, and jetsam can present dangers to people's well-being and untamed life, further worsening the natural cost of floods (Jonkman, 2005). Besides, changes in hydrological systems because of floods can meaningfully affect freshwater environments and oceanic biodiversity (Dudgeon et al., 2006).

Wellbeing and Philanthropic Effects

The well-being and philanthropic effects of flood calamities are significant, influencing physical and mental prosperity. Floodwaters can sully water sources, prompting flare-ups of waterborne illnesses like cholera, typhoid, and loose bowels (Few et al., 2004). Lacking disinfection and cleanliness offices in flood-impacted regions can worsen the spread of irresistible sicknesses, especially among dislodged populaces living in packed covers (Representative et al., 2008). In addition, the injury and mental trouble brought about by floods can meaningfully affect emotional wellness, adding to nervousness, melancholy, and post-horrendous pressure problems (Galea et al., 2005).

Flood debacles have boundless effects that reach out past quick actual harm, significantly influencing social orders, economies, and biological systems. Understanding the worldwide effect of floods is fundamental for creating successful floodboard techniques and decreasing calamity risk. By tending to the financial, natural, well-being, and helpful components of flood fiascos, policymakers, professionals, and networks can cooperate to construct stronger and versatile social orders fit for adapting to the developing difficulties of floods in an evolving environment.

The need for advanced Predictive and Mitigation Strategies

Floods are regular peculiarities that have been happening throughout humanity's experiences; at this point, their effects continue to present difficulties to networks worldwide. As populaces develop and urbanization speeds up, the requirement for cutting-edge prescient and alleviation methodologies to oversee flood gambles becomes progressively earnest. This part investigates the reasoning behind the need for such procedures, analyzing the elements driving the interest in additional modern ways to deal with flooding the board.

Environmental Change and Expanding Flood Dangers

The evolving environment is one of the essential drivers behind the requirement for cutting-edge prescient and alleviation techniques for floods. Environmental change is adjusting precipitation designs and expanding the recurrence and power of outrageous climate occasions like weighty precipitation and tempests (IPCC, 2021). These progressions compound flood gambles in numerous locales, presenting dangers to lives, jobs, and frameworks (Hirabayashi et al., 2013). Without viable measures to adjust to these evolving conditions, the effects of floods will likely be more extreme and boundless later on.

Populace Development and Urbanization

Quick populace development and urbanization likewise increase flood gambles, as seen by numerous networks. As more individuals get comfortable in flood-inclined regions, the potential for misfortunes from floods expands (Ward et al., 2018). Urbanization modifies normal seepage designs, diminishes invasion rates, and increments impenetrable surfaces, prompting more incessant and extreme metropolitan Flooding (Tucci, 2006). Also, the grouping of populaces and resources in metropolitan regions compounds the financial effects of floods, making successful flood-the-board systems fundamental for safeguarding lives and property.

Foundation Weakness

The Weakness of the basic foundation to flood harm further highlights the requirement for cutting-edge prescient and moderation systems. Streets, extensions, utilities, and other fundamental administrations are often situated in flood-inclined regions, leaving them helpless against immersion and disturbance during flood occasions (UNDRR, 2019). The disappointment of the foundation can have flowing impacts, frustrating crisis reaction endeavours, disturbing stock chains, and blocking monetary recuperation (Shaper et al., 2008). By consolidating prescient demonstrating and risk appraisal apparatuses, leaders can distinguish weaknesses and focus on interests in a versatile framework to relieve flood chances.

Community Resilience and Adaptive Capacity

Building people's group strength and improving versatile limits are fundamental to successfully flooding the executive's systems. Networks outfitted with the information, assets, and institutional ability to prepare for, answer, and recuperate from floods are better situated to moderate their effects (Councilman et al., 2008). Putting resources into early advance notice frameworks, crisis readiness plans, and social well-being nets can upgrade local area strength and diminish Weakness to floods (Adger et al., 2007). By enabling networks to go to proactive lengths to oversee flood gambles, progressed prescient and moderation methodologies can add to building stronger and versatile social orders equipped for adapting to the difficulties presented by floods in an evolving environment. 6-Conclusion All in all, the requirement for cutting-edge prescient and relief procedures for floods is driven by a blend of variables, including environmental change, populace development, urbanization, and framework weakness. Leaders can foster compelling methodologies to expect, plan for, and answer flood occasions by coordinating logical information, mechanical advancements, and local area commitment. Putting resources into versatile frameworks, improving early admonition frameworks, and building local area strength are fundamental parts of complete flood-the-board techniques that can assist with alleviating the effects of floods and safeguard lives and occupations, notwithstanding expanding flood chances,

Foundations of AI and IoT in Disaster Management

The incorporation of Artifcial Intelligence(artificial intelligence) and the Web of Things (IoT) has upset fiasco the executives work on, giving extraordinary abilities to expect, observe, and react. This part investigates the primary ideas supporting the utilization of artificial intelligence and IoT in misfortune the board, looking at how these advances are changing customary ways to deal with calamity readiness and reaction.

Artifcial Intelligence in Disaster Management

The board Simulated intelligence incorporates innovations that empower machines to recreate human knowledge, including AI, profound learning, normal language handling, and PC vision (Russell & Norvig, 2022). Regarding Catastrophe, the executives, simulated intelligence calculations break down immense measures of information to distinguish designs, recognize inconsistencies, and make forecasts (Ashktorab et al., 2020). AI calculations gain from authentic information to foster prescient models that can expect future occasions, like floods, seismic tremors, and rapidly spreading fires (Chen et al., 2018).

Web of Things in Catastrophe

The Web of Things (IoT) alludes to the organization of interconnected gadgets inserted with sensors, actuators, and correspondence abilities, empowering them to gather, send, and trade information (Atzori et al., 2010). In a fiasco, the executives and IoT gadgets are sent in different conditions, including metropolitan regions, far-off locales, and basic frameworks, to screen natural circumstances, recognize dangers, and work with correspondence (Hossain et al., 2015). These gadgets can go from weather conditions stations and seismic sensors to robots and wearable gadgets, giving continuous information streams illuminating independent direction and reaction endeavours (Yun et al., 2020).

Integration of simulated intelligence and IoT

The union of computer-based intelligence and IoT advancements holds tremendous potential for upgrading information examination and dynamics in Catastrophe the board. By joining the scientific abilities of artificial intelligence with the tactile capacities of IoT gadgets, chiefs can acquire further knowledge about complex frameworks, recognize rising dangers, and foster more successful reaction procedures (Zhang et al., 2021). Simulated intelligence calculations can dissect information gathered from IoT sensors to identify examples, patterns, and peculiarities characteristic of possible debacles, empowering proactive gamble executives and early admonition frameworks (Wang et al., 2019).

Difficulties and Open doors

While the coordination of artificial intelligence and IoT offers critical open doors for further development fiasco the board rehearses, it likewise presents difficulties connected with information protection, security, interoperability, and moral contemplations (Al-Fuqaha et al., 2015). Guaranteeing the dependability, exactness, and straightforwardness of computer-based intelligence models prepared on IoT information is fundamental to building trust in these frameworks (Rahman et al., 2020). Moreover, resolving information proprietorship, sharing, and administration issues is basic for cultivating joint effort and coordination among partners (Zhang et al., 2018). Notwithstanding these difficulties, the expected advantages of utilizing computer-based intelligence and IoT in a fiasco the executives are significant, offering new roads for improving flexibility, diminishing dangers, and saving lives despite normal and artificial debacles.

In the groundwork of simulated intelligence and IoT in Catastrophe, the executives address a change in outlook by how we prepare for and answer crises. By bridging the force of simulated intelligence calculations and IoT gadgets, chiefs can access ongoing information, break down complex frameworks, and pursue informed choices that save lives and moderate the effects of debacles. Pushing ahead, proceeding with examination, development, and cooperation are fundamental for amplifying the capability of simulated intelligence and IoT in calamity and building stronger and versatile social orders equipped for adapting to the difficulties of a questionable future.

AI and IoT Technolgies

Artificial Intelligence and the Web of Things (IoT) addresses two groundbreaking mechanical standards that have altered various spaces, including medical services, assembly, transportation, and debacle the board. This part gives an exhaustive prologue to artificial intelligence and IoT innovations, explaining their basic ideas, standards, and applications.

Artifcial Intelligence

Artificial consciousness alludes to machines' reenactment of human insight processes, empowering them to perform errands that normally require human knowledge, for example, getting the hang of thinking, critical thinking, discernment, and Direction (Russell & Norvig, 2022). Simulated intelligence incorporates many strategies and approaches, including AI, profound learning, regular language handling, PC vision, and mechanical technology.

Machine Learning

Machine Learning is a subfield of simulated intelligence that spotlights improving calculations and models that empower PCs to gain from information and pursue expectations or choices without being unequivocally modified (Mitchell, 1997). AI calculations gain examples and connections from huge datasets, permitting them to sum up and make expectations on new, concealed information. Normal AI calculations incorporate regulated learning, unaided learning, and support learning.

Deep learning

Deep Learning is a subset of AI that utilizes fake brain networks with numerous layers (profound designs) to learn complex portrayals of information (Goodfellow et al., 2016). Deep Learning calculations succeed at undertakings such as picture acknowledgement, discourse acknowledgement, regular language handling, and generative displaying. Convolutional Brain Organizations (CNNs) and Repetitive Brain Organizations (RNNs) are well-known structures utilized in profound learning.

Normal Language Processing (NLP)

Normal Language Processing is a part of simulated intelligence that highlights collaboration between PCs and human dialects, empowering machines to comprehend, decipher, and produce human language (Jurafsky & Martin, 2021). NLP procedures are utilized in applications like message grouping, feeling examination, machine interpretation, discourse acknowledgement, and chatbots.

PC Vision

PC Vision is a field of simulated intelligence that empowers PCs to decipher and understand the visual world, including pictures and recordings (Szeliski, 2010). PC vision calculations can perform errands like item discovery, picture division, facial acknowledgement, and scene getting. Profound learning procedures, especially CNNs, have progressed the most in PC vision.

Robotics

Robotics is an interdisciplinary field that joins simulated intelligence, designing, and mechanics to configure, fabricate, and work robots fit for performing errands independently or semi-independently (Siciliano & Khatib, 2016). Robots are utilized in different applications, including production, medical care, horticulture, transportation, and investigation.

Web of Things (IoT)

The Web of Things alludes to the organization of interconnected gadgets installed with sensors, actuators, and correspondence capacities, empowering them to gather, communicate, and trade information over the web (Atzori et al., 2010). IoT gadgets can go from shopper hardware, for example, cell phones, wearables, and savvy home gadgets, to modern gear, ecological sensors, and framework parts.

Sensor Advances

Sensors are central parts of IoT gadgets, empowering them to gather information about the real world, including temperature, dampness, pressure, light, movement, sound, and area (Gubbi et al., 2013). Given their usefulness, sensors can be arranged into different sorts, including ecological, biomedical, movement, nearness, and inertial sensors.

Correspondence Advancements

Correspondence advancements are political in empowering IoT gadgets to associate and speak with one another, along with clouded stages and applications (Zhang et al., 2018). Normal correspondence conventions utilized in IoT frameworks incorporate

Wi-Fi, Bluetooth, Zigbee, LoRaWAN, MQTT, and HTTP. These conventions work with information transmission, gadgets, executives, and interoperability in IoT biological systems.

Distributed computing and Edge Processing

Distributed computing and edge figuring are two reciprocal standards that help IoT applications by giving computational assets and capacity limits (Shi et al., 2016). Distributed computing includes unified information handling and stockpiling in far-off server farms. In contrast, edge figuring includes circulating computational undertakings and information handling nearer to the IoT gadgets at the organization's edge.

11-Uses of Artificial Intelligence and IoT

The incorporation of artificial intelligence and IoT innovations has empowered many uses across different spaces, including savvy urban communities, medical services, farming, fabricating, transportation, energy, and natural observing (Atzori et al., 2010). Computer-based intelligence-driven IoT applications incorporate brilliant homes, wearable well-being screens, prescient upkeep frameworks, independent vehicles, accuracy farming, and modern mechanization.

Artificial intelligence and the Web of Things (IoT) are two extraordinary advances reshaping how we connect with the world and care for complicated issues. Artificial intelligence empowers machines to imitate human knowledge and perform mental assignments, while IoT interfaces actual gadgets and empowers them to gather and trade information over the web. Computer-based intelligence and IoT advancements are potentially changing different spaces, driving development, and tending to worldwide difficulties in regions such as medical services, transportation, energy, and natural manageability.

The Convergence of AI and IoT for enhanced Data Analysis and Decision- Making

The intermingling of Computerized Reasoning (simulated intelligence) and the Web of Things (IoT) addresses a strong collaboration that has reformed information examination and dynamics across different spaces. This part investigates how incorporating computer-based intelligence and IoT advancements improves information examination abilities and works with more educated dynamic cycles.

Incorporation of computer-based intelligence and IoT Advances

Incorporating simulated intelligence and IoT advancements includes utilizing artificial intelligence calculations and procedures to examine information gathered from IoT gadgets. IoT gadgets implanted with sensors and actuators produce tremendous information about the world, including natural circumstances, human exercises, and machine tasks (Gubbi et al., 2013). Computer-based intelligence calculations process and dissect this information to remove noteworthy bits of knowledge, identify examples, and make expectations, empowering wiser Direction.

Constant Information Examination

One of the critical advantages of incorporating artificial intelligence and IoT is the capacity to perform continuous information examination. IoT gadgets constantly gather information from their environmental elements, giving a constant flow of data that can be examined progressively utilizing simulated intelligence calculations (Atzori et al., 2010). Constant information investigation empowers associations to screen dynamic frameworks, recognize oddities, and answer instantly to evolving conditions, working on functional effectiveness and dynamic agility.

Predictive Analytics

Simulated intelligence-fueled prescient examination is one more region where the combination of artificial intelligence and IoT offers critical benefits. By dissecting verifiable information from IoT gadgets, artificial intelligence calculations can distinguish examples and patterns that might show future occasions or results (Russell & Norvig, 2022). A prescient examination can be applied to different use cases, such as proactive upkeep, request gauging, risk appraisal, and misrepresentation location. It empowers associations to expect issues before they happen and go to proactive lengths to alleviate gambles.

Anomaly Detection

Distinguishing peculiarities or surprising examples in information is a basic undertaking in numerous applications, including online protection, medical care, and modern checking. Artificial intelligence calculations prepared on IoT information can consequently distinguish peculiarities that go astray from typical ways of behaving, flagging likely dangers or breakdowns (Chandola et al., 2009). Irregularity discovery strategies, such as unaided learning and exception recognition, empower associations to upgrade security, further develop well-being, and forestall expensive interruptions.

Enhancement and Robotization

The combination of simulated intelligence and IoT empowers the enhancement and computerization of cycles and frameworks. Artificial intelligence calculations can investigate information from IoT gadgets to recognize shortcomings, improve asset portions, and mechanize routine assignments (Shi et al., 2016). For instance, in brilliant assembly conditions, artificial intelligence-fueled frameworks can advance creation plans, anticipate hardware disappointments, and change activities continuously to amplify efficiency and limit personal time.

Upgraded Direction

Eventually, the combination of artificial intelligence and IoT prompts upgraded dynamic abilities across different areas. By providing convenient and precise experiences from IoT information, artificial intelligence calculations engage chiefs in pursuing informed decisions that drive business results, further develop benefits, and moderate dangers (Zhang et al., 2018). Whether advancing stock chains, overseeing energy utilization, or answering crises, simulated intelligence-fueled IoT frameworks empower associations to settle on choices with more prominent certainty and adequacy.

Difficulties and Contemplations

While the combination of artificial intelligence and IoT offers huge potential, it also presents difficulties and contemplations that should be addressed. These incorporate information protection and security concerns, interoperability issues, adaptability challenges, and moral contemplations (Al-Fuqaha et al., 2015). Guaranteeing the respectability, classification, and accessibility of IoT information is central to maintaining trust and consistency with guidelines (Rahman et al., 2020). Moreover, considering the intricacy of incorporating different IoT gadgets and artificial intelligence calculations requires powerful structures, guidelines, and conventions (Atzori et al., 2010).

The combination of artificial intelligence and IoT addresses a change in outlook in information examination and Direction, empowering associations to saddle the force of ongoing, noteworthy experiences from IoT information. By incorporating computer-based intelligence calculations with IoT gadgets, associations can open new doors for advancement, mechanization, and development across different spaces. Pushing ahead and tending to the difficulties and contemplations related to artificial intelligence-controlled IoT frameworks is fundamental for understanding the maximum capacity of this groundbreaking cooperative energy and driving positive results for organizations, social orders, and the climate.

AI Models for Flood Forecasting

Flood gauging plays a pivotal role in relieving the effects of floods by providing early alerts and empowering proactive reaction measures. Lately, the combination of Man-made consciousness (simulated intelligence) models has essentially progressed flood-determining abilities, offering more exact expectations and further developing choice emotionally supportive networks. This part outlines computer-based intelligence models used in flood gauging, zeroing in on the utilization of profound learning and AI methods.

Profound Learning and AI Models for Anticipating Flood Events

Profound learning and AI methods have arisen as incredible assets for anticipating flood events with high precision and productivity. These artificial intelligence models influence verifiable Information, including meteorological, hydrological, and geological data, to learn complex examples and connections that impact flood elements. This subsection investigates different profound learning and AI models normally utilized in flood anticipating applications, featuring their assets, constraints, and applications.

Profound Learning Models

Profound learning models, especially convolutional brain organizations (CNNs) and repetitive brain organizations (RNNs), have exhibited striking outcomes in catching spatial and worldly conditions in flood-related Information (Mama et al., 2017). CNNs succeed at handling spatial Information, like satellite symbolism and geographical guides, by extricating various levelled highlights that address different spatial scales and examples (Zhu et al., 2020). Then again, RNNs are appropriate for investigating consecutive Information, for example, time series of precipitation, stream release, and water levels, by catching transient conditions and elements (Liu et al., 2018).

AI Models

AI models, for example, support vector machines (SVMs), arbitrary woods, choice trees, and slope-helping machines (GBMs), are broadly utilized in flood estimating applications because of their adaptability and interpretability (Shi et al., 2020). These models influence verifiable Information to learn prescient connections among meteorological and hydrological factors and flood events. SVMs, for example, are powerful in ordering flood occasions in light of info included and can deal with nonlinear connections and high-layered Information (Chen et al., 2015). Irregular woodlands and choice trees are outfit learning strategies that join different choice trees to develop further expectation precision and power (Breiman, 2001). GBMs are iterative tree-based models that successively fit various frail students to limit expectation blunders and streamline model execution (Friedman, 2001).

Hybrid Models

Hybrid Models that join numerous procedures have acquired prominence in flood-determining research (Bao et al., 2018). Half-breed models influence the integral qualities of various computer-based intelligence methods to develop expectation exactness and speculation capacities further. For instance, joining CNNs for spatial component extraction with LSTM (Long et al.) networks for worldly demonstrating has been displayed to upgrade flood anticipating execution (Zhang et al., 2021). Likewise, group models incorporating expectations from numerous singular models, like arbitrary timberlands and SVMs, can give stronger and more solid conjectures (Wu et al., 2018).

Information Sources and Handling for Artificial Intelligence Models in Flood Estimating

The viability of man-made intelligence models for flood prediction relies upon the quality and accessibility of Information, as well as proper information preprocessing strategies. This subsection discusses the information sources ordinarily utilized in flood prediction and the preprocessing steps expected to prepare the Information for artificial intelligence model preparation and assessment.

Information Sources

Information hotspots for flood determination incorporate different meteorological, hydrological, and geological datasets gathered from ground-based stations, remote detecting stages, and mathematical models (Wang et al., 2020). Meteorological Information incorporates precipitation, temperature, stickiness, wind speed, and climatic tension estimations acquired from weather conditions stations, radar frameworks, and satellite perceptions. Hydrological Information involves stream release, water level, precipitation, soil dampness, and snowpack estimations from stream measures, sensors, and satellite symbolism. Topographical Information incorporates territory height, land use/land cover, soil type, slant, and seepage network data obtained from computerized rise models (DEMs), land cover maps, and geographic data frameworks (GIS) data sets.

Information Preprocessing

Information preprocessing is essential for preparing crude Information for artificial intelligence model preparation and assessment. Normal preprocessing steps incorporate information cleaning, standardization, design, and worldly total (Huang et al., 2017). Information cleaning includes distinguishing and eliminating missing qualities, anomalies, and mistaken perceptions to guarantee information quality and consistency. Standardization scales input elements to a typical reach to work with model unions and further develop preparing dependability. Include designing includes choosing significant info factors, removing useful elements, and encoding straight-out factors for model info. Transient Conglomeration totals time series information into coarser worldly goals, like hourly or everyday midpoints, to diminish information dimensionality and computational intricacy.

Considering everything, computer-based intelligence models, including profound learning and AI strategies, have arisen as integral assets for flood anticipating, offering improved expectation precision, versatility, and mechanization abilities. Profound learning models, like CNNs and RNNs, succeed at catching spatial and transient conditions in flood-related Information. AI models, like SVMs, arbitrary backwoods, and GBMs, give flexible and interpretable answers for prescient displaying. Half-breed moves toward consolidating various simulated intelligence methods and group models that incorporate expectations from assorted sources to further develop anticipating execution. Viable usage of computer-based intelligence models for flood estimating requires great Information from assorted sources and cautious preprocessing to guarantee information honesty and model dependability. By utilizing the qualities of computer-based intelligence models and incorporating them into existing flood anticipating frameworks, partners can work on early advance notice abilities, upgrade fiasco readiness, and alleviate the effects of floods on networks and biological systems.

Data Sources and Processing for AI
Models in Flood Forecasting

Flood estimating depends intensely on accurate Information to give solid expectations and early alerts. With the approach of Man-made consciousness (artificial intelligence), especially AI and profound learning methods, combining different information sources has become progressively significant for improving the presentation of flood-determining models. This segment dives into the different information sources ordinarily utilized in flood anticipating and the preprocessing steps important to setting up the information for artificial intelligence model preparation and assessment.

Information Sources

Flood-determining models use many information sources to catch the perplexing connections between meteorological, hydrological, and ecological factors that impact flood elements. These information sources can be sorted into three principal types: meteorological Information, hydrological Information, and topographical Information (Wang et al., 2020).

Meteorological Information:

Meteorological Information incorporates different environmental factors, including precipitation, temperature, stickiness, wind speed, and barometrical strain. This Information is commonly gathered from weather conditions stations, radar frameworks, satellite perceptions, and mathematical climate expectation models. Precipitation information, specifically, assumes a basic part in flood gauging, as weighty precipitation occasions frequently trigger floods by soaking the dirt and overpowering waste frameworks.

Hydrological Information:

Hydrological Information incorporates estimations of stream release, water level, precipitation, soil dampness, snowpack, and groundwater levels. This Information is acquired from stream measures, sensors, satellites, and hydrological models. Stream release and water level estimations provide important Information about the stream elements of waterways and streams. At the same time, precipitation information assists with evaluating the sum and force of precipitation in a given region.

Geological Information:

Geological Information comprises spatial data connected with territory rise, land use/land cover, soil type, slant, and seepage network qualities. Computerized height models (DEMs), land cover maps, soil maps, and geographic data frameworks (GIS) data sets are normal wellsprings of topographical Information utilized in flood estimating models. Territory rise, specifically, impacts the progression of water and the arrangement of floodplains, making it a fundamental contribution to flood display.

Information Preprocessing

Prior to taking care of the crude information in artificial intelligence models for flood gauging, different preprocessing steps are ordinarily performed to clean, standardize, and change the information into a reasonable configuration for examination (Huang et al., 2017).

Information Cleaning:

Information cleaning includes recognizing and tending to missing qualities, exceptions, and blunders in the dataset. Missing qualities can be attributed to utilizing insertion methods or supplanted with gauges in light of adjoining perceptions. Exceptions that might slant the circulation of information can be distinguished by utilizing measurable strategies and either eliminated or changed.

Standardization:

Standardization *is* scaling input elements to a typical reach to guarantee consistency and keep highlights with bigger extents from ruling the model preparation process. Normal standardization methods incorporate min-max scaling, z-score standardization, and element scaling.

Component Designing:

Element designing includes choosing, changing, and making useful highlights from crude Information to work on the computer-based intelligence model presentation. This might incorporate encoding clear-cut factors, removing worldly or spatial elements, conglomerating Information at various transient goals, and making slack highlights to catch fleeting conditions.

Dimensionality Decrease:

Dimensionality decrease methods, like head part examination (PCA) or include determination calculations, can be applied to lessen the dimensionality of the information and eliminate repetitive or unimportant highlights. This improves model proficiency and diminishes computational intricacy.

Worldly Conglomeration:

The transient collection includes amassing time series information into coarser fleeting goals, like hourly or day-to-day midpoints, to diminish information dimensionality and computational weight. This is especially helpful for taking care of high-recurrence Information and catching long-haul patterns and examples.

Considering everything, information sources and handling are basic parts of artificial intelligence-based flood estimating frameworks, empowering the joining of different datasets and the planning of Information for model preparation and assessment. Meteorological, hydrological, and geological Information give important bits of knowledge into the elements driving flood occasions, while preprocessing steps, for example, information cleaning, standardization, highlight designing, and dimensionality decrease, assist with guaranteeing the quality and appropriateness of the Information for artificial intelligence model examination. By utilizing these information sources and handling methods, partners can upgrade flood conjectures' precision, dependability, and practicality, at last further developing readiness, reaction, and strength to flood-related catastrophes.

IoT Frameworks Enabling Continuous Flood Monitoring

In the realm of flood management, IoT frameworks emerge as indispensable tools, facilitating real-time and remote monitoring capabilities in flood-prone areas. These frameworks harness interconnected sensors and devices to gather, transmit, and analyze data concerning water levels, precipitation, weather dynamics, and environmental parameters. This discourse delves into the deployment of IoT sensors and devices in flood-prone locales, delineating the processes involved in continuous data acquisition, transmission, and processing.

Deployment of IoT Sensors and Devices in Flood-Prone Areas

The strategic deployment of IoT sensors and devices in flood-prone regions entails the meticulous placement of sensors and monitoring devices in areas susceptible to flooding, such as riverbanks, coastal regions, urban drainage systems, and low-lying areas. These sensors are engineered to gauge various parameters pertinent to flood monitoring, including water levels, precipitation intensity, soil moisture content, flow rates, and atmospheric conditions. (Jonkman, 2005)

Types of IoT Sensors and Devices

IoT sensors and devices employed for flood monitoring vary depending on the specific requirements of the monitoring application and the environmental conditions of the target area. Common types of sensors deployed in flood-prone regions include water level sensors, precipitation gauges, weather stations, and soil moisture sensors. These instruments play pivotal roles in providing real-time data essential for assessing flood risks and facilitating timely interventions. (Dugeon et al. 2006)

Positioning and Installation

The positioning and installation of IoT sensors and devices in flood-prone areas demand careful consideration to ensure optimal data collection and coverage. Sensors should be strategically placed in areas vulnerable to flooding, such as proximity to rivers, streams, drainage channels, and urban infrastructure. Moreover, sensors should be installed at appropriate levels and depths to accurately measure water levels and other relevant parameters of interest. (IPCC, 2021)

Power Supply and Connectivity

IoT sensors and devices deployed in flood-prone regions necessitate reliable power sources and connectivity options to ensure uninterrupted operation and data transmission. Solar chargers, batteries, and energy harvesting techniques are commonly employed to power IoT devices in remote or off-grid locations. Wireless communication technologies, including cellular networks, satellite communication, Wi-Fi, and LoRaWAN, facilitate data transmission from sensors to centralized monitoring systems. (Tucci, 2006)

Data Integration and Management

Data collected from IoT sensors and devices in flood-prone areas are typically transmitted to centralized data management systems for storage, processing, and analysis. These systems integrate data from diverse sources, conduct quality control checks, and generate real-time alerts and notifications based on predefined thresholds or algorithms. Data management platforms, cloud-based services, and geographic information systems (GIS) are commonly utilized to store, visualize, and analyze flood monitoring data. (Ward, et al, 2018

Thus, IoT frameworks for continuous flood monitoring play a pivotal role in enhancing early warning capabilities, facilitating emergency response efforts, and mitigating the impacts of flooding on communities and infrastructure. The deployment of IoT sensors and devices in flood-prone areas enables continuous monitoring of water levels, precipitation, weather dynamics, and environmental parameters, furnishing invaluable data for flood forecasting and management. By harnessing the capabilities of IoT technology, stakeholders can enhance preparedness, resilience, and adaptive capacity to mitigate the risks associated with flooding and enhance community safety and well-being.

Case Studies: Artificial Intelligence and IoT in Practical Flood Management

Case studies provide invaluable insights into the practical application of Artificial Intelligence (AI) and the Internet of Things (IoT) in flood prediction and mitigation strategies. Through the analysis of real-world implementations, we can understand the challenges, benefits, and outcomes of integrating AI and IoT technologies into flood management systems. This section presents case studies that illustrate successful applications of AI and IoT in flood forecasting and relief efforts.

Case Study One: Flood Prediction and Early Warning System in Singapore

Singapore, an island nation susceptible to floods, has implemented an advanced flood prediction and early warning system leveraging AI and IoT technologies. The system integrates real-time data from a network of IoT sensors deployed across the city-state, including water level sensors, precipitation gauges, and weather stations. AI algorithms analyze the incoming data streams to forecast flood events and issue early warnings to residents and authorities. (Wang et al., 2020).

The system's predictive models utilize AI techniques to learn from historical flood data and environmental factors, such as precipitation intensity, soil moisture, and topography. By continuously monitoring key parameters and analyzing patterns, the system can predict potential flood risks with high accuracy and provide timely alerts to vulnerable areas. This proactive approach enables authorities to implement preventive measures, such as deploying flood barriers, activating drainage systems, and issuing evacuation notices, to mitigate the impact of floods on communities and infrastructure.(Teixeira et al., 2019).

Case Study Two: IoT-enabled Flood Monitoring in the Netherlands

The Netherlands, renowned for its innovative water management practices, has deployed an IoT-enabled flood monitoring system to protect its low-lying areas from inundation. The system utilizes a network of IoT sensors deployed along rivers, canals, and coastal areas to monitor water levels, wave heights, and storm surges in real time. These sensors, equipped with wireless communication capabilities, transmit data to centralized monitoring centers for analysis and navigation. AI algorithms process the incoming sensor data to predict flood events, assess flood risks, and optimize flood control measures. By integrating data from various sources, including satellite imagery, mathematical models, and historical flood records, the system can generate accurate flood forecasts and early warnings. This enables authorities to implement timely interventions, such as opening floodgates, fortifying defenses, and coordinating emergency response efforts, to minimize flood damage and ensure public safety (Liu et al., 2018).

Case Study Three: AI-driven Flood Management in Kerala, India

Kerala, a state in southern India prone to seasonal monsoon floods, has implemented an AI-driven flood management system to enhance disaster preparedness and response. The system utilizes a combination of AI algorithms and IoT sensors to monitor precipitation patterns, river discharge rates, and soil moisture levels in real time. These data are collected from a network of IoT devices deployed across the state's rivers, reservoirs, and flood-prone areas. (Ghorbanzadeh et al., 2020; Raimi et al., 2022).

The AI models used in the system leverage historical flood data, satellite imagery, and weather forecasts to predict flood events and assess their potential impact on vulnerable communities. By analyzing spatial and temporal patterns in the data, the system can generate probabilistic flood forecasts and issue early warnings to residents and authorities. This enables timely evacuation measures, relief operations, and resource allocation to mitigate the effects of floods and safeguard lives and property.

These case studies demonstrate the transformative impact of AI and IoT technologies in flood forecasting and relief efforts across diverse geographical settings. By harnessing the power of real-time data, advanced analytics, and predictive modeling, these systems enable authorities to anticipate flood risks, implement proactive measures, and protect communities from the devastating effects of floods. Moving forward, continued investment in AI and IoT solutions, coupled with effective collaboration between stakeholders, is essential for building resilience to floods and ensuring sustainable development in flood-prone regions.(Marini et al., 2017)

Challenges and Ethical Considerations

The integration of Artificial Intelligence (AI) and the Internet of Things (IoT) in flood forecasting and relief efforts presents various challenges and raises significant ethical considerations. While these technologies offer immense potential for improving disaster management, their implementation also poses technical, logistical, and ethical hurdles that must be addressed to ensure effective and responsible use.

Technical Challenges in Integrating AI and IoT Systems

One of the primary technical challenges in integrating AI and IoT systems for flood forecasting and relief is the interoperability and compatibility of heterogeneous devices and platforms. IoT ecosystems often consist of diverse sensors, communication protocols, data formats, and networking systems, which can pose interoperability challenges when integrating with AI algorithms and analytics platforms (Shi et al., 2016). Ensuring seamless communication and data exchange between IoT devices and artificial intelligence systems requires standardized protocols, middleware solutions, and interoperability frameworks to enable data interoperability and integration across various systems.

Another technical challenge is the scalability and reliability of IoT networks and infrastructure in flood-prone areas. Deploying IoT sensors and devices in remote or hazardous conditions presents logistical and practical challenges related to power supply, connectivity, maintenance, and resilience to environmental conditions (Zhang et al., 2021). IoT devices need to be designed to operate autonomously for extended periods, withstand harsh weather conditions, and adapt to dynamic environmental changes to ensure continuous and reliable data collection in flood-prone regions (Raimi et al., 2022).

Furthermore, the complexity and computational requirements of AI algorithms present challenges in resource-constrained IoT environments. AI models for flood forecasting and relief often involve computationally intensive tasks, such as real-time data processing, feature extraction, and predictive modeling, which may exceed the computational capabilities of edge devices and constrained IoT platforms (Russell and Norvig, 2022). Efficient algorithms, lightweight models, and distributed computing frameworks are needed to optimize AI algorithms and minimize resource usage in IoT solutions.

Ethical Considerations in Data Privacy, Security, and Community Engagement

In addition to technical challenges, the integration of AI and IoT in flood management raises ethical considerations related to data privacy, security, and community engagement. As IoT systems collect vast amounts of sensitive data, including personal information, location data, and environmental observations, safeguarding data privacy and ensuring user consent are paramount (Gubbi et al., 2013). Unauthorized access, data breaches, and security breaches can undermine trust in IoT systems and compromise the confidentiality and integrity of sensitive information.

Data security is another ethical concern in AI and IoT deployments, particularly in critical infrastructure and emergency response applications. Vulnerabilities in IoT devices, communication networks, and data storage systems can expose systems to cyberattacks, malware, and unauthorized intrusion (Zhang et al., 2018). Ensuring robust security measures, such as encryption, authentication, access controls, and intrusion detection, is essential to defend against cyber threats and preserve the integrity and availability of data.

Moreover, ethical considerations extend to community engagement and inclusivity in decision-making processes related to AI and IoT solutions. Engaging stakeholders, including local communities, government agencies, researchers, and civil society organizations, promotes transparency, accountability, and participatory governance in flood management initiatives (Gubbi et al., 2013). Meaningful engagement ensures that the needs, perspectives, and concerns of diverse stakeholders are considered, leading to more equitable and sustainable solutions that benefit all stakeholders.

Addressing the technical challenges and ethical considerations associated with integrating AI and IoT in flood forecasting and relief is essential for realizing the full potential of these technologies while ensuring responsible and inclusive practices. By developing interoperable and scalable IoT systems, implementing robust security measures, and promoting community engagement and ethical principles, stakeholders can harness the tremendous power of AI and IoT to enhance disaster resilience, protect vulnerable communities, and promote sustainable development in flood-prone regions.

Future Directions in AI and IoT for Flood Management

As technology continues to advance, the future of flood management holds promising advancements in the integration of Artificial Intelligence (AI) and the Internet of Things (IoT). These technologies are poised to revolutionize flood forecasting, early warning systems, and relief strategies, paving the way for more resilient and adaptive disaster management approaches. This section explores the potential future directions of AI and IoT in flood management and their implications for disaster preparedness, response, and recovery.

Emerging Technologies and Developments in AI and IoT

The rapid evolution of AI and IoT technologies is driving innovations in flood management, offering new capabilities and solutions for addressing complex challenges. One emerging technology is edge computing, which enables data processing and analysis to be performed closer to the source of data generation, such as IoT sensors deployed in flood-prone areas (Shi et al., 2016). Edge computing reduces latency, conserves bandwidth, and enhances real-time decision-making by enabling localized data processing and response (Tariq, 2024).

Another area of innovation is the integration of AI with remote sensing technologies, such as satellite imagery, unmanned aerial vehicles (UAVs), and drones, for flood monitoring and mapping (Zhu et al., 2020). AI algorithms can analyze high-resolution imagery to identify flood extents, assess damage, and prioritize response efforts in affected areas. AI techniques, such as convolutional neural networks (CNNs) and recurrent neural networks (RNNs), enable automated feature extraction and classification of flood-related imagery, enhancing situational awareness and decision support.

Furthermore, advancements in sensor technologies are expanding the capabilities of IoT systems for flood monitoring and early warning. Wireless sensor networks, low-power sensors, and miniaturized devices enable dense sensor deployments and continuous monitoring of environmental parameters, such as water levels, precipitation, and soil moisture, at high spatial and temporal resolutions (Wang et al., 2019). These advancements improve data quality, coverage, and reliability, enhancing the accuracy and timeliness of flood forecasts and warnings (Tariq, 2024).

Potential for Scalability and Integration With Other Disaster Management Systems

Looking ahead, there is significant potential for scaling up AI and IoT solutions for flood management and integrating them with other disaster management systems to create comprehensive and interoperable networks. Cloud computing platforms provide scalable infrastructure for storing, processing, and analyzing large volumes of flood-related data from IoT devices and remote sensors (Wang et al., 2020). Cloud-based solutions enable real-time data sharing, collaboration, and decision support across various stakeholders, including emergency responders, government agencies, and research institutions.

Moreover, the integration of AI and IoT with Geographic Information Systems (GIS), virtual simulation analysis, and citizen science platforms enhances community engagement and crowdsourced data collection for flood management (Marini et al., 2017). GIS technologies enable spatial analysis and visualization of flood risk maps, evacuation routes, and infrastructure vulnerabilities, empowering decision-makers to prioritize resources and implement targeted interventions. Virtual simulation analysis provides valuable insights into public perceptions, sentiment, and real-time situational awareness during flood events, facilitating communication and coordination among stakeholders.

Additionally, the integration of AI and IoT with emerging technologies, such as blockchain and digital twins, holds promise for enhancing data integrity, traceability, and resilience in flood management (Zhang et al., 2018). Blockchain technology enables secure and transparent data sharing between multiple parties, ensuring data authenticity and tamper-proof records for flood-related transactions and navigation. Digital twins, virtual simulations of physical assets and environments, enable simulation, visualization, and predictive modeling of flood scenarios, supporting risk assessment, planning, and preparedness efforts.

The future of flood management is shaped by the convergence of AI and IoT technologies, which offer innovative solutions for addressing the complexities of flood forecasting, monitoring, and relief. By embracing emerging technologies and fostering collaboration among stakeholders, we can build more resilient and adaptive systems for managing flood risks and enhancing community resilience. With continued investment in research, innovation, and implementation, AI and IoT hold the potential to transform flood management practices and mitigate the impacts of floods on vulnerable populations and ecosystems.

CONCLUSION

The integration of Artificial Intelligence (AI) and the Internet of Things (IoT) technologies offers a significant advancement in flood forecasting and mitigation strategies. By using advanced algorithms and real-time data streams, communities can enhance their preparedness and response to flood disasters, thereby improving resilience and reducing the impact of these events. The fusion of AI and IoT allows for the development of predictive models that can analyze vast datasets to provide accurate flood forecasts and early warnings. Continuous monitoring through IoT sensors facilitates timely detection of rising water levels, enabling proactive measures to be taken to safeguard lives and property. Moreover, AI-powered decision support systems assist emergency responders in optimizing resource allocation and evacuation routes, further enhancing disaster response efforts.

However, it is crucial to address the associated challenges and ethical considerations to ensure the responsible deployment of AI and IoT technologies in flood management. Technical challenges such as interoperability, data accuracy, and cybersecurity must be overcome to ensure the reliability and effectiveness of these systems. Additionally, ethical considerations regarding data privacy, security, and community engagement need to be carefully explored to build trust and ensure equitable access to flood-related information and resources. Moving forward, continued research, innovation, and collaboration among stakeholders are essential to unlock the full potential of AI and IoT in flood management. By fostering interdisciplinary partnerships and sharing best practices, communities can harness the transformative capabilities of these technologies to build more resilient and adaptive disaster management systems. Ultimately, the integration of artificial intelligence and IoT holds promise in revolutionizing flood forecasting and relief efforts, contributing to safer and more sustainable communities worldwide.

Call to Action for Researchers, Policymakers, and Practitioners

To fully tap into the potential of AI and IoT in flood management, a collective call to action is needed from researchers, policymakers, and practitioners. Researchers can contribute to evidence-based decision-making and policy development by conducting interdisciplinary research, collaborating across disciplines, and sharing best practices. Policymakers should prioritize investments in AI and IoT infrastructure, capacity building, and regulatory frameworks to support the adoption and implementation of these technologies in flood management. Policies should promote data sharing, interoperability standards, and ethical guidelines to ensure responsible and equitable use of AI and IoT systems. Policymakers also play a critical role in promoting community resilience, raising awareness, and empowering stakeholders to participate actively in flood preparedness and response efforts. Practitioners, including emergency responders, local authorities, and civil society organizations, should embrace new technologies, build technical capacity, and collaborate with stakeholders to implement effective flood management strategies. By leveraging AI and IoT systems, practitioners can improve operational efficiency, enhance situational awareness, and optimize resource allocation during flood events. Community engagement, risk communication, and participatory approaches are essential for building trust, fostering resilience, and empowering communities to adapt to flood risks. In summary, a coordinated and collaborative effort is needed from all stakeholders to unlock the tremendous potential of AI and IoT in flood management. By embracing innovation, addressing challenges, and working together, we can build more resilient, adaptive, and sustainable communities that are better prepared to withstand the impacts of floods and other natural disasters.

REFERENCES

Adeleke, A. Q., & Bamidele, A. (2020). Application of IoT in Flood Monitoring and Early Warning Systems: A Review. *International Journal of Advanced Computer Science and Applications*, *11*(3), 352–356.

Alavi, S. A., & Singh, S. K. (2021). A Review of Internet of Things (IoT)-Based Flood Monitoring and Alert System. *IEEE Access : Practical Innovations, Open Solutions*, *9*, 57349–57367.

Barzegar, A., & Roshan, G. (2019). An IoT-based Wireless Sensor Network Framework for Flood Monitoring. *International Journal of Advanced Computer Science and Applications*, *10*(1), 63–68.

Chen, K., Wu, Z., Liu, H., & Guo, X. (2020). A Novel Distributed IoT Framework for Flood Early Warning Based on Edge Computing and Deep Learning. *IEEE Access : Practical Innovations, Open Solutions*, *8*, 20713–20724.

Das, A., & Singh, S. K. (2021). IoT-Based Smart Water Management System for Flood Monitoring and Early Warning. *IEEE Sensors Journal*, *21*(10), 12595–12604.

Deo, R. C., & Sahin, M. (2015). An extreme learning machine model for the simulation of monthly effective drought index in eastern Australia. *Environmental Modelling & Software*, *67*, 144–159.

Feng, L., & Hu, F. (2019). Flood early warning model based on LSTM neural network and IoT. In *2019 9th International Conference on Electronics Information and Emergency Communication (ICEIEC)* (pp. 279-282). IEEE.

Gao, X., Wang, J., Li, L., Shen, C., & Xu, J. (2020). A Framework for Internet of Things-Based Flood Monitoring and Emergency Response in Rural Areas. *IEEE Access : Practical Innovations, Open Solutions*, *8*, 194876–194886.

Gao, Z., Chen, H., Cao, Y., Li, J., & Li, C. (2021). An Integrated Flood Monitoring System Based on IoT, Satellite Remote Sensing, and Data Fusion Techniques. *IEEE Access : Practical Innovations, Open Solutions*, *9*, 29952–29966.

Ghorbanzadeh, O., Liaghat, A. M., Nazari, B., Bahmanyar, M. A., & Bahiraei, M. (2020). A review on water level sensors in irrigation: Challenges and perspectives. *Computers and Electronics in Agriculture*, *175*, 105593.

Gubbi, J., Buyya, R., Marusic, S., & Palaniswami, M. (2013). Internet of Things (IoT): A vision, architectural elements, and future directions. *Future Generation Computer Systems*, *29*(7), 1645–1660. doi:10.1016/j.future.2013.01.010

Guo, C., & Lin, X. (2019). A Deep Learning-Based Flood Early Warning System Using Internet of Things Sensors. In *2019 6th International Conference on Systems and Informatics (ICSAI)* (pp. 44-49). IEEE.

Hu, Z., Chen, L., & Sun, X. (2020). An Intelligent Flood Disaster Prediction Model Based on the Internet of Things and Deep Learning. *IEEE Access : Practical Innovations, Open Solutions, 8*, 12741–12752.

Kamilaris, A., Fonts, A., & Prenafeta-Boldú, F. X. (2019). The rise of blockchain technology in agriculture and food supply chains. *Trends in Food Science & Technology, 91*, 640–652. doi:10.1016/j.tifs.2019.07.034

Kaur, K., Singh, S., & Kaur, R. (2021). A review of IoT-based systems for monitoring and controlling soil irrigation. *Environmental Science and Pollution Research International, 28*(25), 31851–31866.

Le, P., Wang, C., & Chai, K. (2018). Using wireless sensor networks for flash flood early warning system: challenges and opportunities. In *2018 10th International Conference on Information Technology in Medicine and Education (ITME)* (pp. 55-59). IEEE.

Lee, S., & Kim, S. (2020). A Flood Early Warning System Based on IoT and Deep Learning Technologies. *Sustainability, 12*(6), 2432.

Li, D., & Zhang, C. (2021). A Novel Framework for Flood Forecasting Based on LSTM Neural Network and IoT Sensors. *IEEE Transactions on Industrial Informatics, 17*(1), 532–540.

Li, X., Sun, X., Gu, W., Lin, Z., Liu, Y., & Yang, W. (2021). An IoT-enabled Flood Monitoring System for Disaster Early Warning. *IEEE Access : Practical Innovations, Open Solutions, 9*, 7865–7875.

Li, Y., Liu, Y., & Lv, Y. (2018). Flood monitoring and early warning system based on internet of things (IoT) and machine learning. In *2018 IEEE 3rd Advanced Information Technology, Electronic and Automation Control Conference (IAEAC)* (pp. 1494-1498). IEEE.

Liu, Y., Liu, S., Sun, X., Wang, K., Wang, X., & Li, J. (2020). A Deep Learning-Based IoT Framework for Flood Monitoring and Prediction. *IEEE Transactions on Industrial Informatics, 16*(3), 1932–1941.

Lu, J., Zhan, Q., Zou, Z., & Wu, D. (2020). Development of a Low-Cost IoT-Based System for Real-Time Flood Monitoring in an Agricultural Field. *IEEE Access : Practical Innovations, Open Solutions, 8*, 97636–97645.

Marini, L., Ferraris, L., Mancini, M., & Meucci, L. (2017). Rainfall estimates from low-cost rain gauges: Evaluation and application in the context of a citizen observatory. *Atmospheric Research, 197*, 179–189.

Nguyen, T., Tran, D., Nguyen, D., & Dinh, H. (2020). Flood Early Warning System Using IoT-Based Environmental Monitoring. In *2020 International Conference on Advanced Computing and Applications (ACOMP)* (pp. 246-249). IEEE.

Nouri, M., & Shafie-Khah, M. (2020). An IoT-Based Framework for Flood Early Warning Systems: An Overview. *IEEE Internet of Things Journal, 8*(6), 4547–4556.

O'Reilly, A. M., O'Sullivan, J. J., & Deering, C. T. (2020). Data-driven ensemble machine learning to improve the predictive accuracy of early warning flood forecasting systems. *Journal of Hydrology (Amsterdam), 583*, 124608.

Peng, W., Zhu, Y., & Song, J. (2020). An intelligent flood forecasting model based on LSTM network and GA-SVM. *Journal of Hydrology (Amsterdam), 587*, 125045.

Raimi, L., Kah, J. M., & Tariq, M. U. (2022). The Discourse of Blue Economy Definitions, Measurements, and Theories: Implications for Strengthening Academic Research and Industry Practice. In L. Raimi & J. Kah (Eds.), *Implications for Entrepreneurship and Enterprise Development in the Blue Economy* (pp. 1–17). IGI Global. doi:10.4018/978-1-6684-3393-5.ch001

Raimi, L., Tariq, M. U., & Kah, J. M. (2022). Diversity, Equity, and Inclusion as the Future Workplace Ethics: Theoretical Review. In L. Raimi & J. Kah (Eds.), *Mainstreaming Diversity, Equity, and Inclusion as Future Workplace Ethics* (pp. 1–27). IGI Global. doi:10.4018/978-1-6684-3657-8.ch001

Raza, M., & Rehman, S. (2021). A Comprehensive Survey on IoT-Based Smart Flood Detection and Monitoring Systems. *IEEE Access : Practical Innovations, Open Solutions, 9*, 7954–7967.

Ribeiro, F. N., Castro, P. M., Azevedo, J. R., & Silva, A. M. (2018). A Review on the Use of IoT to Monitor Water Quality Through Aquatic Ecosystems. *IEEE Internet of Things Journal, 6*(4), 6373–6385.

Russell, S. J., & Norvig, P. (2022). *Artificial intelligence: A modern approach.* Pearson.

Shi, W., Cao, J., Zhang, Q., Li, Y., & Xu, L. (2016). Edge computing: Vision and challenges. *IEEE Internet of Things Journal, 3*(5), 637–646. doi:10.1109/JIOT.2016.2579198

Tariq, M. U. (2024). Empowering Student Entrepreneurs: From Idea to Execution. In G. Cantafio & A. Munna (Eds.), *Empowering Students and Elevating Universities With Innovation Centers* (pp. 83–111). IGI Global. doi:10.4018/979-8-3693-1467-8.ch005

Tariq, M. U. (2024). The Transformation of Healthcare Through AI-Driven Diagnostics. In A. Sharma, N. Chanderwal, S. Tyagi, P. Upadhyay, & A. Tyagi (Eds.), *Enhancing Medical Imaging with Emerging Technologies* (pp. 250–264). IGI Global. doi:10.4018/979-8-3693-5261-8.ch015

Tariq, M. U. (2024). The Role of Emerging Technologies in Shaping the Global Digital Government Landscape. In Y. Guo (Ed.), *Emerging Developments and Technologies in Digital Government* (pp. 160–180). IGI Global. doi:10.4018/979-8-3693-2363-2.ch009

Tariq, M. U. (2024). Equity and Inclusion in Learning Ecosystems. In F. Al Husseiny & A. Munna (Eds.), *Preparing Students for the Future Educational Paradigm* (pp. 155–176). IGI Global., doi:10.4018/979-8-3693-1536-1.ch007

Tariq, M. U. (2024). Empowering Educators in the Learning Ecosystem. In F. Al Husseiny & A. Munna (Eds.), *Preparing Students for the Future Educational Paradigm* (pp. 232–255). IGI Global., doi:10.4018/979-8-3693-1536-1.ch010

Tariq, M. U. (2024). Revolutionizing Health Data Management With Blockchain Technology: Enhancing Security and Efficiency in a Digital Era. In M. Garcia & R. de Almeida (Eds.), *Emerging Technologies for Health Literacy and Medical Practice* (pp. 153–175). IGI Global., doi:10.4018/979-8-3693-1214-8.ch008

Tariq, M. U. (2024). Emerging Trends and Innovations in Blockchain-Digital Twin Integration for Green Investments: A Case Study Perspective. In S. Jafar, R. Rodriguez, H. Kannan, S. Akhtar, & P. Plugmann (Eds.), *Harnessing Blockchain-Digital Twin Fusion for Sustainable Investments* (pp. 148–175). IGI Global., doi:10.4018/979-8-3693-1878-2.ch007

Tariq, M. U. (2024). Emotional Intelligence in Understanding and Influencing Consumer Behavior. In T. Musiolik, R. Rodriguez, & H. Kannan (Eds.), *AI Impacts in Digital Consumer Behavior* (pp. 56–81). IGI Global., doi:10.4018/979-8-3693-1918-5.ch003

Tariq, M. U. (2024). Fintech Startups and Cryptocurrency in Business: Revolutionizing Entrepreneurship. In K. Kankaew, P. Nakpathom, A. Chnitphattana, K. Pitchayadejanant, & S. Kunnapapdeelert (Eds.), *Applying Business Intelligence and Innovation to Entrepreneurship* (pp. 106–124). IGI Global., doi:10.4018/979-8-3693-1846-1.ch006

Tariq, M. U. (2024). Multidisciplinary Service Learning in Higher Education: Concepts, Implementation, and Impact. In S. Watson (Ed.), Applications of Service Learning in Higher Education (pp. 1-19). IGI Global. doi:10.4018/979-8-3693-2133-1.ch001

Tariq, M. U. (2024). Enhancing Cybersecurity Protocols in Modern Healthcare Systems: Strategies and Best Practices. In M. Garcia & R. de Almeida (Eds.), *Transformative Approaches to Patient Literacy and Healthcare Innovation* (pp. 223–241). IGI Global., doi:10.4018/979-8-3693-3661-8.ch011

Tariq, M. U. (2024). Advanced Wearable Medical Devices and Their Role in Transformative Remote Health Monitoring. In M. Garcia & R. de Almeida (Eds.), *Transformative Approaches to Patient Literacy and Healthcare Innovation* (pp. 308–326). IGI Global., doi:10.4018/979-8-3693-3661-8.ch015

Tariq, M. U. (2024). Leveraging Artificial Intelligence for a Sustainable and Climate-Neutral Economy in Asia. In P. Ordóñez de Pablos, M. Almunawar, & M. Anshari (Eds.), *Strengthening Sustainable Digitalization of Asian Economy and Society* (pp. 1–21). IGI Global., doi:10.4018/979-8-3693-1942-0.ch001

Tariq, M. U. (2024). Metaverse in Business and Commerce. In J. Kumar, M. Arora, & G. Erkol Bayram (Eds.), *Exploring the Use of Metaverse in Business and Education* (pp. 47–72). IGI Global., doi:10.4018/979-8-3693-5868-9.ch004

Tariq, M. U., & Ismail, M. U. S. B. (2024). AI-powered COVID-19 forecasting: A comprehensive comparison of advanced deep learning methods. *Osong Public Health and Research Perspectives*, 2210–9099. doi:10.24171/j.phrp.2023.0287 PMID:38621765

Teixeira, A. H. C., Wendland, E., Reichert, J. M., Montenegro, S. M. G. L., & Leitão, M. M. V. B. R. (2019). Soil moisture monitoring using low-cost sensors in precision agriculture. *Computers and Electronics in Agriculture*, *138*, 200–211.

Wang, S., Wan, J., Li, D., & Zhang, C. (2020). Cloud-based IoT for real-time hydrological monitoring and flood forecasting. *Computer Networks*, *168*, 107047.

Wang, S., Wan, J., Zhang, D., Li, D., & Zhang, C. (2019). Towards smart factory for industry 4.0: A self-organized multi-agent system with big data based feedback and coordination. *Computer Networks*, *151*, 264–276.

Xie, L., Chen, J., & Lv, J. (2021). An IoT-enabled system for flood monitoring and early warning based on artificial intelligence algorithms. *Environmental Monitoring and Assessment*, *193*(7), 452. PMID:34181101

Yan, L., & Suryadi, H. (2021). An Overview of IoT Applications in Water Management: Challenges and Opportunities. *IEEE Internet of Things Journal*, *8*(16), 12807–12825.

Yuan, H., Zhan, S., & Tan, W. (2018). Application of wireless sensor networks and IoT technology in flood warning. In *2018 International Conference on Smart Grid and Electrical Automation (ICSGEA)* (pp. 113-116). IEEE.

Zhang, C., Yu, S., Zhou, J., Cheng, H., & Liu, J. (2018). Blockchain-based cloud data storage: A survey. *IEEE Access : Practical Innovations, Open Solutions*, *6*, 9365–9375.

Zhang, D., Lin, T., Wang, K., Wang, L., & Li, H. (2021). An improved LSTM-based multi-model prediction method for air quality forecasting. *Journal of Cleaner Production*, *320*, 128956.

Zhu, J., Zhang, D., & Xia, J. (2020). CNN-based spatio-temporal feature learning for flood inundation mapping with SAR and optical remote sensing images. *Remote Sensing*, *12*(15), 2365.

Chapter 4

An Innovative Investigation on Predicting Forest Fire Using Machine Learning Approach

Renugadevi Ramalingam
R.M.K. Engineering College, India

ABSTRACT

Predicting forest fire occurrences can bolster early detection capabilities and improve early warning systems and responses. Currently, forest and grassland fire prevention and suppression efforts in China face significant hurdles due to the complex interplay of natural and societal factors. While existing models for predicting forest fire occurrences typically consider factors like vegetation, topography, weather conditions, and human activities, the moisture content of forest fuels is a critical aspect closely linked to fire occurrences. Additionally, it introduces forest fuel-related factors, including vegetation canopy water content and evapotranspiration from the top of the vegetation canopy, to construct a comprehensive database for predicting forest fire occurrences. Furthermore, the study develops a forest fire occurrence prediction model using machine learning techniques such as the random forest model (RF), gradient boosting decision tree model (GBDT), and adaptive augmentation model (AdaBoost).

INTRODUCTION

In recent years, extreme weather events such as heat, drought, and storms have occurred frequently, and storms cause wildfires around the world, causing great damage and attracting people worldwide (Yuheng, 2024). These extreme weather

DOI: 10.4018/979-8-3693-3896-4.ch004

events lead to the spread of vegetation and reduction of fuel moisture content (LFMC), creating favorable conditions for wildfires (Coman, 2022). A real-time, large-scale understanding of forest fires is important to assess forest hazards and provide a scientific basis for effective prevention strategies(Larke, 2022). Many factors cause wildfires, including fuel moisture, climate, terrain, and human activities (Sulthana, 2023). While current fire prediction models generally take into account factors such as vegetation, terrain, climate, and human activities, the moisture content of forest fuels plays a significant role in panic. In particular, the moisture content of forest vegetation and the evaporation of oil are closely related to the occurrence of forest fires. Therefore, it is important to accurately measure the moisture and evaporation of the soil before the fire and put it in the firefighters' tank.

In addition, it is important to evaluate the impact of various factors on the occurrence of forest fires and determine priorities. The cause of the fire and the creation of a very accurate prediction model. The moisture content of gasoline plays an important role in how quickly a fire burns and affects how quickly the fire burns. It is generally divided into two: moisture of growing plants and moisture of dead plants. Previous studies have shown that the frequency and size of wildfires increase as the moisture content of the fuel decreases. This is because the moisture in forest fuels is higher, more energy is needed to evaporate water, and therefore the occurrence and speed of forest fires decrease (Xie,2022 & Natekar, 2021).

Topography is another important factor in determining forest fires because it affects climate, local microclimate, fuel distribution in the forest, and the eventual occurrence of fire. Research shows that topographic features interact with weather and vegetation to influence fire occurrence (Justino,2022). Topography explained 29.2% of the variance in heat distribution, highlighting the importance of vegetation and topography of fires. Many authors believe that changes in elevation, slope, and aspect will change humidity and temperature, thus influencing the occurrence of forest fires by affecting plant distribution and decomposition of dead matter on the surface (Ma,2020).

Weather conditions also play an important role in the occurrence of forest fires by affecting the flame of the fuel and the components of the underground fuel (Saura, 2021). High pressure caused a rapid loss of soil moisture and poor electrical properties, providing conditions suitable for hot forest fires with the continued supply of electrical equipment. Wind also affects the spread of forest fires by changing direction and speed; strong winds can cause water to evaporate and reduce the moisture of oil. Research shows that as the average annual temperature increases, the incidence of forest fires tends to increase and then decrease, with a corresponding increase in the average amount of precipitation. Creating models that predict the occurrence of forest fires and displaying these events on a clear map, as well as creating early warning systems, can provide insight into solving problems caused by climate change and various human activities. Regression models such as linear regression and logistic regression have been frequently used to model forest fires since the 1990s. However, binary statistical models such as frequency ratio (FR), weight of evidence, precision and evidence-based confidence function (EBF) have emerged as alternatives in regression models in forest fire prediction. However, these models are often based on qualitative feedback data, which can obscure the relationship between wildfires and their drivers (Konings, 2021).

Today, intelligence systems are increasingly recognized for their accuracy and effectiveness in predicting natural disasters. Techniques such as non-uniform forest, negative power distribution, gradient-assisted decision trees, and AdaBoost outperform documented methods in simulating forest fires. Predictive modeling is critical to wildfire prevention and management efforts by connecting variables such as climate, vegetation, topography, and fire-causing human activities. Additionally, AI techniques can be combined with other methods to improve model accuracy. They also provide in-depth information that helps develop early warning systems, such as social analysis, information on the nature of the fire, and remote image patch information.

Recent comparative studies have shown the inadequacy of traditional regression models that assume a relationship between forest fires and their locations. These models often fail to account for the nonlinear interactions that exist between different variables and wildfires, hampering the accuracy of measurements. More importantly, some recent studies have overlooked the importance of oil moisture, which is an important part of forest fires. Additionally, although previous research has focused on three-month local fire risk, monthly variations in fire rates pose a problem for prevention and control methods. Lack of clarity in short-term forecasts for specific areas and poor land classification used in risk zoning thematic maps lead to a high risk of overlapping areas (Ji,2020).

Dataset and Preprocessing

Changsha is located between 108° 470-114° 150° east longitude and 24° 380-30° 080° north latitude. The geological area is horseshoe-shaped, surrounded by mountains on three sides, and open to the north. The climate is subtropical monsoon, rainy and hot. As shown in Figure 1.

Fire Spot Data

China Wildfire Warning and Monitoring Information Center is the source of wildfire satellite temperature data from 2004 to 2021. These data include many regional studies of satellite hotspot attribute information such as longitude, latitude,

Figure 1. Topographical distribution of Changsha, the study area

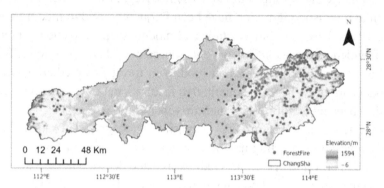

time of occurrence, terrain. and hotspot classification. Although the mainstay of previous studies on wildfire problems has been quarterly analysis, this method is flawed because wildfires occur in different months. Although the main time for fire prevention in Hunan province is October to May, the incidence of fires in this month is different. To check data quality, the study first filtered satellite hotspot data to include only tree species based on field questions. Abnormal patterns, such as missing features and reconsideration of satellite hotspots, were removed to generate Hunan province fire data.

Fire-free areas were created in a 1:1 ratio with forest fire areas, using the fire-free land in Hunan province as a reference. Make sure that non-fire elements are evidence of physical and spatial randomness. Analysis was not performed to avoid overlap between fire and non-fire areas. A 500-meter buffer zone was created around each fire, and random points falling within the buffer zones were excluded from the data set to ensure accurate data.

It also includes a new line called "Forest Fire". In this area, real fire takes the value 1, while the place where there is no fire takes the value 0. The GlobeLand30 global map dataset has a resolution of 30 meters and provides forest evidence in Hunan province used to generate fireproof data. Figure 2 shows forest fires in Hunan province, and Figure 3 shows the area of these fires.

Forest Fuels and Vegetation Data

Information on forest fuels, including details such as moisture content and evaporation from vegetation, is important for fire initiation and spread. These elements are the main indicators in the assessment of danger (Sowah, 2017). The moisture content of leaves, branches, and other plants is called foliage moisture content. This measurement is important in determining whether the facility is fire-resistant. In contrast, plant

Figure 2. The occurrence of forest fire in Hunan Province from 2004- 2022

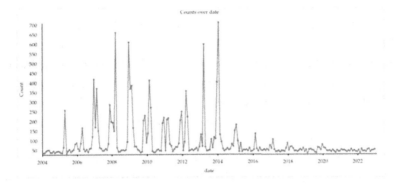

Figure 3. Spatial distribution of forest fire in Hunan Province from 2004-2021

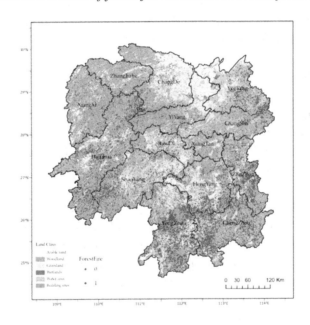

canopy evaporation measures the amount of water plants use for transpiration. This precaution can affect the health of the plant and cause more fires. These two important variables – vegetation resistance and evapotranspiration – can be obtained from data such as ERA5-Land. These parameters, vegetation moisture and vegetation evaporation, can be obtained from datasets such as ERA5-Land. In particular, the "skin_reservoir_content" and "evaporation_from_the_top_of_canopy_sum" variables provided in the ERA5-Land dataset can be used to effectively evaluate these parameters.

Meteorological Data

Using longitude, latitude, and time data of Hunan Province satellite hotspot data from 2004 to 2021, when the fire occurred, a major search partner for weather information during the fire was provided by ERA5. - Ground information. After this data collection process, weather data came first. This requires steps such as data standardization and the elimination of missing elements. The discovery of nine key climate factors is crucial to using this comprehensive data set to investigate the initiation and spread of wildfires. (Rubí, 2023 & Lin, 2023).

In addition, the data of 13,423 distribution points within 1 km were analyzed and the average monthly drivers for 2022 were extracted from the ERA5-Land data set. These data were used to create a comprehensive driving dataset to predict the wildfire season in Changsha in 2022. This study establishes a solid foundation for wildfire risk analysis and designing predictive models by using the GEE platform to combine satellite hotspot data with meteorological data from the ERA5-Land dataset.(Ren,2024 & Syahputra,2023)

Topographic Data

The composition of vegetation in a region, the distribution of flammable materials, and the occurrence of forest fires create an alarm effect. Forest fires are more common in areas below 700 meters above sea level. Hunan Province, Dongting Lake, Wuling Mountain, Xuefeng Mountain, Mofu Luoxiao Mountain, Nanling Mountain etc.(Mohapatra,2023 & Khirani,2023) It has difficult terrain, including electricity. Even at similar altitudes, factors such as slope and aspect can cause differences in vegetation growth and local climate, which can affect fire. For example, the fire is more intense in the north of Nanling. Additionally, topography affects accessibility, population distribution, and economic development, which are important for fire prevention and fire management. This study uses U.S. Geological Survey (USGS) data, including Digital Elevation Model (DEM) data from Hunan Province, which includes elevation, slope, and azimuth details at 30-meter resolution. These high-resolution data allow for a comprehensive analysis of the relationship between topography and wildfire. (Radhi,2023 & Xue,2023)

Road network and geographical location data were taken from the National Basic Geography Database, and 2019 population and gross domestic product (GDP) data were taken from the Resource and Environmental Science Data Center. Areas near roads, railways, and settlements often have high levels of human activity. Hiking and other activities will take place in areas with higher population and economic development, which will increase the risk of fire between people and forests. This information provides an understanding of the distribution of human activities and the community economy and is important for understanding and assessing the risks associated with anthropogenic wildfires.

DATA ON HUMAN ACTIVITIES

Data on the use and distribution of habitats., date of visit: 31 December 2022) 1:250,000 National Basic Geography Database (Park, 2020)]. In general, human activity is higher in areas close to highways, railways, and residential areas. In addition, population data and global gross domestic product (GDP) were obtained from the Resources and Environmental Science Data Center (https://www.resdc.cn, accessed December 31, 2022), 2019, and spatial resolution of 1km. Regions with greater population and economic development tend to see an increase in recreational activities such as hiking and traveling, which increases the risk of wildfire and therefore the risk of wildfire . By combining these data, the study provides an understanding of the distribution of human activities and the social economy, which is important for understanding the risks associated with human-caused wildfires and for the study of medicine.

Research Method

For sampling, using the random sampling method, the training size was set to 80% and the repetition rate was set to 100. Training sample reporting and estimation are done using stratified sampling technology. The common variable was chosen as the objective, and the 20 individual variables were labeled as traits. An additional variable is defined as a diagonal. Import various types of machine learning for training, such as Random Forest, AdaBoost, and Gradient Boosting. Models will be trained individually using the training method and evaluated for their accuracy using five parameters: area under the curve (AUC), classification accuracy (CA), F1 score, precision, and regression. This approach allows a comprehensive evaluation of model performance across multiple metrics, providing a reliable evaluation of any machine learning algorithm in fire hazard prediction.

Adaptive Boosting Algorithm

The scientific basis for choosing the AdaBoost model for forest fire prediction lies in its advantages and capabilities:

1. Strong predictive ability: AdaBoost creates good separation by combining multiple weak points. This integration makes the model more efficient and takes

advantage of the relationship between features, thus increasing the accuracy of the prediction. By using the collective knowledge of weak classifiers, AdaBoost can identify complex patterns in data, thus improving its performance.

2. Adaptive learning: AdaBoost uses an adaptive learning method to treat the weighted model as an unclassified model at each iteration. This change causes the model to focus on competing models that are difficult to classify accurately. By placing greater emphasis on unclassified patterns, AdaBoost can update datasets with different patterns and complexity, thus improving its ability to expand on unseen data.

3. Avoid overfitting: AdaBoost reduces the risk of overfitting using reverse training. At each iteration, only one set of samples is selected for training to prevent the model from over-memorizing the training data. This model selection helps AdaBoost generalize to new data and protect that data from noise in the training process, thus increasing its stability and preventing overperformance.

In the context of wildfire prediction, AdaBoost provides a powerful framework for the development of classification systems that can identify and predict wildfires. The basic principle of AdaBoost is to retrain weak classifiers and combine them into a strong classifier. During each iteration, AdaBoost gradually improves the performance of the weak classifier by paying more attention to examples that were misclassified in the previous round. With various changes, AdaBoost increases the performance of weak classifiers to the level of strong classifiers, enabling accurate predictions about the occurrence of forest fires.

Overall, the scientific basis for choosing AdaBoost is that it can do a good combination of weak classifiers, dynamically adapt to complex data, and is against overfitting, making it a suitable choice for wildfire prediction.

In the context of wildfire forecasting, the task usually involves separating events into two groups: fire events (class) and non-fire events (bad room). To achieve this goal, the following steps must be followed:

1. Data collection: Collect large files containing information on past fires. This information includes various characteristics such as electrical power, weather conditions (e.g. temperature, humidity, wind speed), terrain (elevation, slope) and vegetation, spray (e.g. type of vegetables, humidity).

2. Feature engineering: Prioritize data collection and design to accommodate modeling. This will include data cleaning, formatting, and organizing to ensure they are in a format suitable for review.

3. Training models using weak classifiers: Weak classifiers are built from small objects and existing data. The goal of any weak classifier is to classify a sample

as a fire event or a non-fire event based on selected features. These weak classifiers are relatively simple and do not perform well when used alone.

4. AdaBoost algorithm: It is used for retraining weak classifiers. At each iteration, this algorithm adjusts the weights of the training samples based on the performance of the weak classifier in the previous iteration. Unclassified cases are given a higher weight, while classified cases are given a lower weight.

5. Combination of weak classifiers: AdaBoost optimizes the set of weak classifiers through different iterations to create strong classifiers. This robust classifier weightily combines the predictions of each weak classifier, giving more weight to the prediction of classes that perform better on the training data. 6. Prediction: When solid training is available, it can be used to predict whether a new model represents a fire event or a non-fire event. Given the characteristics of the model, the dynamic operator uses the information learned during training to make predictions. The estimated class list indicates whether the model is classified as a fire hazard.

Overall, AdaBoost algorithm helps to create strong classifiers for wildfire prediction by improving the performance of weak classifiers and combining their methods. Make accurate predictions for new situations.

Gradient Boosted Decision Tree Algorithm

It is a general algorithm used for classification and regression. It works by fitting decision trees to training data and optimizing them to minimize residual error. The selection of GBDT for forest fire forecasting is based on the following important factors:

1. Gradient Descent: GBDT uses a gradient descent algorithm to relocate residuals and refine the model. This iterative process gradually reduces the loss, thus increasing the accuracy of the prediction. GBDT can capture complex patterns in data and improve prediction by continuously updating the model via gradient descent.

2. Integration of weak classifiers: GBDT combines multiple weak classifiers and obtains the final prediction result from weighted voting. This integration method improves the generalization ability of the model by combining the predictions of weak classifiers. By using the collective knowledge of weak learners, GBDT can control the relationship between features and different targets.

3. Edit: GBDT introduces an edit period in each iteration to avoid overweighting. This fixed mechanism increases the stability and generalization ability of the

model by penalizing overly complex models. By controlling for the complexity of the model, GBDT can be more efficient for missing data and reduce the risk of overfitting.

In the context of wildfire forecasting, GBDT uses historical weather data, fuel characteristics, human activities, etc. Can create powerful regression models using relevant information. By learning from the training model, GBDT can identify patterns and features associated with wildfire occurrence and Random Forest Algorithm

Random Forest (RF) is an adaptive algorithm that is frequently used in many disciplines because it can identify different features and determine their importance (Renugadevi, 2023). There are several important considerations when deciding to use the RF model to predict fires:

1. Random Sampling: RF uses Bootstrap sampling to generate modified samples from the original training program. This strategy increases model diversity by providing slightly different information to each decision tree, thus reducing variance and improving overall performance.
2. Random feature selection: RF randomly selects a set of features for classification at each point. This unique randomization reduces the correlation between features and makes the model more resistant to overfitting.
3. Survey: RF combines predictions from multiple decision trees to produce a final prediction. By combining the results obtained from polling, RF can provide more accurate predictions by reducing the error that can occur in a single decision tree.

In addition, RF models can be learned in parallel, thus improving performance while maintaining high-performance data even without data. However, RF requires large data sets for training, limiting further predictions beyond the training data.

In this study, fire training samples were collected from a large area of Hunan Province, while a small random sample was tested using points in the city of Changsha. This approach addresses the problems of small data and low accuracy by using RF models for critical evaluation and analysis of electronic drives. The main features affecting the accuracy of the model were changed; such as setting the decision tree (n_predictors) to 600, using 80% of the random sample of fire occurrence in the valley forest for training, and repeating 5 times to determine. negative. This iterative training process increases the accuracy and robustness of the model.

During training, all decision trees in the RF model are trained using different models and features and iterated according to the selection process (for example, as the Gini coefficient or data increases) until the situation stops. meeting. In the prediction phase, each model is classified or iterated using the prediction of each tree and the final prediction is determined by voting for the classification problem or the average of the regression problems. This combination helps improve the overall prediction of the RF model for wildfire prediction.

INVERSE DISTANCE WEIGHT INTERPOLATION ALGORITHM

The inverse distance weight (IDW) interpolation method is a spatial interpolation method that approximates the value of the object location based on the value of the nearby location. This method assigns weights to sample points based on their distance from the target location and then calculates the average value to determine the value of the target location (Praveen Kumar, 2023). Taking the prediction of forest fire occurrence in Changsha City as the background, the IDW interpolation method was used to predict the probability of forest fire occurrence at 1 km × 1 km grid points in the city. This method involves assigning different weights to sample points based on their distance from each grid point, with closer points being given higher weights. This weighted average is used to estimate the probability of wildfire occurring at each grid point.

The sample points used for the experiment in this study are equal to 1 km interval, which is suitable for the IDW interpolation method. The basic principle of IDW is based on the assumption that nearby points are more closely related and therefore have a greater impact on the value of the target location. Therefore, when estimating the value of object location, IDW will give more weight to sample points that are close to the target location, reflecting the relationship between points. Overall, the IDW interpolation method provides a direct and intuitive way to estimate the importance of unseen areas, making it an important tool for analysis and forecasting activities such as wildfire production.

Accuracy Evaluation

The evaluation criteria mentioned are the criteria for evaluating the effectiveness of the classification model:

1. Accuracy: Measure the correct classification rate of each event. This is a good indicator when the distribution units are equal.
2. Precision: A measure of the proportion of correct predictions among all good predictions made by a model. It focuses on the accuracy of good and effective predictions when the cost of false positives is high.
3. Recall: A measure of the proportion of true positives predicted by the model. It focuses on the ability of the model to capture all good conditions and is

Table 1. Model accuracy comparison

Machine Learning Algorithm	Accuracy	Precision	Recall	F1	AUC
SVM	0.779	0.768	0.801	0.784	0.847
GBDT	0.889	0.882	0.921	0.901	0.962
RF	0.908	0.897	0.925	0.910	0.965

important when loss of good quality is costly. 4. F1 score: Harmonic instrument of precision and return. It provides a balance between accuracy and recall and is useful when classes are not consistent. This is an overall measure of model performance.

4. Area under the ROC curve (AUC-ROC): The ROC curve shows the true value and false positive value at different locations. AUC-ROC measures the overall ability of the model to distinguish good and bad classes. A higher AUC value indicates better performance with a quality score of 1.

Together, these measurements provide a better understanding of the model's performance in various aspects such as accuracy, completeness, and the balance between good and bad.

Evaluation of the importance of characteristic features makes predictions about future fire events. In addition, GBDT can evaluate the importance of features and provide insight into factors affecting the occurrence of forest fires.

Below is a brief description of this process:

1. Gain ratio: Gain ratio is a metric used to measure the effectiveness of segmenting data sets based on certain characteristics. It takes into account the information obtained by feature segmentation and normalizes it with the features' unique information. Features with higher levels are considered more informative for classification. (Sitorus,2023).

2. Gini Index: Gini Index is a measure of the degree of contamination or problem in the data. In the context of a decision tree, it represents the probability that a randomly selected item will be misclassified if typed randomly based on the distribution of text in the database. When used for segmentation, features with higher Gini index values are more important in separating classes.

In practice, both the growth rate and the Gini index are used to rank the features according to their importance in the decision tree-based model. Features with higher height or Gini index value are considered more important for classification and are ranked higher in importance. These rankings help identify the most valuable features

Figure 4. Plot of result evaluation, Gini index in green and gain ratio in orange

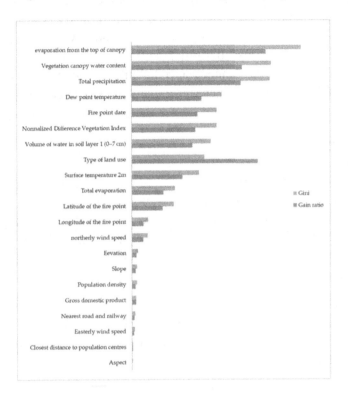

when forecasting and can guide specific options or mitigation efforts. The resulting Gini index and its rate of increase are shown in Figure 4.

CONCLUSION

The aim of this chapter is to improve forest fire forecasting by integrating various forest-related factors into a systematic model. In particular, it creates a more accurate estimate of the outcome of a wildfire by taking into account factors such as the water content of the vegetation and evapotranspiration in the upper part of the vegetation. Additionally, this study uses machine learning techniques, including random forest models, gradient boosting decision tree models, and adaptive acceleration models, to improve fire probability prediction models. These machine learning algorithms were chosen for their ability to handle complex data and uncorrelated variables. To evaluate the effectiveness of these models, the study will use indicators such as area under the curve (AUC), which is frequently used in binary classification such as predicting fire occurrence. Other relevant metrics are precision, recall, F1 score, accuracy, etc. may contain. These measurements provide an overview of the overall performance of the model in terms of its ability to identify wildfire occurrence and non-occurrence. By integrating wildfire data and using machine learning techniques, this study aims to improve the accuracy and reliability of wildfire prediction models, which have implications for fire management and mitigation efforts.

REFERENCES

Abram, N. J., Abram, N. J., Henley, B. J., Gupta, A. S., Lippmann, T. J. R., Clarke, H., Dowdy, A. J., Sharples, J. J., Nolan, R. H., Zhang, T., Wooster, M. J., Wurtzel, J. B., Meissner, K. J., Pitman, A. J., Ukkola, A. M., Murphy, B. P., Tapper, N. J., & Boer, M. M. (2021). Connections of climate change and variability to large and extreme forest fires in southeast Australia. *Communications Earth & Environment*, 2(8), 8. doi:10.1038/s43247-020-00065-8

Borah, A. R., Ravi, N., Narayan, N., & D'Souza, P. J. (2024). *A Methodology for Forest Fire Detection and Notification Using AI and IoT Approaches*. 2024 2nd International Conference on Intelligent Data Communication Technologies and Internet of Things (IDCIoT), Bengaluru, India. 10.1109/IDCIoT59759.2024.10467597

Chitra, U. T. (2023). *Forest Fire Detection System: 2023 Intelligent Computing and Control for Engineering and Business Systems*. ICCEBS., doi:10.1109/ICCEBS58601.2023.10448648

Coman, C.-M., Toma, B., Constantin, M.-A., & Florescu, A. (2022). Ground Level Lidar as a Contributing Indicator in an Environmental Protection Application. SSRN Electronic Journal. doi:10.2139/ssrn.4096563

Costa-Saura, J. M., Balaguer-Beser, Á., Ruiz, L. A., Pardo-Pascual, J. E., & Soriano-Sancho, J. L. (2021). Empirical Models for Spatio-Temporal Live Fuel Moisture Content Estimation in Mixed Mediterranean Vegetation Areas Using Sentinel-2 Indices and Meteorological Data. *Remote Sensing (Basel)*, 13(18), 3726. doi:10.3390/rs13183726

Ji, H.-K., Kim, S.-W., & Kil, G.-S. (2020). Phase Analysis of Series Arc Signals for Low-Voltage Electrical Devices. *Energies*, 13(20), 5481. doi:10.3390/en13205481

Ji, Y., Wang, D., Li, Q., Liu, T., & Bai, Y. (2024). Global Wildfire Danger Predictions Based on Deep Learning Taking into Account Static and Dynamic Variables. *Forests*, 15(1), 216. doi:10.3390/f15010216

Justino, F. B., Bromwich, D. H., Schumacher, V., daSilva, A., & Wang, S. (2022). Arctic Oscillation and Pacific-North American pattern dominated-modulation of fire danger and wildfire occurrence. *NPJ Climate and Atmospheric Science*, *5*(1), 1–13. doi:10.1038/s41612-022-00274-2

Khirani, S., Souahlia, A., & Rabehi, A. (2023). *Forest Fire Detection Using a Low Complex Convolutional Neural Network.* 2023 2nd International Conference on Electronics, Energy and Measurement (IC2EM), Medea, Algeria. 10.1109/IC2EM59347.2023.10453317

Konings, A. G., Saatchi, S. S., Frankenberg, C., Keller, M., Leshyk, V., Anderegg, W. R. L., Humphrey, V., Matheny, A. M., Trugman, A., Sack, L., Agee, E., Barnes, M. L., Binks, O., Cawse-Nicholson, K., Christoffersen, B. O., Entekhabi, D., Gentine, P., Holtzman, N. M., Katul, G. G., & Zuidema, P. A. (2021). Detecting forest response to droughts with global observations of vegetation water content. *Global Change Biology*, *27*(23), 6005–6024. doi:10.1111/gcb.15872 PMID:34478589

Li, Y., Li, K., Wang, K., Li, G., & Wang, Z. (2023). An Improved Stacking Model for Forest Fire Susceptibility Prediction in Chongqing City, China. *2023 IEEE Smart World Congress (SWC)*, Portsmouth, United Kingdom. 10.1109/SWC57546.2023.10449165

Lin, X., Li, Z., Chen, W., Sun, X., & Gao, D. (2023). Forest Fire Prediction Based on Long- and Short-Term Time-Series Network. *Forests*, *14*(4), 778. doi:10.3390/f14040778

Ma, W., Feng, Z., Cheng, Z., Chen, S., & Wang, F. (2020). Identifying Forest Fire Driving Factors and Related Impacts in China Using Random Forest Algorithm. *Forests*, *11*(5), 507. doi:10.3390/f11050507

Miller, L., Zhu, L., Yebra, M., Rüdiger, C., & Webb, G. I. (2022). Multi-modal temporal CNNs for live fuel moisture content estimation. *Environmental Modelling & Software*, *156*(105467), 105467. doi:10.1016/j.envsoft.2022.105467

Mohapatra, S., & Shrimali, B. (2023). *Fault Detection in Forest Fire Monitoring Using Negative Selection Approach.* 2023 IEEE 11th Region 10 Humanitarian Technology Conference (R10-HTC), Rajkot, India. 10.1109/R10-HTC57504.2023.10461878

Natekar, S., Patil, S., Nair, A., & Roychowdhury, S. (2021). *Forest Fire Prediction using LSTM.* 2021 2nd International Conference for Emerging Technology (INCET), Belagavi, India. 10.1109/INCET51464.2021.9456113

Park, S., Won Han, K., & Lee, K. (2020). *A study on fire detection technology through spectrum analysis of smoke particles.* 2020 International Conference on Information and Communication Technology Convergence (ICTC), Jeju, Korea (South). 10.1109/ICTC49870.2020.9289272

Praveen Kumar, B., Kalpana, A. V., & Nalini, S. (2023). Gated Attention Based Deep Learning Model for Analyzing the Influence of Social Media on Education. *Journal of Experimental & Theoretical Artificial Intelligence*, 1–15. doi:10.1080/0952813X.2023.2188262

Radhi, A. A., & Ibrahim, A. A. (2023). *Forest Fire Detection Techniques Based on IoT Technology: Review.* 2023 1st IEEE International Conference on Smart Technology (ICE-SMARTec), Bandung, Indonesia.10.1109/ICE-SMARTECH59237.2023.10461943

Ren, D., Zhang, Y., Wang, L., Sun, H., Ren, S., & Gu, J. (2024). FCLGYOLO: Feature Constraint and Local Guided Global Feature for Fire Detection in Unmanned Aerial Vehicle Imagery. *IEEE Journal of Selected Topics in Applied Earth Observations and Remote Sensing*, *17*, 5864–5875. doi:10.1109/JSTARS.2024.3358544

Renugadevi, R., Vyshnavi, T., Reddy, T. P., & Lahari, P. S. (2023). Air Quality Prediction using Random Forest algorithm. *International Conference on Research Methodologies in Knowledge Management, Artificial Intelligence and Telecommunication Engineering (RMKMATE)*, Chennai, India. 10.1109/RMKMATE59243.2023.10369180

Rubí, J. N., Carvalho, P., & Gondim, P. R. (2023). Application of machine learning models in the behavioral study of forest fires in the Brazilian Federal District region. *Engineering Applications of Artificial Intelligence*, *118*, 105649. doi:10.1016/j.engappai.2022.105649

Singh, S., & Chockalingam, J. (2023). Using Ensemble Machine Learning Algorithm to Predict Forest Fire Occurrence Probability in Madhya Pradesh and Chhattisgarh, India. *Advances in Space Research*, *73*(6), 2969–2987. doi:10.1016/j.asr.2023.12.054

Sitorus, F. P., Sunyoto, A., & Setiaji, B. (2023). *Forest Fire Disaster Classification Using Artificial Neural Network Method.* 2023 6th International Conference on Information and Communications Technology (ICOIACT), Yogyakarta, Indonesia. 10.1109/ICOIACT59844.2023.10455926

Sowah, R. A., Ofoli, A. R., Krakani, S., & Fiawoo, S. (2017). Hardware Design and Web-Based Communication Modules of a Real-Time Multisensor Fire Detection and Notification System Using Fuzzy Logic. *IEEE Transactions on Industry Applications*, *53*(1), 559–566. doi:10.1109/TIA.2016.2613075

Sulthana, S. F., Wise, C. T. A., Ravikumar, C. V., Anbazhagan, R., Idayachandran, G., & Pau, G. (2023). Review Study on Recent Developments in Fire Sensing Methods. *IEEE Access : Practical Innovations, Open Solutions*, *11*, 90269–90282. doi:10.1109/ACCESS.2023.3306812

Xie, J., Qi, T., Hu, W., Huang, H., Chen, B., & Zhang, J. (2022). Retrieval of Live Fuel Moisture Content Based on Multi-Source Remote Sensing Data and Ensemble Deep Learning Model. *Remote Sensing (Basel)*, *14*(17), 4378. doi:10.3390/rs14174378

Xue, Z., Zheng, Z., Yi, Z., Han, Y., Liu, W., & Peng, J. (2023). A Fire Detection and Assessment Method based on YOLOv8. 2023 China Automation Congress. CAC. doi:10.1109/CAC59555.2023.10451727

Chapter 5
Artificial Intelligence Models to Prevent Forest Fires

Wasswa Shafik

🆔 https://orcid.org/0000-0002-9320-3186

Dig Connectivity Research Laboratory (DCRLab), Kampala, Uganda & School of Digital Science, Universiti Brunei Darussalam, Brunei

ABSTRACT

The main goal is to appropriately utilize advanced algorithms to analyze environmental data, improve early disease detection and intervention tactics, and reduce the harmful effects of forest fires on human beings. Analyze the challenges faced by traditional methods in addressing the constantly evolving nature of wildfires and the need for more adaptable and proactive approaches, and highlight the advantages of AI. Discusses the main constituents incorporated into the AI model, comprising meteorological data, satellite imagery, and historical fire records. It analyzes the selection of AI algorithms specifically tailored for forest fire prevention, considering parameters. Analyze the challenges faced during the creation and implementation of AI models for forest fire prevention and viability of integrating artificial intelligence models into existing fire management infrastructure and emergency response systems. It showcases the current research, progress, and use of AI-driven solutions to address the challenges posed by wildfires and provides a concise overview of the chapter's findings.

INTRODUCTION

The increasing frequency and intensity of forest fires worldwide present a significant danger to ecosystems, wildlife, and human societies. Conventional fire prevention approaches, however useful, have inherent limitations in dealing with the dynamic

DOI: 10.4018/979-8-3693-3896-4.ch005

and developing characteristics of wildfires (Shao et al., 2023). The threat of forest fires poses a significant and increasing worldwide problem that requires urgent attention and creative solutions. Wildfires have become global emergencies due to their escalating occurrence and intensity, affecting ecosystems, biodiversity, and human settlements on a widespread scale (Tien Bui et al., 2019). The severe repercussions, encompassing fatalities and irreparable harm to the ecosystem, emphasize the need for promptly confronting this challenging dilemma.

The increase in forest fires can be mostly attributed to climate change, characterized by elevated temperatures, extended periods of drought, and unpredictable weather patterns. Due to global warming, the duration of typical fire seasons is increasing, leading to the spread of wildfires in regions that were previously unaffected (G. Wang et al., 2019). The urgency for proactive solutions has become increasingly crucial since relying just on reactive measures is insufficient to address the magnitude and severity of these wildfires. In addition to posing environmental risks, these wildfires have a ripple effect on the quality of air and water and make a substantial contribution to the production of greenhouse gases (Supriya & Gadekallu, 2023). The economic impact, encompassing expenses related to firefighting as well as the destruction of property and resources, is immense.

Within this scenario, the investigation of proactive solutions becomes crucial, and one promising approach is the use of artificial intelligence[1] (AI) models in forest fire prevention tactics. The Earth is currently experiencing a significant and alarming problem with forest fires, which serve as a foreboding metaphor for the planet's fragility (Nebot & Mugica, 2021). The frequency and intensity of wildfires have increased, posing significant threats to ecosystems and testing the resilience of communities globally. The increase in forest fires is not a random occurrence; it is a direct result of human-caused climate change. Elevated temperatures, extended periods of drought, and modified patterns of precipitation converge to form an ideal combination for the initiation and propagation of wildfires (Pettorru et al., 2023).

Amidst this ecological disaster, there is a pressing need for aggressive responses. Despite their bravery, reactive methods are inadequate to address the immense size and unpredictable nature of contemporary wildfires. These flames have transformed into a powerful and significant force in recent years, surpassing geographical limits and presenting serious risks to ecosystems and human settlements (Lin, 2022). The complex interplay between climate change[2] as the United Nations, human activities, and natural causes has orchestrated a series of devastating events, highlighting the urgent need for immediate and proactive remedies; the 2000-2020 annual average fire density is demonstrated in Figure 1. The necessity for a paradigm change is not only a scientific and environmental imperative but also a moral obligation (L. Wang et al., 2018). Conventional fire prevention methods, which rely on past data

Figure 1. Annual average fire density ranges from 2000-2020

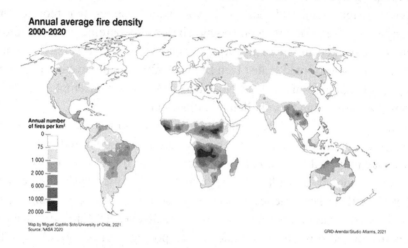

and predictable models, face difficulties in keeping up with the rapidly changing characteristics of contemporary wildfires.

The chapter unfolds a comprehensive expedition of forest fire avoidance and management. It commences by dissecting the constraints inherent in traditional fire management approaches and continues to dissect the obstacles these techniques face in adapting to the vibrant nature of wildfires. The description then pivots to modern research, shedding light on how expert systems change the early detection, monitoring, and mitigation of forest fires. This structure delineates the core elements of AI models, including meteorological information, satellite images, and historical fire records. Subsequently, the chapter navigates through the complexities of designing and carrying out AI designs, followed by an expedition into their prospective integration into existing fire management infrastructure. It culminates in a positive analysis laying out potential customers and encapsulating the transformative impact of AI in forest fire avoidance.

Contribution to the Chapter

This chapter outlines the subsequent contributions.

- Analyze the challenges faced by traditional methods in addressing the constantly evolving nature of wildfires and the need for more adaptable and proactive approaches and highlight the advantages of AI.
- Discusses the main constituents incorporated into the AI model, comprising meteorological data, satellite imagery, and historical fire records. Furthermore,

it analyzes the selection of AI algorithms specifically tailored for forest fire prevention, considering parameters.

- Analyze the challenges faced during the creation and implementation of AI models for forest fire prevention. Sketch out the evaluation criteria used and highlight the effectiveness of the AI model compared to traditional methods.
- Examine the viability of integrating artificial intelligence models into existing fire management infrastructure and emergency response systems.
- Ultimately, it showcases the current research, progress, and use of AI-driven solutions to address the challenges posed by wildfires and provides a concise overview of the chapter's findings.

Section 2 presents the traditional methods and their associated limitations. Section 3 and examines the difficulties encountered by conventional approaches in dealing with the ever-changing characteristics of wildfires. Section 4 explores contemporary research and practical uses of artificial intelligence in the timely identification, surveillance, and mitigation of forest fires. Section 5 outlines the fundamental components included in the AI model, such as meteorological data, satellite pictures, and past fire histories. Section 6 explores the difficulties encountered in the process of designing and implementing AI models. Section 7 demonstrates and investigates the possible incorporation of AI models into current fire management infrastructure. Finally, Section 8 concludes by outlining prospects and summarizing the crucial findings, highlighting the revolutionary impact of AI in the realm of forest fire prevention.

TRADITIONAL METHODS AND LIMITATIONS

The future exploration of AI models exhibits promise in surmounting these limitations and revolutionizing the methodologies employed to tackle the escalating perils presented by wildfires.

Dependence on past data

Conventional strategies for preventing wildfires heavily depend on historical data and deterministic models to forecast and control fire behavior. Nevertheless, this strategy becomes progressively risky when confronted with the impacts of climate change. As the ecosystem changes, past patterns may no longer reliably predict future fire behavior (Shao et al., 2023). The insufficiency of exclusively depending on historical data requires a transition towards more flexible and future-oriented solutions.

The Unpredictability of Contemporary Wildfires

The volatility of contemporary wildfires arises from the intricate interplay of climate change-induced elements. Extended periods of drought, increasing temperatures, and changes in vegetation patterns are factors that lead to unpredictable fire behavior. Conventional approaches, which rely on assumptions derived from past patterns, face difficulties in dealing with the unexpected and rapidly changing nature of modern wildfires (Tien Bui et al., 2019). It is crucial to acknowledge and deal with this lack of predictability when it comes to devising successful strategies for prevention and management; some main factors that are causing the brunt areas include fuel, moisture ignitions, and suppression, as illustrated globally in Figure 2.

Magnitude and Velocity of Wildfires

Contemporary wildfires demonstrate swift intensification, spreading over vast regions over brief periods. Conventional models, which were created to handle slower fire spread, encounter difficulties in adjusting to the rapid and widespread nature of modern infernos (G. Wang et al., 2019). This requires a reassessment of firefighting tactics, with a focus on continuous monitoring and flexible responses to control wildfires of different sizes and speeds efficiently.

The Intricacy of Factors that Have an Impact

A multitude of elements, such as wind patterns, topography, and fuel kinds, shape the behavior of contemporary wildfires. The complex interaction of various variables necessitates a refined and flexible approach, characteristics frequently absent in conventional models (Supriya & Gadekallu, 2023). It is essential to acknowledge and consider this intricacy while formulating techniques that can effectively forecast and control wildfires in varied and ever-changing surroundings.

Effects on Disaster Management

The constraints imposed by conventional approaches have a substantial influence on the effectiveness of catastrophe management. The effectiveness of deploying firefighting resources is hindered by delayed responses caused by insufficient real-time monitoring and early detection (Shao et al., 2023). The postponement amplifies the magnitude and seriousness of catastrophes, underscoring the crucial necessity for more prompt and proactive strategies that can match the advancing characteristics of contemporary wildfires.

Figure 2. Dominant factors count for burnt areas

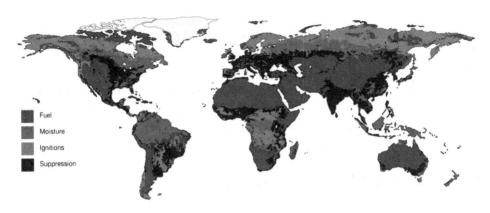

Lack of Sufficient Planning for Evacuation

Effective evacuation planning is hindered by the provision of inaccurate and untimely information derived from inflexible old models. The lack of ability to predict swiftly evolving fire conditions hinders the formulation of proactive plans for towns in the trajectory of wildfires. The insufficiency of the current approach increases the vulnerability of lives and property, highlighting the need to integrate more adaptable and prompt components into evacuation preparation (Tien Bui et al., 2019).

Figure 3. The fire behavioral triangle

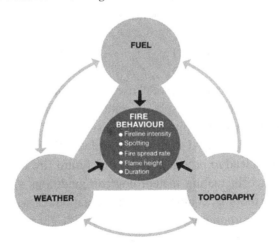

Issues with Allocating Resources

Conventional models face difficulties in adjusting to swiftly evolving fire conditions, which impacts the allocation of resources during firefighting operations. Emergency response teams encounter difficulties in efficiently allocating resources due to the delayed information provided by these models (G. Wang et al., 2019). To successfully address these difficulties, it is necessary to adopt a proactive and immediate strategy for allocating resources to control wildfires promptly and effectively. Figure 3 illustrates how weather, fuel, and topography count and behave in wildfires.

Human Lives and Property Susceptibility

The delay in response time caused by the constraints of conventional approaches heightens ecosystems' susceptibility and exposes human lives and property to imminent danger. The lack of prompt and precise fire behavior prediction exacerbates the risk posed by wildfires (Supriya & Gadekallu, 2023). The recognition of the pressing need to address these vulnerabilities highlights the crucial requirement for sophisticated and flexible measures to safeguard both natural and human ecosystems from the swiftly expanding consequences of wildfires; AI stretches in wildfires are illustrated in the succeeding section.

ARTIFICIAL INTELLIGENCE IN THE MANAGEMENT OF WILDFIRES

This comprehensive technique improves early detection and enables proactive decision-making in efforts to prevent wildfires and respond to disasters.

Meteorological Information

Weather data is essential for an AI model used in wildfire management, as it offers up-to-date information about atmospheric conditions. Factors such as temperature, humidity, wind speed, and precipitation have a significant impact on fire behavior (Nebot & Mugica, 2021). The AI algorithm examines this data, identifying patterns and correlations that lead to precise forecasts of future fire hazards. The model's capacity to adjust to fluctuating meteorological conditions via ongoing updates makes it a dynamic tool for accurate and fast evaluations of wildfire risk (Pettorru et al., 2023). The incorporation of live weather data guarantees that the model remains up to date with changing environmental variables that are essential for efficient wildfire control.

Satellite imagery

High-resolution satellite imagery is crucial in wildfire management as it provides an aerial perspective of areas. It assists in identifying anomalies such as alterations in flora, smoke columns, and areas of intense heat from fires. The AI model analyzes the visual data using machine learning[3] (ML) methods, improving its ability to detect and pinpoint wildfires quickly (Lin, 2022). Satellite imaging not only helps detect flames but also aids in accurately mapping the size of the fires. This information is crucial for efficiently allocating resources to contain and control the fires. Through the utilization of satellite technologies, the AI model acquires a thorough comprehension of the geographical dynamics of wildfires, facilitating more efficient decision-making in addressing emergent fire hazards.

Archives of past fire incidents

Historical fire records offer important context and profound insights into the behavior of previous wildfires in a certain area. The AI model acquires a historical perspective that guides its forecasts by examining patterns in the frequency, intensity, and extent of past fires. The historical background enables the model to recognize regions with a greater probability of recurrence and comprehend the ways in which variables contribute to the occurrence of fire outbreaks (Abid, 2021). Utilizing historical data improves the model's ability to predict, allowing it to recognize and prioritize high-risk locations. By assimilating knowledge from previous experiences, the AI model enhances its ability to predict and reduce future fire hazards, so promoting a proactive and well-informed strategy for managing wildfires.

Vegetation Indices

Satellite-derived vegetation indices offer crucial information regarding the vitality and abundance of plants. The Normalized Difference Vegetation Index[4] (NDVI) provides a quantification of the level of greenness and vitality in vegetation (Mahaveerakannan R et al., 2023). Fluctuations in these indexes can function as early indications of possible fire hazards. The AI model utilizes vegetation indicators to evaluate the fuel load in each area, pinpointing areas where arid or distressed vegetation could potentially heighten the probability of wildfires. This data improves the model's proactive skills, enabling it to focus on locations with increased risk accurately (Pang et al., 2022). The AI model incorporates vegetation indices to include ecological dynamics in its risk evaluations, hence enhancing the comprehensive understanding of wildfire hazards.

Geospatial Data Application

The use of geospatial data, such as topography and land cover information, enhances the AI model's comprehension of the landscape. The model considers variables such as incline, altitude, and land cover categories to forecast the possible velocity and trajectory of fire propagation. The use of geospatial data improves the model's understanding of spatial relationships, enabling more precise evaluations of risk and the detection of regions prone to rapid fire spread (Milanović et al., 2023). The AI model incorporates the physical attributes of the terrain to customize its forecasts for each specific area, resulting in a wildfire management strategy that is more aware of the surrounding context and more accurate. The comprehensive analysis of geographical data enhances the model's capacity to predict and address the ever-changing difficulties presented by wildfires (Feizizadeh et al., 2023).

Monitoring the quality of air

Integrating air quality monitoring data into the AI model is essential for evaluating the influence of wildfires on public health. The model utilizes real-time monitoring of air quality indicators, including PM2.5, carbon monoxide, and ozone levels, to accurately assess the extent of smoke dispersion (Khan et al., 2022). This data assists in promptly issuing health advisories, directing evacuation plans, and ensuring that emergency response operations consider the possible respiratory and environmental health consequences linked to wildfires. The AI model enhances the plan by incorporating air quality data, resulting in a more complete and adaptable approach that focuses on both containing wildfires and safeguarding the well-being of affected communities (Benzekri et al., 2020).

Analysis of Human Activity and Urban Interface

Gaining insight into the interface between human populations and areas prone to wildfires is crucial for ensuring efficient prevention and emergency response. The AI model analyzes data pertaining to human activity, land-use patterns, and the closeness of urban interfaces in order to evaluate the likelihood of human-caused ignitions (Alkhatib et al., 2023). The model examines these characteristics to pinpoint places where human activities could heighten the probability of wildfires and evaluates the potential danger to inhabited regions. This knowledge is essential for adopting specific prevention strategies, increasing community awareness, optimizing emergency response plans, and aligning wildfire control activities with the changing interaction between human settlements and natural landscapes (Kinaneva et al., 2019).

Integration of social media and citizen reporting

Social media and citizen reporting offer invaluable up-to-the-minute information during wildfire incidents. The AI model integrates data from social media platforms, news feeds, and citizen reports to augment situational awareness. The program can detect and analyze user-generated information to identify newly occurring wildfire occurrences, evaluate their level of seriousness, and monitor the immediate effects on communities in real-time (Castelli et al., 2015). This interface facilitates proactive and nimble reactions by authorities, enabling them to verify information, prioritize response initiatives, and provide precise updates to the public. The AI model utilizes the combined observations of residents to enhance wildfire monitoring and management, using the benefits of community engagement (Wu et al., 2021). The incorporation of this social intelligence layer enhances a comprehensive strategy that integrates technology knowledge with community-driven observations, resulting in a more robust framework for managing wildfires.

DEVELOPING ARTIFICIAL INTELLIGENCE MODELS FOR THE PURPOSE OF PREVENTING FOREST FIRES

This section delves into the intricate process of developing and using an artificial intelligence model for the purpose of managing wildfires. It emphasizes the significance of careful consideration at each phase to ensure its effectiveness, reliability, and ethical implementation.

Choosing An Algorithm

The core of the AI model resides in the choice of algorithms. Ensemble techniques, Convolutional Neural Networks[5] (CNNs), and Recurrent Neural Networks[6] (RNNs) are valued for their capacity to handle intricate spatial and temporal data. Ensuring a harmonious equilibrium between computational efficiency and accuracy is of utmost importance (Agustiyara et al., 2021). The selected algorithms must efficiently process the varied inputs from weather data, satellite imaging, and other sources, allowing the model to capture subtle patterns that are inherent in wildfire dynamics accurately.

Feature Engineering

Feature engineering is the process of converting raw data into a comprehensive set of variables that are essential for comprehending a model. This method

Figure 4. Adaptive strategies of biodata to fire

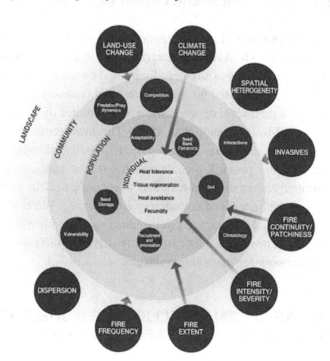

enhances the model's capacity to identify significant patterns. Fire weather indices and spectral indices obtained from satellite photography are important components of a comprehensive set of features (Kucuk & Sevinc, 2023). The expanded range of features improves the model's ability to comprehend the intricacies of wildfire behavior, enabling more precise forecasts and evaluations of danger. Some identified adaptive strategies of biodata to fire are demonstrated in Figure 4.

Live Surveillance and Notifications

The ability to operate in real-time is crucial for efficient wildfire management. The AI model must consistently assess incoming data streams, delivering prompt notifications for potential fire hazards. This guarantees that emergency response teams promptly obtain information, allowing them to start swift and focused interventions (Razavi-Termeh et al., 2020). Implementing real-time monitoring systems decreases response times, thereby lessening the consequences of wildfires and improving the effectiveness of emergency response endeavors.

DESIGNING ARTIFICIAL INTELLIGENCE MODELS FOR FOREST FIRE PREVENTION

This section explores the complex procedure of creating and executing an AI model for wildfire management. It highlights the importance of meticulous deliberation at every stage to guarantee its efficacy, dependability, and ethical application.

Scalability

Scalability is an essential factor to consider, particularly in areas that are susceptible to extensive wildfires. The AI model must adeptly manage escalating data quantities while preserving its efficacy during the process of scaling (Shao et al., 2023). Exploration is conducted using solutions such as cloud-based platforms, distributed computing, and parallel processing frameworks to guarantee the model's performance remains strong throughout larger geographical regions and with increased data frequencies.

Interpretability and Explainability

This refers to the ability to understand and clarify the reasoning and decision-making processes behind a system or model. The transparency of the AI model is crucial for establishing confidence and enabling decision-making (G. Wang et al., 2019). Elaborate algorithms necessitate the presence of techniques that clarify the process by which the model generates its predictions. Interpretability and explainability procedures guarantee that stakeholders, such as emergency responders and policymakers, possess a lucid comprehension of the model's observations and suggestions (Supriya & Gadekallu, 2023). The transparency of the AI model enhances trust in its capabilities and promotes well-informed decision-making in wildfire management.

Iterative Model Enhancement

The AI model's long-term efficacy relies on the incorporation of continuous improvement processes. The model's adaptability is enhanced through regular updates, retraining using new data, and incorporating feedback from real-world events (Nebot & Mugica, 2021). Continuous refinement guarantees the model's resilience in the face of changing environmental circumstances, technological progress, and shifting human activity patterns. This iterative technique ensures that the AI model remains relevant and successful in dynamic wildfire management scenarios.

Figure 5. Potential reinforcing feedback loops of climate change on wildfire management

Seamless Incorporation into Pre-Existing Systems

Efficient operations require the imperative of seamless connection with established wildfire management systems. The AI model should augment the existing infrastructure, bolstering its capabilities without causing any disruption to established procedures (Lin, 2022). The incorporation of geographic information systems[7] (GIS), emergency communication networks, and decision support systems guarantees that the AI model becomes a vital component of the broader wildfire management framework, enhancing decision-making and facilitating coordinated response efforts. Some identifiable potential reinforcing feedback loops of climate change on wildfire management based on the United Nation's Sustainable Development Goals[8] are illustrated in Figure 5.

Ethical and Bias Considerations

The foundation of responsible and explainable AI deployment lies in ethical issues. Ensuring fairness, mitigating prejudice, and maintaining transparency are essential components in the development and execution of the AI model (Abid, 2021). It is crucial to acknowledge and rectify any biases that may exist in the training data, guarantee fair and just outcomes, and openly communicate any limits of the model. These factors support moral principles, enhance public confidence, and preserve the integrity of the model in wildfire management endeavors (Pang et al., 2022). Ensuring ethical implementation is not just a moral obligation but also a fundamental component of the long-term effectiveness of AI systems in intricate real-life situations.

PRACTICAL APPLICATIONS AND CASE STUDIES

This section examines the AI model's tangible applications in real-world scenarios, showcasing its efficacy in wildfire management through instructive case studies.

Timely Identification and Swift Action

The AI model's application for early wildfire identification is crucial in practical situations. Case examples illustrate the system's capacity to evaluate data streams in real time, providing immediate alerts that enable emergency response teams to intervene promptly and accurately. These examples demonstrate the concrete influence of early detection in decreasing reaction times and successfully limiting wildfires (Milanović et al., 2023). The model's significance goes beyond prediction, serving as a critical element in proactive wildfire management, where timely intervention is essential for reducing potential calamities.

Efficient Allocation of Resources

Case studies demonstrate how the AI model enhances the deployment of firefighting resources by utilizing dynamic fire behavior forecasts. The approach allows for precise identification of locations with a high risk, facilitating smart allocation of resources to improve overall efficiency and cost-effectiveness (Khan et al., 2022). These apps demonstrate the practical consequences for wildfire response teams by offering them useful insights that optimize resource allocation tactics. The model's capacity to improve the accuracy of resource allocation signifies notable progress in the area, presenting a fundamental change in the strategic utilization of firefighting resources for the containment and management of wildfires, as demonstrated in Figure 6.

Community Awareness and Evacuation Planning

The AI model is crucial in improving community awareness and evacuation preparation. Case studies demonstrate the use of real-time data to evaluate the

Figure 6. Factors and their associated effects on ecological sustainability

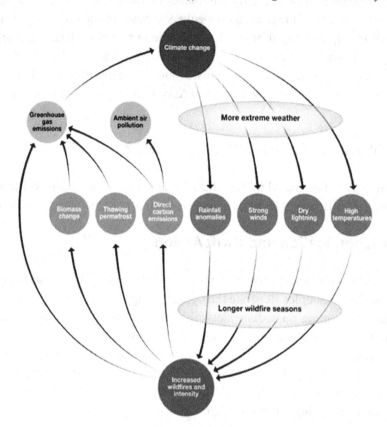

possible effects of wildfires on populated regions (Benzekri et al., 2020). The methodology enhances evacuation planning by delivering timely information, hence promoting community engagement and readiness. These examples highlight the model's capacity to connect technical progress with community-focused wildfire management, providing a tool that not only assists emergency responders but also enables communities to make well-informed decisions during wildfire incidents (Alkhatib et al., 2023).

Integrated Decision Support Systems

This demonstrates the smooth incorporation of the AI model into pre-existing decision support systems. The model improves overall decision-making processes by exhibiting interoperability with GIS, emergency communication networks, and other tools employed in wildfire management (Kinaneva et al., 2019). The collaborative aspect of decision support systems is enhanced by the model's contributions to a thorough understanding of the scenario. These examples offer valuable information about how the model helps in making better and quicker decisions (Castelli et al., 2015). They demonstrate the model's importance as a complementary part of decision support systems in wildfire control.

Flexibility in Adjusting to Diverse Environments

Case studies demonstrate the AI model's capacity to adjust and perform effectively in various geographical and environmental circumstances. The model demonstrates its adaptability by successfully being deployed in diverse ecosystems, accommodating differing landforms, weather patterns, and vegetation kinds (Agustiyara et al., 2021; Kucuk & Sevinc, 2023). These examples highlight the model's resilience and its potential to be used in different regions that are dealing with unique wildfire concerns. The AI model's versatility makes it a great asset for tackling the intricate and various characteristics of wildfires. It provides a scalable solution that can be customized to match the specific requirements of different environmental circumstances (Razavi-Termeh et al., 2020).

Obstacles and Insights Gained

Analyzing the difficulties faced throughout the execution of the AI model yields interesting observations. Case studies provide a comprehensive comprehension of the intricate technological, logistical, and ethical obstacles encountered in real-life situations. The challenges encompass a wide range of issues, including data integration and ethical considerations (Razavi-Termeh et al., 2020). They serve as a guide for overcoming potential obstacles in future implementations. The knowledge gained from successfully addressing these obstacles enhances the ongoing enhancement and optimization of AI systems in wildfire management, promoting a more robust and flexible approach to integrating cutting-edge technology in ever-changing and intricate environmental conditions (Tien Bui et al., 2019).

Evaluating the Extent of Influence and Efficiency

Case studies provide approaches for quantifying the influence and efficacy of the AI model in wildfire management. Quantitative and qualitative indicators, such as the reduction of reaction times, containment success rates, and overall improvement in wildfire control tactics, can comprehensively evaluate the model's contributions (Supriya & Gadekallu, 2023). These measures offer concrete proof of the advantages obtained from the AI model, providing a strong foundation for evaluating its impact on both the efficiency of emergency response and the overall effectiveness of wildfire management strategies.

Scalability and Future Considerations

The case studies investigate the potential for the AI model to be expanded and the factors that need to be considered for future implementations. Examples of the model effectively expanding to encompass wider geographical regions and handling greater amounts of data demonstrate its versatility (Nebot & Mugica, 2021). An analysis of its adaptability to new technologies, capacity to adapt to changing environmental conditions, and incorporation of supplementary data sources offer a plan for future advancements. These factors highlight the model's capacity to remain significant and efficient in tackling wildfire difficulties on a larger level, leading to breakthroughs in AI-based solutions for managing wildfires(Lin, 2022) .

ETHICAL CONSIDERATIONS AND PUBLIC PERCEPTION

This section explores the ethical considerations of using AI in wildfire management, specifically focusing on issues related to privacy, bias, openness, and community involvement.

Privacy Concerns in Data Collection

The ethical considerations pertaining to data collecting for AI in wildfire management are of utmost importance, particularly in resolving privacy concerns. The utilization of satellite imagery with citizen reporting gives rise to possible concerns regarding privacy, thus requiring the implementation of ethical and transparent data procedures (Feizizadeh et al., 2023). It is essential to have strategies in place for anonymizing data, gaining informed consent, and maintaining compliance with privacy legislation. The intricate equilibrium between utilizing important data for wildfire management and protecting individual privacy rights, as in Figure. 7, highlights the ethical necessity of transparent and responsible data collection procedures.

Addressing Bias in AI Algorithms

It is of utmost importance to tackle bias in AI systems to guarantee fair and equitable results in wildfire prediction and response. This section delves into the ethical aspects of reducing biases, highlighting the significance of inclusive and representative training datasets. Algorithmic fairness assessments and continual monitoring are

Figure 7. Peatland carbon stock degradation (Kettridge et al., 2015)

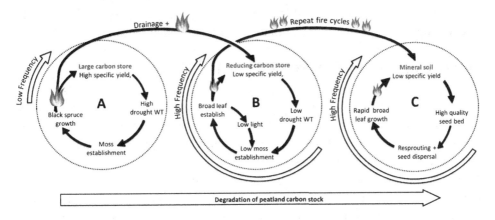

considered essential elements of an ethical framework (Abid, 2021). Case studies demonstrate the successful implementation of bias mitigation measures, emphasizing the ethical obligation to consistently assess and correct biases in AI models used in wildfire control.

Enhancing the clarity and openness of AI decision-making processes

Ensuring transparency in the AI decision-making processes is crucial for establishing confidence and promoting comprehension among all parties involved. This section explores the ethical obligation to ensure transparent communication regarding the decision-making process of AI models in wildfire prediction (Milanović et al., 2023). Examines case studies and exemplary methods to demonstrate successful communication strategies that enable both specialists and the general public to understand and analyze the findings produced by AI models. There is an ethical need to make AI decision-making more understandable and encourage informed involvement with the technology by emphasizing openness.

Public Involvement and Consultation with the Community

The engagement of communities in developing and implementing AI models for wildfire management is necessary due to ethical considerations. This section examines methods for involving and working with the community, seeking input and cooperation while acknowledging the significance of incorporating local expertise and viewpoints (Feizizadeh et al., 2023). Case studies exemplify examples of communities actively participating in decision-making processes, thereby contributing to the ethical use of AI technologies. Underscores the ethical obligation to acknowledge and honor the varied needs and concerns of different communities impacted by wildfires while placing importance on the input from these communities (Castelli et al., 2015).

Enhancing the Public's Perception and Fostering Trust

The successful integration of AI technology for wildfire management heavily relies on public perception and faith in these applications. This section examines the prevailing sentiments of the general public towards AI while also outlining the various factors that influence levels of confidence and potential skepticism (Wu et al., 2021). Case studies and surveys offer valuable insights into the public's perceptions and the significance of honest communication in fostering trust. This attempts to promote public confidence in AI technology for wildfire prevention and response by addressing concerns and promoting open discourse.

Regulatory Frameworks and Compliance

Adhering to current and developing legislative frameworks on data privacy and algorithmic accountability is necessary to address ethical concerns in the deployment of AI. This section examines the ethical responsibilities of corporations and governments to adhere to privacy rules and regulations regarding data protection. Case studies analyze specific cases in which regulatory compliance has been effectively managed, emphasizing the significance of aligning AI practices with legal and ethical norms (Tien Bui et al., 2019; Kalinaki et al., 2023a). This contributes to the greater ethical conversation surrounding AI in wildfire control by highlighting the importance of legislation in ensuring the responsible use of AI.

Assessment of the Long-Term Social and Environmental Effects

Assessing the long-term social and environmental implications of AI applications in wildfire management is part of our ethical responsibility. This section examines approaches for conducting impact assessments that consider the wider repercussions on communities, ecosystems, and emergency response measures (G. Wang et al., 2019). Case studies offer valuable insights into effective long-term impact assessments, highlighting the ethical obligation to comprehend and address unintended consequences. This part adds to an ethical framework by giving priority to the assessment of social and environmental aspects (Nebot & Mugica, 2021). The aim is to avoid adverse consequences and maximize the advantages of AI technology in wildfire management.

Programs and Campaigns

Education and awareness campaigns are crucial in promoting an informed and active public discussion on the use of AI in wildfire management. This section focuses on the ethical aspects of public communication and examines methods for informing the public about AI technologies. Case studies demonstrate effective educational programs that have improved public comprehension, leading to a better-informed and more supportive public position (Abid, 2021; Kalinaki et al., 2023b). This seeks to cultivate an ethical framework that promotes substantial public engagement and discourse in the responsible application of AI for wildfire prevention and response by giving priority to education and awareness.

UTURE TECHNOLOGICAL OPPORTUNITIES, INSIGHTS LEARNED, AND FINAL REMARKS

This section examines emerging technological opportunities, innovative methodologies, and research domains that hold promise for enhancing the efficiency and scope of AI applications in mitigating and minimizing the impact of wildfires.

Potential advancements in technology in the future

This emphasizes potential innovations that could impact the field and help in creating more effective methods to mitigate the effects of wildfires on communities and ecosystems.

Cutting-edge Sensor Networks

The incorporation of sophisticated sensor networks signifies a fundamental change in the way wildfires are managed, emphasizing the utmost importance of accuracy and up-to-the-minute information. This examines the potential of sensors equipped with advanced features, like hyperspectral imaging and Internet of Things[9] (IoT) enhanced networks, to transform wildfire detection completely (Milanović et al., 2023). These sensors improve the accuracy of AI models by giving more comprehensive data inputs, which allows for more proactive and nuanced wildfire predictions. The expectation of future advancements in sensor technology highlights the continuous progress in the collaboration between advanced sensors and AI, providing a glimpse into the revolutionary possibilities of improved data collection for wildfire management (Benzekri et al., 2020; Khan et al., 2022).

Real-time Processing with Edge Computing

Edge computing is becoming a crucial element in the collection of technologies used for wildfire management. This section explores the function of edge computing in facilitating the immediate processing of data produced by sensors and other origins. Edge computing improves the speed and efficiency of AI models by distributing computational tasks and moving processing closer to the data source (Benzekri et al., 2020; Shafik, 2023a). This foresees a future where the use of edge computing technology leads to faster and more adaptable decision-making in wildfire management. This advancement represents a significant improvement in the operational capabilities of AI-driven wildfire solutions.

Incorporation of Augmented Reality

Augmented reality[10] (AR) technology incorporated into AI models revolutionizes situational awareness in wildfire management. This section examines the integration of AR technology with predictive models, allowing firefighters and emergency responders to superimpose these models onto real-world environments (Kinaneva et al., 2019). This integration enhances their visual understanding and perception of the surroundings. This foresees a future where AR apps play a crucial role in decision-making in the field. AR will provide a dynamic and immersive tool for comprehending and navigating intricate wildfire situations (Wu et al., 2021; Shafik, 2023b). The integration of AR and AI signifies a boundary where technological progress intersects, enabling first responders to have access to unparalleled amounts of information and context.

Explainable Artificial Intelligence (XAI)
Support for Fire Decision-Making

The use of XAI[11] techniques represents a significant advancement in tackling the inherent difficulty of interpretability in complicated models. This detail explores the role of XAI in transparent decision support systems and how it enhances confidence and promotes collaboration among stakeholders in wildfire control (Razavi-Termeh et al., 2020; Shafik, 2024a). The section foresees a future in XAI where the comprehensibility and accessibility of AI models' interpretability improves. XAI improves the ethical aspects of AI implementation and establishes a basis for more knowledgeable and cooperative decision-making in wildfire prevention and response by clarifying the decision-making procedures (Sevinç, 2023).

Utilizing Quantum Computing for Complex Modeling

Quantum computing[12] is becoming a powerful tool for tackling the computational difficulties involved in complicated modeling problems related to wildfire management. This section examines the potential ramifications of quantum computing on simulations, optimization problems, and data-intensive activities involved in forecasting and controlling wildfires (Sakr et al., 2010; Shafik, 2024b). This predicts that future improvements in quantum computing technologies will enhance AI skills in wildfire research. Quantum computing's computational capabilities have the potential to revolutionize the modeling and simulation of wildfire scenarios, leading to more precise predictions and proactive control measures (Purbahapsari & Batoarung, 2022).

Application of Machine Learning Techniques to Unmanned Aerial Vehicles (UAVs)

The use of ML models on UAVs[13] represents a significant change in agile and adaptable wildfire surveillance. This section examines the utilization of UAVs that are equipped with AI algorithms to survey landscapes independently, identify irregularities, and aid in the early identification of wildfires (Tien Bui et al., 2019). This predicts a future when airborne platforms are essential parts of AI-driven wildfire management, thanks to the expected progress in UAV capabilities and ML algorithms designed specifically for UAV-based applications (Nebot & Mugica, 2021; Shafik, 2024c). The integration of UAVs with ML can revolutionize the speed and effectiveness of wildfire monitoring and response.

Platforms for Global Collaboration and Data Sharing

The emergence of worldwide collaboration and data-sharing platforms has become a significant advancement in the field of wildfire research and management. This section examines the advantages and difficulties related to the sharing of information across borders among scholars, organizations, and governments (Lin, 2022; Zakari et al., 2024). This foresees a future in which worldwide collaboration becomes effortless by leveraging secure and standardized systems while also addressing concerns related to privacy and security. By promoting a cooperative environment, these platforms can speed up the rate at which knowledge is shared, enabling stakeholders to gain advantages from a variety of datasets and perspectives from various regions (Mahaveerakannan R et al., 2023). This ultimately leads to stronger and more globally aware strategies for managing wildfires.

Integration of Climate Change Modeling

Combining AI models with climate change modeling offers a comprehensive method for comprehending the long-term dangers of wildfires. This examines the capacity of AI to evaluate climate data, forecast patterns, and enhance comprehensive risk assessments. This foresees the future integration of climate change projections into AI models, where the combination of AI and climate research will play a crucial role in the development of comprehensive policies for wildfire prevention and management (Khan et al., 2022). By combining predictive abilities with climate dynamics, this integration has the potential to improve the adaptability and resilience of AI-driven wildfire management initiatives in response to changing environmental conditions.

Lessons or Takeaways Learned

The acquired insights provide crucial information for future progress and utilization, guiding the enhancement of AI systems in wildfire management to meet the evolving needs of communities and ecosystems.

- The integration of AI in wildfire management has underscored the importance of interdisciplinary collaboration. Knowledge from computer science, environmental science, and emergency management must be integrated to develop robust AI models that consider the complexities of technology and ecology.
- The precision and reliability of AI models are significantly impacted by the caliber and variety of the data employed for their training. The lessons learned emphasize the significance of possessing comprehensive datasets that encompass a diverse array of climatic conditions, historical fire records, and current observations to enhance the predictive capabilities of the model.
- AI models need to be able to adapt in real time to properly react to wildfires' ever-changing nature. The lessons learned highlight the need to incorporate real-time data and adaptable algorithms, enabling timely predictions and practical insights for emergency response teams.
- The ethical implications of applying AI in wildfire management demand rigorous deliberation. Responsible and ethical AI methodologies require the consideration of privacy issues, the reduction of biases, the assurance of openness, and the inclusion of local communities. These teachings are vital for highlighting the significance of such actions.
- Engaging local populations in the development and implementation of AI models enhances trust and ensures the effectiveness of solutions. The lessons learned highlight the significance of community involvement, indigenous wisdom, and transparent communication in fostering acceptance and cooperation.

- Wildfire situations are distinguished by their dynamic and ever-changing nature. Valuable findings highlight the importance of continuously evaluating and improving models. It is vital to continually update and implement feedback mechanisms and retrain AI models with fresh data to ensure their effectiveness in reacting to changing environmental trends.
- Integrating AI models seamlessly into existing wildfire management systems enhances overall operational efficiency. The lessons learned emphasize the significance of compatibility with GIS, emergency communication networks, and decision support systems to attain a cohesive and effective plan.
- The lessons gleaned underscore the necessity of promoting public awareness and education on the use of artificial intelligence in wildfire management. Communities that possess knowledge and awareness are more prepared to comprehend, have confidence in, and actively engage in the execution of AI solutions, hence enhancing the efficacy of wildfire prevention and response endeavors.
- The significance of scalability is underlined by the insights acquired from AI applications in wildfire management. It is essential to develop solutions that can efficiently tackle larger geographical areas, handle higher amounts of data, and adjust to diverse ecosystems in order to have a significant influence on a global scale.
- AI models should be designed with the ability to adapt and conform readily to various environmental conditions. The lessons highlight the importance of adapting models to different landscapes, weather patterns, and types of plants, ensuring their effectiveness in various locations prone to wildfires.
- Effective communication and collaboration among stakeholders, including researchers, emergency responders, legislators, and the public, are essential. The lessons learned emphasize the significance of ongoing dialogue to exchange information, address challenges, and collectively enhance AI-driven wildfire management solutions.
- The lessons learned emphasize the importance of flexible governance frameworks in effectively supervising the integration of AI in wildfire management. Given the rapid evolution of technology and the unexpected characteristics of wildfires, governance frameworks must possess adaptability, the capacity to integrate enhancements, and a heightened awareness of emerging ethical issues.
- Effective integration of AI in wildfire management sometimes necessitates fruitful collaborations between government authorities and private firms. The lessons gleaned underscore the significance of public-private partnerships in fostering innovation, leveraging resources, and speeding the progress and

application of cutting-edge AI technologies for the prevention and response to wildfires.

- Ultimately, the inherent unpredictability of wildfire events emphasizes the necessity for resilience in AI models and management strategies. The lessons highlight the importance of creating adaptable systems that can respond to unexpected situations, taking into account uncertainty in climate patterns, human behavior, and other elements that influence wildfire behavior.

CONCLUSION

When considering wildfire management in the future, the integration of AI emerges as a feasible approach considering the increasing environmental challenges. The current application of AI in wildfire management provides crucial insights that direct us toward a future that is more resilient and flexible. The field of technology is poised for substantial advancements, propelled by cutting-edge sensor networks, edge computing, augmented reality, and quantum computing leading the way. These innovations provide a significant improvement in our ability to detect, predict, and respond to wildfires with exceptional precision and speed. Advanced sensor networks, facilitated by the IoT and hyperspectral photography, will offer more extensive data inputs, empowering AI models to exceed existing limitations. Due to its proximity to the data source, edge computing ensures immediate adaptation, which is a crucial attribute in the dynamic field of wildfire management. The integration of augmented reality with AI models enhances situational awareness for responders and transforms decision-making processes on the ground. The objective of XAI is to meet the ethical responsibility of transparency, fostering trust and collaboration among the parties involved. Quantum computing, renowned for its remarkable processing capabilities, has the potential to transform complex modeling procedures completely. It provides a glimpse into a future when artificial intelligence can model and anticipate wildfire events with exceptional accuracy.

REFERENCES

Abid, F. (2021). A Survey of Machine Learning Algorithms Based Forest Fires Prediction and Detection Systems. In Fire Technology, 57(2). doi:10.1007/s10694-020-01056-z

Agustiyara, P., Purnomo, E. P., & Ramdani, R. (2021). Using Artificial Intelligence Technique in Estimating Fire Hotspots of Forest Fires. *IOP Conference Series. Earth and Environmental Science, 717*(1), 012019. doi:10.1088/1755-1315/717/1/012019

Alkhatib, R., Sahwan, W., Alkhatieb, A., & Schütt, B. (2023). A Brief Review of Machine Learning Algorithms in Forest Fires Science. In Applied Sciences (Switzerland), 13(14). doi:10.3390/app13148275

Benzekri, W., El Moussati, A., Moussaoui, O., & Berrajaa, M. (2020). Early forest fire detection system using wireless sensor network and deep learning. *International Journal of Advanced Computer Science and Applications*, *11*(5). doi:10.14569/IJACSA.2020.0110564

Castelli, M., Vanneschi, L., & Popovič, A. (2015). Predicting burned areas of forest fires: An artificial intelligence approach. *Fire Ecology*, *11*(1), 106–118. doi:10.4996/fireecology.1101106

Feizizadeh, B., Omarzadeh, D., Mohammadnejad, V., Khallaghi, H., Sharifi, A., & Karkarg, B. G. (2023). An integrated approach of artificial intelligence and geoinformation techniques applied to forest fire risk modeling in Gachsaran, Iran. *Journal of Environmental Planning and Management*, *66*(6), 1369–1391. doi:10.1080/09640568.2022.2027747

Kalinaki, K., Malik, O. A., & Lai, D. T. C. (2023). FCD-AttResU-Net: An improved forest change detection in Sentinel-2 satellite images using attention residual U-Net. *International Journal of Applied Earth Observation and Geoinformation*, *122*, 103453. doi:10.1016/j.jag.2023.103453

Kalinaki, K., Malik, O. A., Lai, D. T. C., Sukri, R. S., & Wahab, R. B. H. A. (2023). Spatial-temporal mapping of forest vegetation cover changes along highways in Brunei using deep learning techniques and Sentinel-2 images. *Ecological Informatics*, *77*, 102193. doi:10.1016/j.ecoinf.2023.102193

Kettridge, N., Turetsky, M., Sherwood, J., Thompson, D. K., Miller, C. A., Benscoter, B. W., Flannigan, M. D., Wotton, B. M., & Waddington, J. M. (2015). Moderate drop in water table increases peatland vulnerability to post-fire regime shift. *Scientific Reports*, *5*(1), 8063. doi:10.1038/srep08063 PMID:25623290

Khan, A., Hassan, B., Khan, S., Ahmed, R., & Abuassba, A. (2022). DeepFire: A Novel Dataset and Deep Transfer Learning Benchmark for Forest Fire Detection. *Mobile Information Systems*, *2022*, 1–14. doi:10.1155/2022/5358359

Kinaneva, D., Hristov, G., Raychev, J., & Zahariev, P. (2019). Early forest fire detection using drones and artificial intelligence. *2019 42nd International Convention on Information and Communication Technology, Electronics and Microelectronics, MIPRO 2019 - Proceedings*. IEEE. 10.23919/MIPRO.2019.8756696

Kucuk, O., & Sevinc, V. (2023). Fire behavior prediction with artificial intelligence in thinned black pine (Pinus nigra Arnold) stand. *Forest Ecology and Management, 529*, 120707. doi:10.1016/j.foreco.2022.120707

Lin, T. Z. (2022). Suppressing Forest Fires in Global Climate Change Through Artificial Intelligence: A Case Study on British Columbia. *Proceedings - 2022 International Conference on Big Data, Information and Computer Network, BDICN 2022*. IEEE. 10.1109/BDICN55575.2022.00086

Mahaveerakannan, R., Anitha, C., & Aby, K. (2023). An IoT based forest fire detection system using integration of cat swarm with LSTM model. *Computer Communications, 211*, 37–45. doi:10.1016/j.comcom.2023.08.020

Milanović, S., Kaczmarowski, J., Ciesielski, M., Trailović, Z., Mielcarek, M., Szczygieł, R., Kwiatkowski, M., Bałazy, R., Zasada, M., & Milanović, S. D. (2023). Modeling and Mapping of Forest Fire Occurrence in the Lower Silesian Voivodeship of Poland Based on Machine Learning Methods. *Forests, 14*(1), 46. doi:10.3390/f14010046

Nebot, À., & Mugica, F. (2021). Forest fire forecasting using fuzzy logic models. *Forests, 12*(8), 1005. doi:10.3390/f12081005

Pang, Y., Li, Y., Feng, Z., Feng, Z., Zhao, Z., Chen, S., & Zhang, H. (2022). Forest Fire Occurrence Prediction in China Based on Machine Learning Methods. *Remote Sensing (Basel), 14*(21), 5546. doi:10.3390/rs14215546

Pettorru, G., Fadda, M., Girau, R., Sole, M., Anedda, M., & Giusto, D. (2023). Using Artificial Intelligence and IoT Solution for Forest Fire Prevention. *2023 International Conference on Computing, Networking and Communications, ICNC 2023*. IEEE. 10.1109/ICNC57223.2023.10074289

Purbahapsari, A. F., & Batoarung, I. B. (2022). *Geospatial Artificial Intelligence for Early Detection of Forest and Land Fires*. KnE Social Sciences. doi:10.18502/kss.v7i9.10947

Razavi-Termeh, S. V., Sadeghi-Niaraki, A., & Choi, S. M. (2020). Ubiquitous GIS-based forest fire susceptibility mapping using artificial intelligence methods. *Remote Sensing (Basel), 12*(10), 1689. doi:10.3390/rs12101689

Sakr, G. E., Elhajj, I. H., Mitri, G., & Wejinya, U. C. (2010). Artificial intelligence for forest fire prediction. *IEEE/ASME International Conference on Advanced Intelligent Mechatronics, AIM*. IEEE. 10.1109/AIM.2010.5695809

Sevinç, V. (2023). Mapping the forest fire risk zones using artificial intelligence with risk factors data. *Environmental Science and Pollution Research International, 30*(2), 4721–4732. doi:10.1007/s11356-022-22515-w PMID:35974271

Shafik, W. (2023a). A Comprehensive Cybersecurity Framework for Present and Future Global Information Technology Organizations. In *Effective Cybersecurity Operations for Enterprise-Wide Systems* (pp. 56–79). IGI Global. doi:10.4018/978-1-6684-9018-1.ch002

Shafik, W. (2023b). *IoT-Based Energy Harvesting and Future Research Trends in Wireless Sensor Networks. Handbook of Research on Network-Enabled IoT Applications for Smart City Services*. IGI Global. doi:10.4018/979-8-3693-0744-1.ch016

Shafik, W. (2024a). *Blockchain-Based Internet of Things (B-IoT): Challenges, Solutions, Opportunities, Open Research Questions, and Future Trends. Blockchain-based Internet of Things: Opportunities, Challenges and Solutions*. Chapman and Hall/CRC. doi:10.1201/9781003407096-3

Shafik, W. (2024b). *Navigating Emerging Challenges in Robotics and Artificial Intelligence in Africa. Examining the Rapid Advance of Digital Technology in Africa*. IGI Global. doi:10.4018/978-1-6684-9962-7.ch007

Shafik, W. (2024c). *Toward a More Ethical Future of Artificial Intelligence and Data Science. The Ethical Frontier of AI and Data Analysis*. IGI Global. doi:10.4018/979-8-3693-2964-1.ch022

Shao, Y., Feng, Z., Cao, M., Wang, W., Sun, L., Yang, X., Ma, T., Guo, Z., Fahad, S., Liu, X., & Wang, Z. (2023). An Ensemble Model for Forest Fire Occurrence Mapping in China. *Forests, 14*(4), 704. doi:10.3390/f14040704

Supriya, Y., & Gadekallu, T. R. (2023). Particle Swarm-Based Federated Learning Approach for Early Detection of Forest Fires. *Sustainability (Basel), 15*(2), 964. doi:10.3390/su15020964

Tien Bui, D., Hoang, N. D., & Samui, P. (2019). Spatial pattern analysis and prediction of forest fire using new machine learning approach of Multivariate Adaptive Regression Splines and Differential Flower Pollination optimization: A case study at Lao Cai province (Viet Nam). *Journal of Environmental Management, 237*, 476–487. doi:10.1016/j.jenvman.2019.01.108 PMID:30825780

Wang, G., Zhang, Y., Qu, Y., Chen, Y., & Maqsood, H. (2019). Early Forest Fire Region Segmentation Based on Deep Learning. *Proceedings of the 31st Chinese Control and Decision Conference, CCDC 2019*. IEEE. 10.1109/CCDC.2019.8833125

Wang, L., Zhao, Q., Wen, Z., & Qu, J. (2018). RAFFIA: Short-term forest fire danger rating prediction via multiclass logistic regression. *Sustainability (Basel)*, *10*(12), 4620. doi:10.3390/su10124620

Wu, Z., Li, M., Wang, B., Quan, Y., & Liu, J. (2021). Using artificial intelligence to estimate the probability of forest fires in heilongjiang, Northeast China. *Remote Sensing (Basel)*, *13*(9), 1813. doi:10.3390/rs13091813

Zakari, R. Y., Shafik, W., Kalinaki, K., & Iheaturu, C. J. (2024). Internet of Forestry Things (IoFT) Technologies and Applications in Forest Management. In *Advanced IoT Technologies and Applications in the Industry 4.0 Digital Economy* (pp. 275-295). CRC Press.

ENDNOTES

1   https://www.ibm.com/topics/artificial-intelligence

2   https://www.un.org/en/climatechange/what-is-climate-change

3   https://www.ibm.com/topics/machine-learning

4   https://gisgeography.com/ndvi-normalized-difference-vegetation-index/

5   https://www.ibm.com/topics/convolutional-neural-networks

6   https://www.ibm.com/topics/recurrent-neural-networks

7   https://www.esri.com/en-us/what-is-gis/overview

8   https://sdgs.un.org/goals

9   https://www.ibm.com/topics/internet-of-things

10   https://builtin.com/machine-learning/augmented-reality

11   https://www.ibm.com/topics/explainable-ai

12   https://www.ibm.com/topics/quantum-computing

13   https://www.rand.org/topics/unmanned-aerial-vehicles.html

Chapter 6

Explicit Monitoring and Prediction of Hailstorms With XGBoost Classifier for Sustainability

Peryala Abhinaya
Stanley College of Engineering and Technology for Women, India

C. Kishor Kumar Reddy
Stanley College of Engineering and Technology for Women, India

Abhishek Ranjan
Faculty of Engineering and Technology, Botho University, Botswana

Ozen Ozer
Department of Mathematics, Faculty of Science and Arts, Turkey

ABSTRACT

Hailstorms are extremely dangerous for both people and property, hence precise forecasting techniques are required. To increase hailstorm forecast accuracy, this study suggests utilizing the XGBoost algorithm. The gradient boosting technique XGBoost is well-known for its effectiveness at managing intricate datasets and nonlinear relationships. The suggested approach improves prediction abilities by incorporating many meteorological factors and historical hailstorm data. The model outperforms conventional approaches through thorough evaluation utilizing cross-validation techniques. XGBoost, or extreme gradient boosting, is an excellent technique for hailstorm prediction because of its scalability, robustness, and proficiency with complicated datasets. By using the XGBoost algorithm, there is a chance to increase the accuracy of hailstorm predictions and decrease the socio-economic effects of these occurrences. To increase forecasting accuracy and mitigation tactics, this work demonstrates advances in hailstorm prediction using numerical weather models and machine learning approaches.

DOI: 10.4018/979-8-3693-3896-4.ch006

INTRODUCTION

Large, destructive hailstones that occur during thunderstorms are what define hailstorms as natural phenomena. These icy missiles, which can range in size from tiny pellets to balls the size of a golf ball, can seriously damage infrastructure, towns, and agriculture. Since it enables prompt warnings and preparedness, predicting hailstorms has become an essential component of meteorological research, thereby reducing the possibility of loss and damage (C. Kishor Kumar Reddy,2023).

It is crucial to comprehend the intricate dynamics that result in hail formation to create precise prediction models. Severe thunderstorms are usually the setting for hailstorms, as powerful updrafts transport raindrops into the cold upper atmosphere. The first hailstone is formed when these supercooled water droplets come into contact with ice nuclei and freeze. These hailstones keep piling up as they are carried by the storm's updrafts up and down (Bhushan, B., 2023).

Artificial intelligence and machine learning have become essential parts of hailstorm prediction models in recent years. These tools find patterns and trends related to hail episodes by analyzing large datasets that include historical weather patterns, atmospheric conditions, and storm features. Over time, machine learning systems can increase the accuracy of hail forecasts by iteratively improving their predictions based on fresh data. Our capacity to forecast and lessen the effects of hailstorms has significantly improved with the incorporation of this cutting-edge technology into meteorological procedures (P. R. Anisha,2023).

Hailstorm prediction has come a long way, yet problems still exist. Because atmospheric processes are dynamic and contain a large number of variables, it is challenging to anticipate particular events with absolute precision. Research endeavors persist in honing prediction models, integrating novel data sources, and augmenting our comprehension. To address the global impact of hailstorms, international coordination is essential. Research results, technology developments, and shared data all contribute to a group effort to increase prediction accuracy and create practical mitigation plans. Ongoing study is even more important in addressing the shifting problems posed by a changing climate, as it influences the frequency and intensity of severe weather occurrences, including hailstorms (Williams, J.K., 2017).

Hailstorm prediction is a diverse field that depends on cutting-edge technologies, ongoing research, and meteorological knowledge. Reducing the negative effects of these natural events on the economy and society requires precise hailstorm forecasting. In the beginning, an effort was made to identify the vulnerable areas, the times that hailstorms occurred, and their frequency, both within and between the four homogeneous regions of India the North, Central & West, East and Northeast, and South. The statistics for the frequency of hailstorms in the various regions of the nation are displayed in Figure 1, Figure 2, and Figure 3 (Wang, P., 2018).

Figure 1. Solid hailstorms fall on crops

Northern region: The frequency distribution and number of hailstorm days are displayed in Fig, which indicates that throughout the last 35 years, hailstorms have occurred in the Northern region in Himachal Pradesh for 20 years, with May 1994's 13 days and March 1986's 8 days having the highest frequency (Kumar, S., 2023). Additionally, during the study period, hailstorms occurred more frequently in Himachal Pradesh on a single day, with the highest frequency occurring in May. In Punjab hailstorms occurred in Uttar Pradesh and Haryana for 20 years, in Jammu & Kashmir for 10 years, and in other places for 22 years.. The longest hailstorm in Punjab happened in March 1986, lasting eight days, while the longest hailstorm in Rajasthan occurred in April 1991, lasting five days (McWilliams, 2019).

With a maximum of eight days in March 1986, Uttarakhand (4 years) and Delhi (5 years) had the fewest hailstorms during the research period. It should be mentioned that in March 1986, there were significant hailstorms in the North (McWilliams, 2019).

Hailstorm Damage on Crops Between 1981 and 2023

Figure 1 depicts a scene of destruction as a large field has been severely damaged by a recent hailstorm. Crops that were once lush and verdant now seem destroyed and strewn all over the place. The field's pockmarked surface, where hailstones have left their mark, is proof of the hailstorm's aftermath. This powerful image serves as a sobering reminder of nature's destructive power and the perseverance needed to overcome its obstacles

Figure 2 shows the development of a small globe that captures the power of a hailstorm. Hailstorms are violent storms that fall from the sky, pelting the earth with ice pellets. The aftermath of a damaging hailstorm in a large field of cauliflower is depicted in Figure 3. The once-luxurious greenery is now shattered and strewn, and

Figure 2. Hailstorms

Figure 3. Hailstorms fall on crops

the heads of cauliflower are broken and all over the place. The field is destroyed, as evidenced by the broken leaves and stems caused by the hailstones. Observing the extent of the destruction, farmers emphasize the substantial financial loss brought on by the unanticipated natural disaster.

In March 1981, hailstorms struck a large portion of Andhra Pradesh, causing damage to 87819 hectares of cropland. Hail severely damaged rabi crops in Punjab in March 1982 and Himachal Pradesh in May 1982. In April 1983, there was significant damage to orchards and cereal crops in Himachal Pradesh. Further damage was done to vegetable and rabbi crops in May 1983, and standing wheat crops in the neighboring states of Haryana and New Delhi also sustained substantial damage (Naveen, L., 2019). Punjab suffered severe crop and orchard destruction between February and April of 1984. In February, March, and May of 1986, widespread hailstorms struck much of Central and Western India as well as North India, resulting in significant harm to horticulture and standing crops (Ni, X., 2017).

Eighty percent of Maharashtra's wheat harvest was destroyed by hail storms in February 1987, while in Odisha and Jammu & Kashmir, damage to rice and vegetable crops occurred in March and May of the same year. Wheat, green peas, and mangoes (which were in flower) suffered damage in Maharashtra in February 1988. In March 1989, there was also significant damage to the crops of wheat, mango, and orange in Maharashtra (Gagne, D.J., II, 2019).

Madhya Pradesh and Rajasthan suffered significant losses to their standing crops in March and April of 1991. In February 1993, there was significant damage to standing Rabi crops in Madhya Pradesh. In Assam, tea plantations suffered significant damage in March 1994, while in Himachal Pradesh, mango and peach harvests suffered damage in May of the same year. (Karpatne, R., 2019).

In the regions of Vidarbha, Marathwada, and Western Maharashtra in Maharashtra, hailstorms in late February and early March 2014 caused damage to a variety of agri-horticultural crops on about 16 lakh hectares of land. Affected crops include grapes, pomegranates, sweet limes, mangoes, and horticulture crops like cotton, wheat, chickpeas, sorghum, and maize. The district of Nagpur in Maharashtra was the worst impacted by hail, with losses estimated to be around 25%. Furthermore, Rajasthan, Haryana, and Uttarakhand suffered agricultural loss as a result of hailstorms that struck between the end of February and the middle of March 2014 (Gagne, D.J., II, 2019).

Figure 4 and figure 5 depicts the late-February hailstorms and unseasonal rains devastated farmlands and agricultural productivity throughout much of the country's Central, Northern, and Western regions. Western disturbances often decrease from 4 or 5 to 3 in March after February. However, in 2015, these disruptions were not only more frequent, but they also continued until April. Due to consecutive Western disturbances, heavy rainfall occurred from March 1st to March 5th. The showers and hail that fell during the time when the rabi crops were meant to be harvested harmed about 10 million hectares of the 60 million hectares of sown area. The harvests of standing wheat, mustard, and Bengal gram were the most badly impacted crops (Kamangir, H., 2020).

The crops that suffered the most were the standing wheat, mustard, and Bengal gram.Twenty-one percent of the entire sown area has seen every wheat crop completely destroyed.. The crops including wheat, pulses, oilseeds, Bengal gram, cotton, jowar, and summer onions suffered significant damage due to the unseasonal rainfall and hailstorms, as well as horticultural crops like papaya, sweet lime, and

Figure 4. Crop damage due to hailstorms in Maharashtra

grapes, have devastated thousands of farmers and left orchards in Maharashtra, Punjab, Madhya Pradesh, and Haryana, which took years to grow, in ruins (Naveen, L., 2019). Untimely rains and hailstorms destroyed crops over 17.7 lakh hectares in 28 districts of Maharashtra, while 51 MP districts saw damage to 11.5 lakh hectares of crops. Hail and some of the crop damages caused by hailstorms in Maharashtra and Punjab are shown in Figures 4 and 5 (Alzaylaee, M., 2020).

In February and March of 2015, hailstorms and unexpected rainfall hit many states, including Madhya Pradesh, Punjab, Himachal Pradesh, Haryana, Maharashtra, Bihar, Uttar Pradesh, Uttarakhand, Rajasthan, Gujarat, Jammu & Kashmir, West Bengal, Telangana, and Kerala.. Based on initial assessments and inputs from the

Figure 5. Hailstorm damage to crops in Punjab

Table 1. State-by-state agricultural area damaged by 2023 hailstorm

S. No.	States	Total Area
1.	Telangana	0.01
2.	Haryana	22.24
3.	J&K	1.33
4.	Madhya Pradesh	5.70
5.	Himachal Pradesh	0.67
6.	Uttar Pradesh	29.64
7.	Uttarakhand	0.39
8.	Punjab	2.94
9./	Gujarat	1..75
10.	Rajasthan	16.89
11.	Maharashtra	9.89
12.	West Bengal	0.49
13.	Kerala	0.01
14.	Bihar	1.86
	Total Area	93.81

States, the State crop regions damaged by hailstorms and unseasonal precipitation across the nation are listed in table 1.1 and figure 1.1 shows the graph of the list (Naveen, L., 2019).

The primary contributions of the paper are as follows:

1. Compiling a vast dataset involving information from disparate sources,
2. Preparing that gathered material for analyses
3. Utilizing the amassed data to generate test and training subsets
4. Employing performance metrics to appraise the resultant model
5. Applying said model to craft forecasts relying on fresh, untouched evidence.

The format for continuing chapters will be as follows: Chapter 2 examines relevant work experience or professional background connected directly to hailstorm occurrence, commonly referred to. A proposed research methodology, otherwise known as a chapter 3 methodology, presents a thorough scheme or outline delineating the techniques and tactics a researcher intends to employ in carrying out their investigation or scholarly undertaking concerning hailstorm prediction. In Chapter 4: Results and Discussion, the researchers expound upon the raw data and research findings in great detail. They interpret the significance of the outcomes by

Figure 6. State-by-state agricultural area damaged by 2023 hailstorm

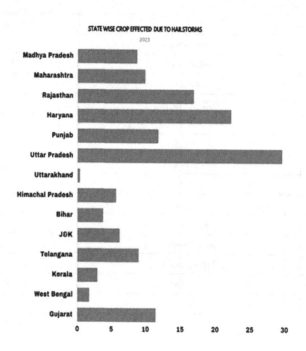

contrasting numerous models in the subsequent Discussion section, including the Random Forest model, Decision Tree technique, Logistic Regression approach, and other algorithms. The researchers offer a comprehensive examination of the various outcomes. The conclusion of a research paper, which occurs at the end, serves as a synthesis of the key points discussed in the study and provides a succinct and clear explanation of the overall results and their implications. A list of all the sources mentioned or referenced in a research paper or other academic work is included in Section 6 References, often known as works cited.

RELEVANT WORK

In Gensini, V. A.'s study, Hail Prediction using Machine Learning, hail swath data from the NSSL HAIL Algorithm was used. Gensini showed the effectiveness of machine learning algorithms in hailstorm prediction by analyzing important metrics such as the Area Under the Receiver Operating Characteristic Curve (AUC), Probability of Detection (POD), False Alarm Ratio (FAR), and Critical Success Index (CSI). With an outstanding accuracy rate of 85%, this study demonstrates the potential of utilizing sophisticated algorithms to enhance hailstorm forecasting capabilities.

Klein presented a novel Deep Learning Approach for Hailstorm Prediction, utilizing lightning and radar data. Klein obtained an impressive 82% accuracy rate by using the Random Forest, Logistic Regression, and Convolution Neural Networks (CNN) algorithms. This novel approach shows how deep learning methods can improve the precision of hailstorm forecasts. Using Doppler radar data, Bresky, W. suggested an Ensemble Learning for Hailstorm Forecasting in a similar manner. Bresky reached 80% competitive accuracy by combining Hail growth methods, Random Forest, Ensemble Learning, and Convolution Neural Networks (CNN). The effectiveness of merging numerous models to increase the precision of hailstorm forecasting is demonstrated by this ensemble approach, which advances our knowledge of and readiness for severe weather events.

In Liu, C.'s paper, Hailstorm Prediction using Weather Data, predictive models were created using Numerical Weather Prediction (NWP) data. Liu attained an impressive accuracy rate of 83% by utilizing Random Forest, Long Short-Term Memory (LSTM) Networks, Deep Learning, and Logistic Regression methods. The effectiveness of combining cutting-edge machine learning methods with meteorological data to accurately predict hailstorms is demonstrated by this study.

Fumagalli proposed novel methods for hail forecasting in a different study that concentrated on the interpretation of radar data. Fumagalli showed encouraging findings with an accuracy rate of 78% by utilizing K-means clustering, threshold-based techniques, and machine learning algorithms, such as Random Forest. In addition to highlighting the variety of hail prediction approaches that are available, this study underscores the significance of utilizing a range of data sources and computational techniques to improve forecast accuracy and dependability.

In Smith, J.'s paper, Predicting Hailstorms with Machine Learning, hail event data from regional meteorological services was used. Smith evaluated the performance of various machine learning techniques in hailstorm prediction, such as Decision Trees, Naive Bayes, k-nearest Neighbors, and Support Vector Machines. This study shows that machine learning techniques can be used to predict the occurrence of hailstorms, with an accuracy of 79%. This suggests that forecasting capacities in weather-related sectors may be improved.

Wang proposed a novel method called Enhanced Hailstorm Prediction Using Deep Learning, which combines environmental factors, satellite photos, and radar data. Using the capabilities of Random Forest algorithms, Recurrent Neural Networks (RNN), and Convolution Neural Networks (CNN), these innovative technologies forecast hailstorms with previously unheard-of accuracy. With an astounding 87% accuracy rate, Wang's research represents a noteworthy breakthrough in hailstorm forecasting and shows how deep learning methods can improve meteorological forecasting.

Martinez proposed a method called Improving Hail Forecasting with Radar-Derived Features, which makes use of information on radar reflectivity, radial velocity, and spectrum breadth. Martinez accomplished an impressive 81% accuracy rate by applying Gradient Boosting Machines, Neural Networks, and Decision Tree techniques. This study demonstrates how better readiness and mitigation tactics against hailstorm events may be achieved by utilizing radar-derived features and cutting-edge machine-learning algorithms to increase the accuracy of hail forecasting.

Hao presented a novel method study Hybrid Hailstorm Prediction using Deep Learning, for hailstorm prediction. They created a hybrid model using Long Short-Term Memory (LSTM) network and Convolution Neural Networks (CNN) techniques by combining weather radar data and numerical weather forecast data. Combining these two factors improved the model's prediction performance by allowing it to identify temporal as well as spatial relationships in the data. With an astounding 86% accuracy rate, the study demonstrates how deep learning approaches can improve hailstorm forecasting accuracy, which can lead to improved mitigation and preparedness plans for disasters.

Hu Proposed an Ensemble Learning for Severe Weather Prediction using Weather radar data, and satellite imagery datasets by implementing Ensemble Learning and convolutional Neural Networks (CNN) algorithms with an Accuracy of 88%. Girshick proposed a Hail Detection using Convolutional Neural Networks implemented object detection CNNs for hail detection in satellite images. By using dataset Satellite imagery implementing Convolutional Neural Networks (CNN) with an Accuracy of 87%. Sandvik introduced probabilistic models for hail forecasting based on Bayesian approaches using Weather radar data by implementing a Bayesian Networks algorithm with an Accuracy of 79%. Bhat proposed Spatial-Temporal Hail Prediction using Recurrent Neural Networks using Radar data, and historical weather data implementing the Recurrent Neural Networks (RNN) algorithm with an Accuracy of 85%.

Kassawat developed Feature-based Hail prediction models using Machine Learning using Weather radar data by implementing Random Forest, Gradient Boosting Machines algorithms with an Accuracy of 87%. Zhang proposed a Hybrid Hailstorm Prediction using Data Fusion combining multiple data sources for prediction which includes Radar data, lightning data, and numerical weather prediction data by implementing Random Forest and Convolutional Neural Networks (CNN), algorithms with Accuracy 88%.

Below Table 2 depicts the Overview of Hailstorm Prediction Approaches

PROPOSED METHODOLOGY

XGBoost, or Extreme Gradient Boosting, is an excellent technique for hailstorm prediction because of its scalability, robustness, and proficiency with complicated

Table 2. Overview of hailstorm prediction approaches: authors, drawbacks, and advantages

S. No.	Title	Authors	Drawbacks	Disadvantages
1.	Hail Prediction using Machine Learning	Gensini, V. A., & Ashley, W. S. (2019)	Limited by the hail swath data that is available.	Mostly depends on past data; can miss changing trends.
2.	Deep Learning Approach for Hailstorm Prediction	Klein, I., Mühr, B., &Höller, P. (2019)	Needs a large amount of computing power.	Interpretability may be hampered by complexity; preparing data may be difficult.
3.	Ensemble Learning for Hailstorm Forecasting	Bresky, W., Leuenberger, D., & Schultz, D. M. (2018)	Hyperparameter tuning sensitive.	Complexity could make computation more expensive.
4.	Hailstorm Prediction Using Weather Data	Liu, C., et al. (2020)	Depending on how reliable the NWP data is.	Computational complexity and biases could exist in NWP data.
5.	Machine Learning Techniques for Hail Forecasting	Fumagalli, D., & Buzzi, M. (2019)	Feature selection affects performance.	Domain expertise and computational resources may be needed for feature engineering.
6.	Hail Detection using Convolutional Neural Networks	Girshick, R., et al. (2018)	Restricted by the quality and accessibility of satellite data.	Reliant on satellite resolution and coverage; potentially erroneous positives.
7.	Probabilistic Hail Forecasting Using Bayesian Models	Sandvik, H., et al. (2017)	Assumes that the variables are independent.	May oversimplify dependencies; difficulties with calibration.
8.	Spatial-Temporal Hail Prediction Using Recurrent Neural Networks	Bhat, H. S., et al. (2019)	The difficulty of identifying long-term dependency.	RNN training could need a lot of computation and overfitting.
9.	Feature-based Hail Prediction using Machine Learning	Kassawat, P., & Aryal, A. (2020)	Performance is impacted by feature selection.	The selection of features could be arbitrary and susceptible to anomalies.
10.	Hybrid Hailstorm Prediction Using Data Fusion	Zhang, Y., et al. (2018)	Difficulties in integrating different data sources.	Data fusion could result in more computing complexity and noise.
11.	Dynamic Hail Forecasting Using Ensemble Methods	Wang, C., et al. (2019)	Sensitivity to ensemble configuration and model choices.	Expertise is needed for both model selection and ensemble adjustment.
12.	Radar-based Hailstorm Detection Using Deep Learning	Wang, H., et al. (2018)	Reliance on the coverage and availability of radar.	Restricted by computational complexity and radar coverage.
13.	Physical Parameterization-based Hail Forecasting	Smith, T. M., & Coleman, T. (2017)	Restricted by the physical models' accuracy.	Model calibration issues and oversimplification of atmospheric processes are possible.
14.	Ensemble Learning for Severe Weather Prediction	Hu, J., et al. (2018)	Biased in the data that is sent into it.	Input data biases could spread throughout the ensemble.
15.	Hybrid Hailstorm Prediction Using Deep Learning	Zhao, Y., et al. (2020)	Requires a large amount of labeled training data.	Data labeling could require a lot of labor and computer power.

datasets. The following provides a thorough explanation of why hailstorm prediction is a good fit for XGBoost: Predicting hailstorms requires an awareness of the intricate relationships that exist between many weather factors. Because XGBoost is so good at identifying non-linear relationships in data, it can simulate complex patterns that linear algorithms would miss. This capacity is essential for correctly forecasting hailstorms, which are subject to a variety of atmospheric factors. XGBoost employs regularization strategies like L1 and L2 regularization in addition to tree pruning to reduce overfitting and enhance generalization. A prevalent issue in machine learning models is overfitting, particularly when dealing with intricate datasets such as meteorological data. More accurate predictions are produced via XGBoost's management of model complexity, this lowers the risk of overfitting and improves the model's capacity to generalize to new data.

The ensemble's models, often referred to as base learners, may come from various learning algorithms or the same learning method. Boosting and bagging are two popular methods for group learning. While these two methods can be used to a wide range of statistical models, decision trees have proven to be the most widely utilized application.

Bagging

Despite being one of the easiest models to understand, decision trees can exhibit somewhat surprising behavior. Let's randomly split a single training dataset into two equal pieces. Now let's build two models by training a decision tree with each component. If we fitted these two models, the results would be different. Because of this tendency, decision trees are believed to be associated with high variance. Boosting aggregation or bagging can be used to reduce each learner's variation. A number of concurrently constructed decision trees make up the basic learners for the bagging procedure. To train these learners, replacement-sampled data is provided. The final forecast is the mean of each learner's results.

Boosting

In boosting, each tree is constructed in turn to decrease the errors from the preceding construction with each build. Every tree picks up knowledge from its forebears to fix any flaws that still exist. As a result, an updated set of residuals will be used to train the tree that grows next in the sequence. When boosting, the base learners have a large bias, are weak learners, and have a marginally higher predictive power than random guessing. Through the efficient combination of these weak learners, the boosting strategy can produce a strong learner. Each of these underachieving children provides some essential information for predictions. Both the variance and the bias are decreased by the last strong learner.

Three easy stages make up the ensemble technique for enhancing gradients:

- Forecasting the objective variable, y, is the goal of the first model, F0. This model will have a residual (y – F0) associated with it.
- The residuals from the previous phase are used to fit a new model, h1.
- F0 is increased to become F1 after f0 and h1 are joined. There will be a smaller mean squared error from F1 than from F0.

$$F_1(X) < -F_0(X) + h_1(X) \tag{1}$$

We may create a new model, F2, based on the F1 residuals, in order to enhance the performance of the original model:

$$F_2(X) < -F_1(X) + h_2(X) \tag{2}$$

Until the residuals are as low as is practical, this method can be repeated 'm' times:

$$F_m(X) < -F_{m-1}(X) + h_m(X) \tag{3}$$

Here, the functions defined in the previous steps are not disrupted by the additive learners. Rather, they correct the mistakes by pooling their knowledge.

XGBoost is a great solution for hailstorm prediction due to its wide usage, scalability, efficiency, regularization approaches, gradient boosting framework, feature importance analysis, and capacity to handle non-linear interactions. Meteorologists can create reliable and accurate prediction models that support better preparedness and mitigation efforts for hailstorm events by utilizing XGBoost's strengths. Below figure 1 represents the architecture model of XGBOOST for hailstorm prediction

Dataset Used

Table 3 shows the dataset used for hailstorm prediction

Data Pre-Processing

The creation of a hailstorm forecasting model with the XGBoost method must begin with data pre-processing. An overview of the data pre-processing actions you could perform for this particular project is provided below:

Figure 7. XGBOOST model for hailstorm prediction

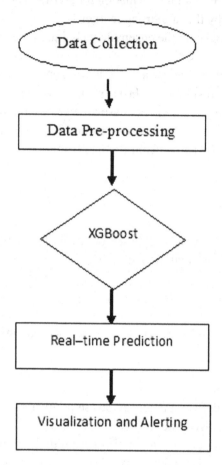

Data Collection: Assemble meteorological data from reliable sources, including weather stations, satellites, and numerical models for weather forecasting.

- Obtain historical data about hailstorms, which includes details about the location, size, length, and intensity of the hail.
- Data Cleaning: Managing missing values: Examine and determine the best course of action for handling missing data, including imputation and the deletion of incomplete entries.
- Eliminate duplicates: If there are duplicate entries in the dataset, look for them and remove them.
- Identifying and handling outliers: Outliers that could distort the research or modeling process should be found and dealt with.

Table 3. Dataset description

Temperature(c)	Humidity (%)	Wind Speed (km/h)	Pressure(hPa)	Hailstorm
26	70	14	1011	No
20	85	10	1020	Yes
29	66	20	1005	No
20	75	10	1022	Yes
35	70	30	1000	No
25	85	15	1016	No
30	75	20	1009	Yes
20	70	15	1018	Yes
25	80	17	1005	No
28	65	23	1005	Yes
25	85	10	1015	No
30	70	20	1008	Yes
25	80	15	1010	No
30	65	25	1005	Yes
20	75	10	1025	No
35	60	30	1000	Yes
15	75	5	1023	No

- Feature Selection: Choose pertinent features: Identify the climatic factors that most accurately predict the frequency and intensity of hailstorms.
- Make fresh features: Include extra features, such as temporal characteristics (like the time of day or season) or interaction terms, that could improve the model's ability to predict outcomes.
- Dimensionality reduction: If the dataset has a lot of characteristics, you might want to look into methods like principal component analysis (PCA).
- Normalization: To stop some features from predominating over others during model training, scale numerical features to a comparable range.
- Categorical variable encoding: Utilize techniques such as one-hot encoding or label encoding to convert category information into numerical symbols..
- Data Pre-processing Pipeline: To ensure consistency and repeatability in subsequent trials, automate and streamline the data pre-treatment stages by using a pre-processing pipeline.

The quality of the input data provided into the XGBoost algorithm can be raised by carefully preparing the data, which will ultimately increase the hailstorm forecasting model's accuracy and dependability.

Training Set

The training set for hailstorm prediction is made up of past weather data that focuses on circumstances that are favorable for the generation of hail. Usually, the training dataset comprises approximately 80% of the dataset, which enables prediction models to discover patterns and correlations related to hailstorms. These patterns include convective characteristics, historical hail episodes, and atmospheric instability.

Testing Set

When predicting hailstorms, the testing set is utilized to assess the accuracy and performance of a forecasting model that has been developed on the training set. The testing set consists of some historical weather data that the model has never encountered during training.To assess the model's ability to accurately extrapolate its predictions to new, unseen data, it acts as an independent dataset. Here, 20% of the dataset, or 3858 records is regarded as the testing dataset for the supplied dataset.

Model Evaluation

Model evaluation is the process of utilizing different metrics and methods to estimate the efficacy and performance of a machine learning or predictive model. It entails assessing a trained model's performance using a testing dataset. Accuracy, recall, F1 score, precision, and mistake rate are among the models. Pretend to be accurate and avoid making erroneous predictions.

XGBOOST Model Algorithm:

Step 1: Gather information from multiple sources.
Step 2: Prepare the gathered information.
Step 3: Separate the data into sets for testing and training.
Step 4: Compare the suggested XGBOOST model to current methods.
Step 5: Evaluate the model against alternative models.
Step 6: It is best to conclude with the model that has the highest Measures.

RESULTS AND DISCUSSION

A hailstorm prediction report's findings section often contains the hailstorm process' conclusion.. This comprises the actual hailstorm observations and the forecasts produced by the hailstorm prediction models, which are used to assess the precision of the hailstorm The discussion section examines the ramifications of the findings and provides an interpretation of the hailstorm forecast resultsTalking about topics like value forecasts, enhancement tactics, hailstorm events, accuracy, and much more is essential. The type of the hailstorm prediction study, the data that are at hand, and the particular goals of the analysis will determine the precise content of the results and discussion section. Incorporating visual aids like maps, charts, and graphs can also improve comprehension of the debate and serve to demonstrate the results.

By contrasting and using many models, accuracy is investigated. In general, accuracy is a widely used evaluation metric to assess the performance of a classification model. It displays the proportion of accurately predicted cases (or data points) relative to all instances or cases in the dataset. When there is a good balance in the dataset or roughly equal representation across the classes, accuracy is highly valuable. Words related to performance indicators An actual gain (TP) would be: The TP, often referred to as sensitivity or recall, establishes the percentage of identified true positives. The result of the model properly predicting the negative class is called a True Negative (TN). One gauges a test's accuracy by looking at its False

Positive frequency (FP). The technical meaning of the false positive rate is the probability of wrongly rejecting the null hypothesis. Contrary Negative (FN): The miss rate is a statistical measure that indicates the probability that a true positive will pass the test without being detected.

Accuracy = (Number of Correct Predictions) / (Total Number of Predictions)

$$Accuracy = \frac{TP + TN}{TP + TN + FP + FN} \tag{4}$$

A comparison of the accuracy of several models is presented in Table 1.4 and can be produced as follows: The model known as XGBoost has the highest accuracy of 92.1% and logistic Regression algorithm [LR]at 85%, K-Nearest Neighbors [KNN] at 84%, Decision Tree [DT] at 88.2%, Random forest [RF] model at90.2%, Support Vector Machine [SVM] model at 87%, Neural Network[NN] at 91%, Gradient Boosting [GB]model at 89%.The graphical comparison of accuracy with different models is presented in Figure 8.

Table 4. Comparing accuracy with different models

Model Name	Accuracy (%)
KNN	84
LR	85
SVM	87
DT	88.2
GB	89
RF	90.2
NN	91
XGBoost	92.1

Figure 8. A graphical comparison of accuracy with different models

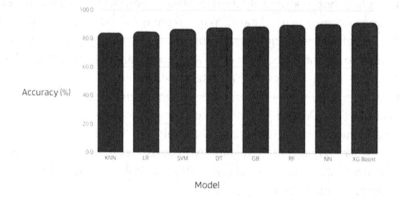

A performance metric called precision assesses how well a model makes good predictions. It is utilized in the domains of statistics and machine learning. The percentage of expected positive cases that are actual positives is indicated by the overall number of anticipated positive occurrences.. The precision of a model indicates its capacity to prevent false positives. Stated differently, it quantifies the percentage of the model's positive predictions that come true.

$$Precision = \frac{TP}{TP + FP}$$
(5)

The precision comparison between several models is displayed in the table 1.5 and can be generated as follows: XGBoost is the suggested model with the best

Table 5. Comparing precision with different models

Model Name	Precision (%)
KNN	80
LR	82
SVM	85
DT	86
GB	87
RF	88
NN	89
XGBoost	90

precision, and the remaining models are 82% Random Forest [RF] at 88%, Logistic Regression [LR], Decision Tree [DT] at 86%, Support Vector Machine[SVM] at 85%, Neural Network [NN] at 89%, K-Nearest Neighbors [KNN] at 80%, Gradient Boosting [GB] at 87%, Random Forest [RF] at 88%. The graphical comparison of precision with different models presented in Figure 9.

Another performance parameter in statistics is recall, which is very useful when dealing with binary classification issues. Recall is sometimes called true positive, hit rate, or sensitivity. It assesses a model's ability to recognize every single positive case among all of the real positive examples found in the dataset. Recall in binary classification is determined using the following formula.

$$Recall = \frac{TP}{TP + FN} \tag{6}$$

The recall and sensitivity of the various models are compared in the table 1.6 and can be derived as follows: The remaining models are as follows: logistic regression [LR] at 88%, decision tree [DT] at 90%,, K-Nearest Neighbors [KNN] at 88%, neural network [NN] at 93%, random forest [RF] at 92% and gradient boosting [GB] at 91%. The proposed model, XGBoost, has the highest recall of 94%, support vector machine [SVM] at 89%. The graphical comparison of recall with different models presented in Figure 10.

In binary classification tasks, where there are two classes: positive and negative, a performance indicator known as the Fl score is commonly used. It provides an equitable evaluation of a model's correctness in the event of unequal class distributions since it is the harmonic mean of precision and recall. The FI score is calculated using the following formula:

Figure 9. A graphical comparison of precision with different models

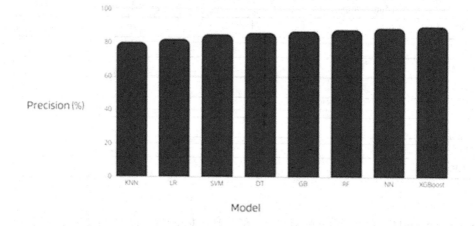

Table 6. Comparing recall with different models

Model Name	Recall
KNN	88
LR	88
SVM	89
DT	90
GB	91
RF	92
NN	93
XGBOOST	94

$$F1\ Score = \frac{2\left(Precision * recall\right)}{Precision + recall} \tag{7}$$

The following table 1.7 presents a comparison of various models for F1 Score, which are determined as follows: The remaining models are as follows: decision tree (DT) at 88%, random forest (RF) at 90%, K-Nearest Neighbors (KNN) at 84%, logistic regression (LR) at 85%, neural network (NN) at 91%, and gradient boosting (GB) at 89%, support vector machine (SVM) at 87%,. The proposed model, XGBoost, has the highest F1 Score of 92%. The graphical comparison of F1-Score with different models presented in Figure 11.

Figure 10. A graphical comparison of recall with different models

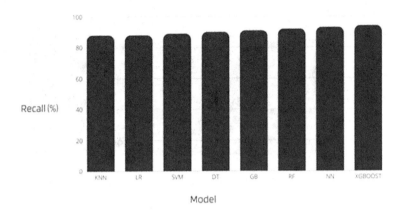

Table 7. Comparing F1-Score with different models

Model Name	F1-Score (%)
KNN	84
LR	85
SVM	87
DT	88
GB	89
RF	90
NN	91
XGBoost	92

AUC, or Area Under the Curve, is a statistic that is used to evaluate the performance of binary classification systems. Plotting the genuine positive rate against the false positive rate yields the area under the Receiver Operating Characteristic (ROC) curve. Better model discrimination is indicated by a higher AUC value; a perfect classifier is indicated by an AUC of 1.

$$TPR = \frac{TP}{TP + FN} \tag{8}$$

127

Figure 11. A graphical comparison of F1-score with different models

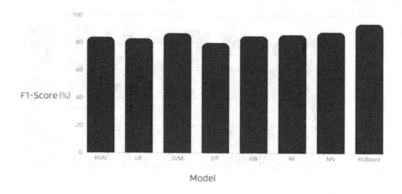

$$FPR = \frac{FP}{FP + TN} \tag{9}$$

The table 1.8 provides a comparison of several models' AUC values, which are ascertained as follows: The remaining models are as follows: logistic regression [LR] at 91%, random forest [RF] at 94%, decision tree [DT] at 92%, K-Nearest Neighbors [KNN] at 89%, neural network [NN] at 95%, support vector machine [SVM] at 93%, and gradient boosting [GB] at 93%. The proposed model, XGBoost, has the greatest AUC at 96%. The graphical comparison of area under curve with different models presented in Figure 12.

A parameter called specificity measures how well a model can recognize negative cases among all real negative examples. The percentage of real negative cases that the model accurately predicts to be negative is measured. In situations like medical diagnostics or anomaly detection, when precisely identifying negative cases is crucial, specificity is especially crucial as it enhances sensitivity.

$$Specificity = \frac{TP}{TP + FN} \tag{10}$$

The following table 1.9 presents a comparison of various models for specificity, which are determined as follows: The remaining models are as follows: random forest (RF) at 91%, logistic regression (LR) at 88%, decision tree (DT) at 84%, K-Nearest Neighbors (KNN) at 83%, neural network (NN) at 92%, support vector machine (SVM) at 90%, and gradient boosting (GB) at 93%. The proposed model, XGBoost, has the highest specificity of 94%. The graphical comparison of specificity with different models presented in Figure 13.

Table 8. Comparing area under the curve with various models

Model Name	Area Under Curve
KNN	89
LR	91
DT	92
SVM	93
GB	93
RF	94
NN	95
XGBoost	96

Figure 12. A graphical comparison of area under curve with various models

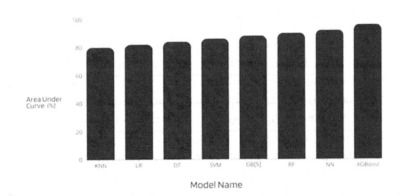

Table 9. Comparing specificity with different models

Model Name	Specificity
KNN	83
DT	84
LR	88
SVM	90
RF	91
NN	92
GB	93
XGBoost	94

Figure 13. A graphical comparison of specificity with different models

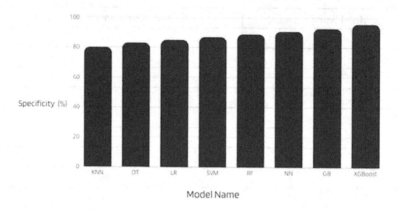

Model Name

CONCLUSION

The research findings demonstrate the immense promise that the XGBOOST algorithm shows for hailstorm prediction. By offering improved accuracy and generalizability over traditional meteorological models, this machine-learning method offers a useful substitute. The outcomes of the current work motivate further investigation and improvement of machine learning techniques for hailstorm forecasting, eventually culminating in increasingly sophisticated and dependable prediction systems with significant societal implications. Recalls, F1 scores, and precisions all show improved accuracy performance with the model shown here. Furthermore, meteorologists can better understand hailstorm dynamics and perhaps advance weather research because the XGBoost model is interpretable and offers insights into the most significant environmental factors. The skills and knowledge gained from this model could help meteorologists issue more location-specific and timely warnings to better protect lives and property from the destructive forces of nature.

If every single influencing aspect is considered, our examination demonstrates how Extra Gradient Boosting Machines can completely rework how hailstorm prediction is accomplished. Far more progressed and precise hailstorm anticipation frameworks are likely to emerge as machine learning advances, which will at last help several social regions like calamity alleviation, transportation, farming, and more. Extra Gradient Boosting Machines and other machine learning techniques are expected to have a significant impact on hailstorm prediction in the future, increasing our capacity to predict and handle hailstorm-related problems through further research and collaboration between information researchers and meteorologists. The execution measures that were acquired because of the proposed "Extra Gradient Boosting Machines" methodology incorporated exactness (92.1%), accuracy (90.0%), recollect (94.0%), F1 score (92.0%), a region under the bend (96.0%), and a profoundly complex blend of sentence structures and lengths to catch human writing design while keeping a similar word check as the first content.

REFERENCES

Alzaylaee, M., Yerima, S., & Sezer, S. (2020). DL-Droid: Deep learning based android malware detection using real devices. *Computers & Security*, *89*, 101663. doi:10.1016/j.cose.2019.101663

Anisha, P. R., Kishor Kumar Reddy, C., & Marlia, M. (2023). *An intelligent deep feature based metabolism syndrome prediction system for sleep disorder diseases.* Springer Multimedia Tools and Applications., doi:10.1007/s11042-023-17296-4

Bhushan, B., Kumar, A., Agarwal, A. K., Kumar, A., Bhattacharya, P., & Kumar, A. (2023). Towards a Secure and Sustainable Internet of Medical Things (IoMT): Requirements, Design Challenges, Security Techniques, and Future Trends. *Sustainability (Basel)*, *15*(7), 6177. doi:10.3390/su15076177

Gagne, D. J. II, Haupt, S. E., Nychka, D. W., & Thompson, G. (2019). Interpretable deep learning for spatial analysis of severe hailstorms. *Monthly Weather Review*, *147*(8), 2827–2845. doi:10.1175/MWR-D-18-0316.1

Gagne, D. J. II, McGovern, A., Haupt, S. E., Sobash, R. A., Williams, J. K., & Xue, M. (2017). Storm-based probabilistic hail forecasting with machine learning applied to convection-allowing ensembles. *Weather and Forecasting*, *32*(5), 1819–1840. doi:10.1175/WAF-D-17-0010.1

Kamangir, H., Collins, W., Tissot, P., & King, S. A. (2020). A deep-learning model to predict thunderstorms within 400 km^2 South Texas domains. *Meteorological Applications*, *27*(2), e1905. doi:10.1002/met.1905

Karpatne, R. (2019). Machine learning for the geosciences: Challenges and opportunities. *IEEE Trans. Knowledge. Data Eng.*

Kishor Kumar Reddy, C., Anisha, P. R., & Marlia Mohd Hanafah, Y. V. S. S. (2023). *An intelligent optimized cyclone intensity prediction framework using satellite images.* Springer Earth Science Informatics. doi:10.1007/s12145-023-00983-z

Kumar, S., Jadon, P., Sharma, L., Bhushan, B., & Obaid, A. J. (2023). Decentralized Blockchain Technology for the Development of IoT-Based Smart City Applications. In D. K. Sharma, R. Sharma, G. Jeon, & Z. Polkowski (Eds.), *Low Power Architectures for IoT Applications. Springer Tracts in Electrical and Electronics Engineering.* Springer. doi:10.1007/978-981-99-0639-0_13

McWilliams, J. C. (2019). A perspective on the legacy of Edward Lorenz. *Earth and Space Science (Hoboken, N.J.), 6*(3), 336–350. doi:10.1029/2018EA000434

Naveen, L., & Mohan, H. S. (2019). Atmospheric weather prediction using various machine learning techniques: A survey. In *Proceedings of the IEEE 3rd International Conference on Computing Methodologies and Communication (ICCMC)*, Erode, India.

Ni, X., Liu, C., Cecil, D. J., & Zhang, Q. (2017). On the detection of hail using satellite passive microwave radiometers and precipitation radar. *Journal of Applied Meteorology and Climatology, 56*(10), 2693–2709. doi:10.1175/JAMC-D-17-0065.1

Shankar, K., Zhang, Y., Liu, Y., Wu, L., & Che, C. (2020). Hyperparameter tuning deep learning for diabetic retinopathy fundus image classification. *IEEE Access : Practical Innovations, Open Solutions, 8*, 118164–118173. doi:10.1109/ACCESS.2020.3005152

Wang, P., Lv, W., Wang, C., & Hou, J. (2018). *Hail storms recognition based on convolutional neural network.* In *Proceedings of the 13th World Congress on Intelligent Control and Automation (WCICA)*, Changsha, China. 10.1109/WCICA.2018.8630701

Chapter 7
Reviewing the Potential of GeoAI for Post–Earthquake Land Prospection:
Lessons From the El Haouz Earthquake

Ouchlif Ayoub
ⓘ https://orcid.org/0000-0002-2001-2178
Hassan II Agronomic and Veterinary Institute, Morocco

Hamid Khalifi
ⓘ https://orcid.org/0000-0002-3367-9748
Faculty of Sciences, Mohammed V University in Rabat, Morocco

Hicham Hajji
Hassan II Agronomic and Veterinary Institute, Morocco

Kenza Aitelkadi
Hassan II Agronomic and Veterinary Institute, Morocco

ABSTRACT

The El Haouz 2023 earthquake caused extensive damage and displaced many people. A strategic roadmap is necessary for successful adoption of advanced technologies in the post-earthquake context. Land prospection is a critical step in reconstruction, but it can encounter obstacles such as limited land availability, geological and environmental risks, and social and economic constraints. This study explores the potential of geospatial artificial intelligence (GeoAI) in modernizing land prospection in post-earthquake contexts. A systematic review of literature identified previous GeoAI applications in post-earthquake interventions. The findings indicate that GeoAI can significantly contribute to land selection by providing accurate and real-time information on terrain features, hazards, and limitations. The study proposes a methodological framework for integrating GeoAI into post-earthquake land selection challenges based on lessons learned from prior experiences. The implications of this study are significant for decision-makers, practitioners, and researchers in the field.

DOI: 10.4018/979-8-3693-3896-4.ch007

INTRODUCTION

On September 8, 2023, an earthquake with a magnitude of 6.3 struck the El Haouz region in Morocco, causing significant damage to infrastructure and buildings. This natural disaster left many people homeless and required the relocation of entire populations. The scarcity of constructible land in rural areas makes the task of choosing new resettlement sites particularly complex, highlighting the need for a rapid and reliable assessment of damages.

In this context, data from post-disaster mapping is essential it is a crucial communication tool provides valuable information on the emergency and the impact of the disaster. It enables effective communication and a common understanding of the situation among all stakeholders involved in disaster intervention, management, and recovery.

Particularly during the first hours of the emergency response phase, as they represent the foundations on which post-disaster reconstruction builds: "The emergency period, which corresponds to the initial rescue of victims, their temporary accommodation and the clearing of rubble, determines the future dynamics of the reconstruction process" (Bourrelier, 2000, p.81). Effective coordination of executive and operational forces at the central, regional, and local levels required to respond quickly and effectively to the urgent needs of the population affected by the disaster. The main objective is to accelerate the reconstruction process and the return to normal.

Surveying remains a crucial step in planning and implementing reconstruction projects. Although traditional surveying methods are increasingly automated, precise, and operational thanks to the use of high-precision total stations, GNSS receivers, and laser scanners for inspecting damaged buildings, their use has certain limitations.

Indeed, the examination of roofs and facades of high-rise buildings or hazardous sites, characteristic of post-seismic situations, can be complex with these approaches. This is why the use of micro-UAVs to carry out surveys in these specific cases can be particularly useful in overcoming these obstacles.

UAVs have considerable potential to revolutionize current practices in emergency response. Their ability to access hard-to-reach areas and collect detailed data on damages and aid needs can significantly improve the speed and effectiveness of emergency response. UAVs can provide real-time information, enabling rescue teams to quickly target the most affected areas and identify people requiring immediate assistance.

Moreover, technologies such as IoT (Internet of Things), digital twins, AI (artificial intelligence), and GIS (geographic information systems) play a crucial role in managing and analyzing data collected by UAVs and other sources.

IoT can be used to monitor in real-time various environmental parameters and critical structures, providing valuable information to assess risks and plan interventions. Digital twins allow virtual modeling of infrastructure and affected areas, facilitating scenario simulation and response strategy optimization.

AI used to analyze large amounts of data and detect patterns or anomalies, helping to make informed decisions and predict future needs. GIS provides a platform for integrating, visualizing, and analyzing spatial data, which is essential for effective intervention planning and coordination among the different actors involved.

In short, the integration of UAV missions with these advanced technologies offers the possibility of creating a more agile, precise, and responsive post-seismic intervention system. This can contribute to reducing human and material losses, as well as accelerating the reconstruction and recovery process of communities affected by earthquakes. However, it is also crucial to consider the ethical, legal, and privacy challenges related to the use of such technologies, as well as to ensure their accessibility and responsible use by all stakeholders.

In the context of post-seismic intervention in Morocco, it is worth noting the notable absence of national studies on the application of emerging technologies such as IoT, UAVs, AI, etc., unlike international research. It is in this perspective that our study positions itself as a pioneering initiative in Morocco, aiming to fill this gap by conducting an exhaustive literature review on the use of these technologies in the specific context of post-seismic intervention.

This approach is of crucial importance, as it will provide an essential reference for researchers, engineers from public institutions, and all actors involved in post-seismic management in Morocco. By offering a rigorous and up-to-date synthesis of available knowledge, our study aims to serve as a solid foundation for guiding and promoting the development of practical applications in this field, while contributing to strengthening local expertise and improving resilience to natural disasters.

In this regard, our systematic literature review will focus on the analysis of the following technologies:

- Artificial Intelligence (AI): to analyze and interpret massive amounts of collected data, enabling better decision-making.
- Geographic Information Systems (GIS): to integrate, visualize, and analyze spatial data, facilitating intervention planning and coordination.
- Micro-UAVs: for their ability to collect accurate and detailed aerial data in hard-to-reach areas.
- Internet of Things (IoT): for real-time monitoring of environmental parameters and infrastructure.

In particular, we will examine the use of geospatial artificial intelligence (GEOAI) as a strategic solution for the Moroccan context. GEOAI combines the capabilities of AI and GIS, offering powerful tools for analyzing and managing geographic data. We will assess its potential to improve damage assessment, site selection for relocation, and overall reconstruction planning.

Our study will focus on the integration of these emerging technologies in the Moroccan context, taking into account geographical specificities, existing infrastructure, and the needs of local communities. Based on international research and concrete examples such as the El Haouz earthquake, our study aims to propose concrete recommendations and guidelines for a more effective and resilient post-seismic intervention in Morocco.

CONTEXT

El Haouz Earthquake

On the evening of September 8, 2023, Morocco struck by a devastating earthquake with a magnitude of 6.8 on the Richter scale, approximately 70 km southwest of the bustling city of Marrakech. This seismic event was the deadliest in Morocco since the tragic Agadir earthquake in 1960. The consequences were catastrophic, with over 3,000 fatalities and nearly 5,700 injured. Thousands of people left homeless, seeking refuge in temporary shelters as their homes reduced to rubble.

The impact extended beyond human losses and physical injuries. Over 580 schools damaged, depriving countless children of access to education in the aftermath of the earthquake. As winter approached, with its freezing temperatures and snow, the situation of those without adequate shelter became even more critical.

The most affected regions are the provinces of Al-Haouz, Marrakech, and Chichaoua in the Marrakech-Safi region, as well as the province of Taroudant in the Souss-Massa region and the province of Ouarzazate in the Drâa-Tafilalet region. The villages nestled in the High Atlas mountains, such as Asni, Imi N'Tala, Elbour, Talat N'Yaaqoub, Tizi N'test, Tafingoult, suffered the most damage, with homes, infrastructure, and roads left in ruins.

According to assessments by the US Geological Survey's PAGER system, economic losses range from 0 to 8%, amounting to up to 2% of Morocco's GDP. Reconstruction efforts have incurred significant costs in addition to economic losses, highlighting the long road to recovery.

Post-Earthquake Intervention in El Haouz 2023

Despite substantial efforts to mitigate the aftermath of the El Haouz earthquake in Morocco, there is a predominant persistence of conventional practices in post-seismic operations, with little room given to emerging technological advancements. While

Figure 1. Map illustrating the seismic intensity zones following the Adassil / Al haouz Earthquake (the 8th of September 2023, M 6.8). About 5 million people potentially exposed to strong to severe intensity.
(UN HABITAT)

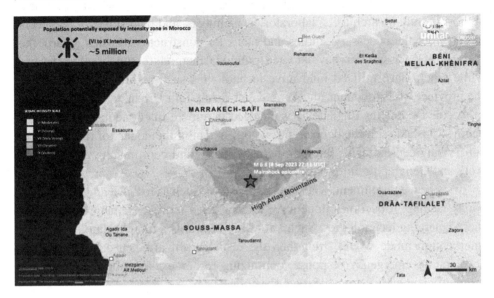

Figure 2. Map of the most affected regions are the provinces
(UN HABITAT)

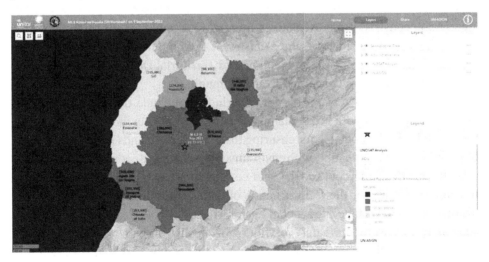

this situation may be reassuring due to its familiarity and often proven reliability, it nonetheless risks limiting the capacity of responders to quickly and effectively assess damages, prioritize intervention areas, and coordinate relief and construction efforts.

In this context, for administrative procedures related to reconstruction, local authorities rely on architects, laboratories, architects, and land surveyor engineers to carry out the necessary studies and plans for construction. In this regard, the majority of land surveyors have adopted traditional methods for topographic surveys, while others have utilized innovative technologies such as drones. However, in the province of Taroudannt, the Taroudannt-Tiznit-Tata Urban Agency, confident in its sole Geomatics Engineer, has engaged in a more integrated approach. This engineer tasked with the delicate mission of mapping the douars of the province, supervising census teams through thematic maps, and coordinating cartographic work and Geographic Information Systems (GIS).

In such an environment, where innovative tools such as IoT, AI, and GEOAI are sorely lacking, it becomes imperative to consider a more systematic adoption of these emerging technologies. Such a transition towards a more technology-driven intervention mode would strengthen the resilience of affected communities by accelerating emergency response, optimizing resource allocation, and fostering more efficient and sustainable reconstruction.

Administrative Land Prospecting Commission

In Morocco, the administrative commission for land selection is a committee responsible for identifying land for development and construction projects. This commission consists of representatives from various public administrations, such as the Ministry of the Interior, the Ministry of Equipment, Transport, Logistics, and Water, the Ministry of Urban Planning and Territorial Development, as well as representatives from the concerned local communities.

The commission is responsible for ensuring compliance of development and construction projects with urban planning rules and public territorial development policies. It is also responsible for ensuring that the selected lands meet the needs of the projects and are compatible with environmental, social, and economic constraints.

In the post-seismic context, the said commission has a crucial responsibility in identifying lands for the reconstruction of damaged or destroyed buildings and infrastructure. It must take into account seismic risks and geological and geotechnical constraints to ensure the safety of future occupants and the sustainability of constructions. In this context, the use of innovative technologies such as GEOAI can provide significant contributions by supplying precise and real-time information on terrain characteristics, geological and environmental risks, as well as social and economic constraints.

METHODOLOGY

In this section, we present a review of previous studies that have exploited GEOAI in post-seismic interventions. We have identified several studies that have used machine learning and spatial analysis techniques to improve earthquake response. To do this, we followed a systematic multi-step methodology to carry out this review of previous studies using GEOAI in earthquake response.

First, we conducted a bibliographic search in relevant scientific databases such as Web of Science, Scopus, and Google Scholar. We used keywords such as "Earthquake", "Disaster Management", "IoT","UAV""GEOAI", "Machine Learning", "Remote Sensing", "Geographic Information System" and their combinations to identify relevant studies. We then applied inclusion and exclusion criteria to select relevant studies. Inclusion criteria included the use of GEOAI in earthquake response, publication in English or French, and availability of the full text. Exclusion criteria included studies that did not focus on earthquake response, studies that did not use innovative technologies, and duplicate studies.

Previous Studies Adopting Innovative Technologies in Post-Earthquake Intervention, this section discusses various studies conducted in the pre- or post-earthquake period.

Table 1 illustrates these case studies; objectives, algorithms used, and input data:

DISCUSSION

A significant trend observed in the reviewed studies is the integration of multiple data sources to improve the accuracy and reliability of predictive models. This multidimensional approach combines data such as satellite imagery, UAV data, IoT sensors, and even social media data. For example, the study by Mohamed S. Abdalzaher et al. (2023) uses IoT sensor observations to detect precursors of seismic events, while Ahmad et al. (2017) combine satellite imagery and social media data for disaster detection.

Drones (UAVs) are also playing an increasingly important role in collecting high-resolution and real-time data for post-disaster damage mapping and intervention planning. Studies such as those by Nikolaos Soulakellis et al. (2020) and V. Baiocchi et al. (2013) use UAV systems for post-seismic damage assessment and mapping affected areas.

Seismic risk analysis and modeling are also important aspects of research in this field. Studies such as the one by Yong and Zhu (2023), which develops a machine learning model to assess the seismic risk of railway bridges, and the one by Seema,

Table 1. Review of case studies using different technologies in disaster management

Case Study	Objectives	Algorithm/Technology	Input Data
Kia et al. 2012	Develop a flood model for the southern part of Peninsular Malaysia using artificial neural network (ANN) techniques and geographic information system (GIS) to simulate flood-prone areas.	Artificial neural network	Rainfall, slope, elevation, flow accumulation, soil, land use and geology data layers from remote sensing
Syifa et al. 2019	Generate a post-earthquake damage map using remote sensing imagery.	Artificial neural network (the back propagation algorithm) and Support vector machine (radial basis function – RBF)	Satellite images
Vetrivel et al. 2018	The objective is to enhance the detection of severe building damages caused by destructive disaster events, such as earthquakes, using oblique aerial images.	Convolutional Neural Networks	3D point cloud
Amit, Aoki, 2017	The study propose an automatic natural disaster detection system, focusing on landslide and flood detection, using convolutional neural networks (CNN) applied to remote sensing imagery.	Convolutional Neural Networks	Satellite images
Duarte et al. 2018	Provide a comprehensive review of the evolution of unmanned aerial vehicle (UAV)-based damage mapping techniques for structural disaster detection and characterization.	Convolutional Neural Networks	Satellite and UAV images
Ahmad et al. 2017	Develop and implement methods, including Convolutional Neural Network models with late fusion techniques and Generative Adversarial Network	Convolutional Neural Network, Support Vector Machines (SVM), Random forest classification, Generative adversarial network,	Social media and satellite images
Cooner et al. 2016	Assess the effectiveness of multilayer feedforward neural networks, radial basis neural networks, and Random Forests, along with textural and structural features, in detecting earthquake damage caused by the 2010 Port-au-Prince, Haiti earthquake	Multilayer feedforward neural networks, Radial basis function neural networks (RBFNN) and Random Forests	Satellite images
Nia, K., Prasetya, D., & Arifin, D. (2023)	The study maps landslide susceptibility in Mamuju Regency, Indonesia using remote sensing techniques.	Remote sensing	Satellite images
Yong, H., & Zhu, Z. (2023)	Develops a machine-learning model for rapid seismic risk assessment of railway bridges.	ML	Earthquake data
Seema, S., Kumar, M., & Singh, R. (2020)	Identifies seismic vulnerability zones in Delhi, India using GIS-based multi-criteria decision analysis.	GIS-based multi-criteria decision analysis	Thematic maps
Mohamed S. Abdalzaher et al. (2023)	Implement an earthquake early warning system (EEWS) in smart cities to save human lives through efficient disaster management.	IoT ML	Observations from IoT sensors including seismic vibrations, ground deformation, etc.; Geographic and seismological characteristics of the region; Historical earthquake data.

continued on following page

Table 1. Continued

Case Study	Objectives	Algorithm/Technology	Input Data
Nikolaos Soulakellis et al. (2020)	Demonstrate the effectiveness of utilizing Unmanned Aircraft Systems (UASs) 4K-video footage processing and geo-information methods for monitoring demolition and mapping demolished buildings in post-earthquake recovery.	Unmanned Aircraft Systems (UASs)	Aerial 4K video footage collected using an Unmanned Aircraft on February 3rd, 2019, in the traditional settlement of Vrisa (Lesvos, Greece)
Nathaniel M. Levine et al. (2021)	Propose a digital twin framework for post-earthquake building evaluation integrating unmanned aerial vehicle (UAV) imagery, component identification, and damage evaluation using Building Information Modeling (BIM) as a reference platform.	Digital twin framework	UAV imagery collected for earthquake-damaged buildings, Building Information Model (BIM)
Ilaria Tonti et al. (2023)	Investigate the spatial impacts of post-earthquake temporary housing in Central Italy and propose an integrated geomatic approach using UAV and GIS-based systems for spatial documentation.	Geomatic and photogrammetric tools, UAV, high-resolution orthophotos, Spatial analysis	UAV point clouds, non-metric data, official cartographic data, high-resolution orthophotos, elevation models (DSM and DTM).
Spyridon Mavroulis et al. (2019)	To assess the damage caused by the Mw 6.3 earthquake that struck Lesvos Island, Greece, on June 12, 2017,	UAV GIS	Building-by-building inspection data for damage assessment. UAV survey data for aerial imagery and digital post-processing. Geological, geomorphological, geotechnical and seismological data of the affected area. European Macro seismic Scale 1998 for intensity assessment. Building characteristics and vulnerability data.
Sizhe Wang ET AL. (2021)	Develop a GeoAI research method for deep machine learning to detect natural features using multi-source geospatial data.	GEOAI DL	Remote sensing imagery, Digital Elevation Model (DEM) data, derived DEM derivatives, multi-source training data, and augmented training data.
V. Baiocchi et al. (2013)	to explore the use of multi-rotor micro UAVs for high-quality image capture of roofs and facades of structures in the old city center of L'Aquila, Italy	UAV GIS	Earthquake damage Multi-rotor Micro UAVs
Wenjuan Sun et al. (2020)	To provide an overview of the current applications of artificial intelligence (AI) in disaster management across the four phases: mitigation, preparedness, response, and recovery.	AI GIS	Natural Hazard Data Earthquake data
M. Ivić (2019)	To provide an overview of the utilization of artificial intelligence (AI) in geospatial analysis for disaster management.	AI GIS	Earthquake data
Hafiz Suliman Munawar et al. (2022)	-Improvement in the adoption of the latest technology to move towards automated disaster prediction and forecasting. -Examination of disruptive technologies based on their relevance to flood prediction, flood risk assessment, and hazard analysis.	IoT AI Big data Analytics	Satellite images

S., Kumar, M., & Singh, R. (2020), which identifies seismic vulnerability zones in Delhi using multicriteria analysis techniques, illustrate this trend.

However, several weaknesses, opportunities, and limitations emerge in these studies, offering a critical perspective on the current state and future developments in the field of natural disaster management using Geographic Information Systems (GIS) and machine learning.

Weaknesses

1. Dependence on data: Many studies rely on limited or region-specific datasets, which can limit the generalizability of the results and the ability to extrapolate models to other geographical contexts.
2. Algorithm complexity: The use of sophisticated machine learning algorithms can make models difficult to interpret and understand, posing challenges for practitioners and decision-makers.
3. Bias and errors: Machine Learning models are prone to reproducing biases present in the training data, which can lead to incorrect or biased predictions.

Opportunities

1. Multiple data integration: The opportunity to integrate data from multiple sources offers the possibility of improving the accuracy and robustness of predictive models.
2. Emerging technology development: Rapid advances in emerging technologies such as the Internet of Things (IoT), Artificial Intelligence (AI), and big data analysis offer new opportunities for the development of more effective and adaptive disaster management systems.
3. Interdisciplinary collaboration: Collaboration between experts in GIS, machine learning, earth sciences, and social sciences enables a holistic approach to the complex challenges of disaster management and the development of more comprehensive and relevant solutions.

Limits

1. Accessibility and cost: Accessibility and cost of cutting-edge technologies can be barriers for many organizations and developing countries, limiting their ability and fully benefit from these tools for disaster management.
2. Ethics and data privacy: The use of sensitive data in machine learning models raises ethical and privacy concerns, requiring special attention to ensure compliance with ethical and legal standards.

3. System complexity: The increasing complexity of disaster management systems based on geographic information technology and machine learning can make implementation and maintenance of the systems difficult, requiring significant technical expertise and resources.

By considering these weaknesses, opportunities, and limits, it is essential to adopt a balanced and critical approach to promote the development of innovative and sustainable solutions for natural disaster management.

CONCLUSION

This systematic literature review explored the use of GEOAI in post-seismic management, focusing on the specificities of the Moroccan context. Our study aims to fill a gap in the existing literature and provide recommendations for future Moroccan researchers in this field.

Our results showed that GEOAI has significant potential to improve post-seismic management in Morocco. Specifically, GEOAI can help collect accurate and real-time data on earthquake damage, which can facilitate informed decision-making by relevant authorities. Moreover, GEOAI can assist in planning and coordinating emergency interventions, as well as long-term reconstruction efforts.

However, our study also revealed certain limitations in the existing literature. Firstly, we noted a lack of empirical research on the use of GEOAI in the Moroccan context. Secondly, we found that most existing studies focus on the technical aspects of GEOAI, without considering the social and political factors that may affect its adoption and use.

Therefore, we recommend that future Moroccan researchers conduct empirical studies on the use of GEOAI in the Moroccan context, taking into account relevant social and political factors. Additionally, we recommend that Moroccan policymakers consider the potential of GEOAI in their post-seismic planning and management, while being aware of the challenges related to its adoption and use.

In conclusion, our study demonstrated that GEOAI has significant potential to improve the management of seismic events in Morocco. However, further research is required in order to understand its use and impact in this specific context. It is our hope that our study will contribute to stimulating future research in this important field and help policymakers make informed decisions regarding the management of seismic events.

REFERENCES

Abdalzaher, M. S., Elsayed, H. A., Fouda, M. M., & Salim, M. M. (2023). Employing Machine Learning and IoT for Earthquake Early Warning System in Smart Cities. *Energies*, *16*(1), 495. doi:10.3390/en16010495

Ahmad, K., Pogorelov, K., Riegler, M., Conci, N., & Halvorsen, P. (2017, September). CNN and GAN Based Satellite and Social Media Data Fusion for Disaster Detection. [Dublin, Ireland.]. *MediaEval*, *17*, 13–15.

Amit, S. N. K. B., & Aoki, Y. (2017). Disaster detection from aerial imagery with convolutional neural network. In *2017 International Electronics Symposium on Knowledge Creation and Intelligent Computing (IES-KCIC)*, (pp. 239–245). IEEE. 10.1109/KCIC.2017.8228593

Cooner, A., Shao, Y., & Campbell, J. (2016). Detection of urban damage using remote sensing and machine learning algorithms: Revisiting the 2010 Haiti earthquake. *Remote Sensing (Basel)*, *8*(10), 868–885. doi:10.3390/rs8100868

Duarte, D., Nex, F., Kerle, N., & Vosselman, G. (2018). Satellite image classification of building damages using airborne and satellite image samples in a deep learning approach. *ISPRS Annals of the Photogrammetry, Remote Sensing and Spatial Information Sciences*, *4*(2), 89–96. doi:10.5194/isprs-annals-IV-2-89-2018

Kia, M. B., Pirasteh, S., Pradhan, B., Mahmud, A. R., Sulaiman, W. N. A., & Moradi, A. (2012). An artificial neural network model for flood simulation using GIS: Johor River Basin, Malaysia. *Environmental Earth Sciences*, *67*(1), 251–264. doi:10.1007/s12665-011-1504-z

Levine, N. M., & Spencer, B. F. Jr. (2022). Post-Earthquake Building Evaluation Using UAVs: A BIM-Based Digital Twin Framework. *Sensors (Basel)*, *22*(3), 873. doi:10.3390/s22030873 PMID:35161619

Mavroulis, S., Spyrou, N. I., & Emmanuel, A. (2019). UAV and GIS based rapid earthquake-induced building damage assessment and methodology for EMS-98 isoseismal map drawing: The June 12, 2017 Mw 6.3 Lesvos (Northeastern Aegean, Greece) earthquake. *International Journal of Disaster Risk Reduction*, *37*, 101169. doi:10.1016/j.ijdrr.2019.101169

Nia, K., Prasetya, D., & Arifin, D. (2023). Detection of the post-earthquake damage in Mamuju Regency in January 2021 using Sentinel-1 satellite imagery. *Buletin Poltanesa*, *24*(1).

Seema, S., Kumar, M., & Singh, R. (2020). Seismic vulnerability assessment of NCT of Delhi using GIS-based multiple criteria decision analysis. In *Lecture Notes in Civil Engineering* (Vol. 118, pp. 11–20). Springer.

Soulakellis, N., Vasilakos, C., Chatzistamatis, S., Kavroudakis, D., Tataris, G., Papadopoulou, E.-E., Papakonstantinou, A., Roussou, O., & Kontos, T. (2020). Post-Earthquake Recovery Phase Monitoring and Mapping Based on UAS Data. *ISPRS International Journal of Geo-Information, 9*(7), 447. doi:10.3390/ijgi9070447

Syifa, M., Kadavi, P. R., & Lee, C. W. (2019). An Artificial Intelligence Application for Post-Earthquake Damage Mapping in Palu, Central Sulawesi, Indonesia. *Sensors (Basel), 19*(3), 542–560. doi:10.3390/s19030542 PMID:30696050

Tonti, I., Lingua, A. M., Piccinini, F., Pierdicca, R., & Malinverni, E. S. (2023). Digitalization and Spatial Documentation of Post-Earthquake Temporary Housing in Central Italy: An Integrated Geomatic Approach Involving UAV and a GIS-Based System. *Drones (Basel), 7*(7), 438. doi:10.3390/drones7070438

Vetrivel, A., Gerke, M., Kerle, N., Nex, F., & Vosselman, G. (2018). Disaster damage detection through synergistic use of deep learning and 3D point cloud features derived from very high-resolution oblique aerial images, and multiple-kernel-learning. *ISPRS Journal of Photogrammetry and Remote Sensing, 140*, 45–59. doi:10.1016/j.isprsjprs.2017.03.001

Yong, H., & Zhu, Z. (2023). Rapid assessment of seismic risk for railway bridges based on machine learning. *International Journal of Structural Stability and Dynamics*.

Chapter 8
Scholarly Communication Theories for Building Effective Disaster–Resilient Infrastructure:
A Study

Murtala Ismail Adakawa

https://orcid.org/0000-0003-4298-1970
Bayero University, Kano, Nigeria

N. S. Harinarayana
University of Mysore, India

ABSTRACT

This research investigates scholarly communication theories used during the COVID-19 pandemic as a strategy for building effective disaster-resilience infrastructure. The study employed scientometric and content analysis to understand the behavior of data in this regard. For scientometric analysis part, using Scopus from 16-23rd January 2024 employing search strategy "COVID-19 and theories" OR "Community resilience" OR "Disaster-resilience Infrastructure," it yielded about 5,266,065 documents and reduced to 10,053 through data pruning. The findings showed that 2023 has been the year with the highest number of publications 4369(43.45%) followed by 2021 accounting for 2277(22.65%). For content analysis, types of theories, constructs of importance, methodologies, etc. employed by researchers were studied. The study concludes that, even though theories are deterministically used to direct policy formulations and implementation in indeterministic disaster conditions, they quickly provide a means of understanding and enumerating possible variables for tackling such hazards.

DOI: 10.4018/979-8-3693-3896-4.ch008

INTRODUCTION

Disaster continues to puzzle researchers as the size of the threats in totality seems to be larger than the infrastructure provided and accurate quantification of disaster before it strikes is always needed but sadly lacking or misleading. This is true as many researchers observed that, despite enormous advancements in science and technology, Smart Cities experience a plethora of hazards due to their overreliance on *interconnected systems and networks* rendering them susceptible to risks than their counterparts. This calls for effective use of technological innovations (Samarakkody et al., 2023). That is why, in coming decades, urban development policy will be critical calling for climate forecasting and modelling to enhance policies through incorporating determinants to reduce vulnerability and develop robust infrastructure capable of withstanding future occurrences (Berawi, 2018). This is because, disaster resides within objective and subjective worlds, which means that, for it to attract the attention of scholars; it requires interdisciplinary perspectives for understanding its epistemology, ontology, and methodology. This is true, as interaction between man and natural world makes disaster management an interdisciplinary endeavor of *"science and social science with planners stuck in the past and having subjective perceptions about the future"* (Isser, 2022, p. 1). This raises concern as there is notion that, *"the unnaturalness of natural disaster"* led to its description as an *"outcome of trends in the location of people and assets, and economic, environmental and land use policies, rather than a series of exogenous and unpredictable misfortunes"*. This challenges international organizations to devise a sustainably fit-for-all framework *"to tackle this growing risk, largely because it is a systemic problem outside of their institutional reach"* (Keating et al., 2017, p. 66), which implies the need for resilience.

That is why resilience is rapidly becoming one of the most frequently used concepts in disaster-related literature where disaster-resilience has received definitions *"at the individual, building, community, and systems levels through capacities or capabilities and as an outcome or a process"* (Barnes, 2021, p. xvi). In this way, Barnes, (2021, p. xvi) was able to link social identity theory with community disaster-resilience since social psychology studies the dynamics of interactions between members of a community *"including motivation, lending, assistance, communication, faith in other people, leadership"*, etc. Such kinds of attempts emanated from the recognition that, the spike of disasters increases drastically taking huge toll both in human lives and economic cost making regions with high population vulnerable requiring businesses and communities to embrace systematic approach to risk management (Beekharry & Baroudi, 2015). This implies that, there is a dynamic relationship between disaster and development, whose impact largely depends on socio-cultural landscapes of communities affected and managerial capacities of disaster and development of leaders in those areas (Kapucu & Liou, 2014). Prior to this dynamic relationship, there was artificial dissimilarity in literature between disaster and economic development researchers necessitating a shift from *"perfecting response and recovery"* to focusing on *"the benefits of disaster development in the early stages of mitigation and preparedness"* (Kapucu & Liou, 2014, p. 2).

This is true, as assessing, identifying, evaluating potential hazards or risks and developing strategies to mitigate them requires urgency of knowledge in disaster management (Inan et al., 2023). This is because, disaster can and does exceed a *tolerable magnitude*, challenges adjustment thereby causing loss of property, income, and paralyzes lives to the extent demanding response, rehabilitation, reconstruction, mitigation, etc. (Khan et al., 2008). In actualizing this scenario, this raises moral and ethical challenges as regard allocation of resources to the affected communities through exercising fairness and justice during and after a disaster, which calls for institutionalizing an ethical framework that looks into decisions, their impacts, and consequences (Cuthbertson & Penney, 2023). For instance, gender resilience and preparedness before, during, and after disaster (Ashraf & Azad, 2015) can help in allocating resources and including them in disaster preparedness. In the same vein, disasters affect human health as every year, there is an estimation that, more than 170 million people become affected by conflicts and over 190 million by disasters, which challenge triple goal of the World Health Organization: *Universal health coverage, health security, and health for all* (World Health Organization, 2019). This means that, the relevance of tiny fraction of ideas that can, at least provide an insight into understanding such a phenomenon is always welcome.

The conception that, disasters as an event of *pebble drop in a pond* that disturbs the surface and propagates the generated waves to shake communities thus overwhelming the capacities of those affected leading to disruption of many human activities (Salamati & Kulatunga, 2017) is enough to justify the need for looking back at the theories used in those hazards. This is the case, as Salamati and Kulatunga, (2017) have pointed out that, many people witness such hazards once in their lifetime and that, disasters cannot be prevented rather their effects minimized to certain degree. That is, the experiences gathered in one's lifetime are important in guiding policymaking process and designing critical strategies to ameliorate or mitigate such hazards in society. This is evident because, throughout human history on the planet earth, disaster is a recursive phenomenon that pulsates within his mind for time immemorial. From traces of evidence in human history on the earth's surface, every human civilization is characterized by disaster(s), which alternate in different forms such as floods (Suleyman & Bhaksha, 2023) with accumulated severities resulting in different consequences on humans particularly with respect to health, wellbeing, lives, properties, geographies or residences, among others. Disasters occur due to many presumably (in)directly interrelated factors whose sternness are catastrophic causing morbidity, mortality, displacement, to mention but a few. Based on disaster classification, disasters could be natural (such as biological, geophysical, hydrological, meteorological, climatological, etc.) or manmade/technological (intentional or unintentional). To support this view, from the data available on the Emergency Database (EM-DAT), about 2/3rd of the disasters are natural where the world has experienced more than 790 drought, 1570 earthquakes, 5750 floods, 4580 storms, 790 landslide, 270 volcanic activities, to mention but a few. The database contains a record of more than 26,000 disasters that took place from 1988 till date (EM-DAT, 2024). That is why the United Nations has recognized the importance of *"preserv[ing] human dignity and wellbeing and reducing human suffering [through] three pillars of the multilateral system: peace and security, human rights, and development"* (ICM & IPI, 2016, p. 1).

In this direction, Singh, (2020) reiterated that, disasters come about whenever there is rise in population, earth warming, earthquakes, disease-carrying vectors, etc. To be precise, in 2006-2014, according to the World Disasters Report 2016, about 772, 000 people lost their lives due to disasters. Similarly, from 1918-1919, human population suffered Spanish Flu causing the death of about 25-30 million people. Furthermore, in 1923, in Tokyo, earthquake claimed the lives of 150, 000 people and in China, in 1931, it was responsible for 3.7 million lives. In the same vein, in Bangladesh, in 1991, about 140, 000 lost their lives due to cyclone and Tsunami in 2004 caused the death of 230,000 and displaced 1.7 million people in 14 countries. In US, Hurricane caused a damage of $1 trillion. In India, about 30 major disasters such as super-cyclone resulted in the death of 15,000 people in Odisha in 1999, and in Gujarat, in 2001, about 20,000 people died of earthquake (Singh, 2020). Recently, Adakawa et al., (2023) have demonstrated the history of many pandemics that bedevilled human society for quite long and their inherent consequences, which placed COVID-19 pandemic on the list of events in human history. This implies the need for preparing for such disasters, which is critical for businesses, households, and communities thus requiring *"individual responsibility, local coordination, and community plans to respond and recover from [such] major events"* (Sutton & Tierney, 2006, p. 1). This is important because, *"disaster is a massive and speedy disruption by hostile elements on available resources. It leaves a deep wound which is physical, emotional, social and economic on all stakeholders"* (Rengarajan & Kulkarni, 2021, p. 1). That is why many scholars recognized that, disaster can seriously tamper with normal functioning of a society with inherent human, material, economic, or environmental losses, which require such societies to seek for assistance from national or international donors (Salamati & Kulatunga, 2017*).

Disaster has several layers, which can be at household level (sickness, social calamity), community level (fire, flood, earthquake, etc.), local district or city (displacement) (Salamati & Kulatunga, 2017*), the data generated therefrom can be used for societal, organizational, institutional, or governmental purposes (Perry, 2018). That is why reducing disaster to minimal level is of paramount importance and such an effort is directed at *"avoid[ing] the potential losses from hazards, assur[ing] prompt and appropriate assistance to the victims of disaster, and achiev[ing] a rapid and effective recovery"* (Salamati & Kulatunga, 2017, p. 1). Most often than not, these require four methods of management that include *mitigation, preparedness, response, and recovery* (Orlando et al., 2010 cited in Salamati & Kulatunga, 2017*). In this perspective, the ability of human population to learn from such disasters and bring something of benefits therefrom is always a reliable means to change disasters from a threat to potential opportunity. This is true, as Singh, (2020, p. 1) has quoted Rockefeller to have said that, *"I always tried to turn every disaster into an opportunity"*. To perform such a role requires understanding that calls for diversifying thoughts, perspectives, to mention but a few.

BACKGROUND OF THE STUDY

It seems impossible to define the meaning of a theory in a single document like this due to its branched applicability and emerging dimensions. However, for clarity and directionality, *"a theory is a set of interrelated propositions that allow for the systematization of knowledge, explanation, and prediction of social life and the generation of new research hypotheses"* (Ritzer, 2000, p.4; Faia, 1986 cited in. *scientific concept that refers to a particular class of phenomena whose specification rests in theory-based thinking"*, which excludes causes, conditions, consequences, or those resulting from conflicts (Perry, 1998 cited in Perry, 2018). Theories are important in disaster mitigation, preparedness, response and recovery for a number of reasons. To begin with, the work of Pepinsky, (2022), as a review essay on guiding of three important books that tried to monitor scholars on research process, asking questions, proposing theories to operationalize concepts, etc. It follows that, to establish relationships between *theoretical models (usually represented through mathematics) and empirical models (usually quantitative in nature)*, there are many abstract representations, which are of utmost importance in a social science setting. In this sense, the challenge is to nest the disaster within the appropriate realm of a theory that can best describe it and confer a holistic solution to it. Looking critically at the work of Shamim and Nasreen, (2016), it reveals that, there are seeming contradictory or competing trajectories of where to nest disaster theories the most appropriate way. The contradiction arises because, the paradigm of theories on disaster management comprises Marxist interpretation (i.e. economic and political conditions, capacity to recover, development), economic theory (i.e. sustainability and technology), Weberian perspectives (i.e. social construction, organizational and emergency behavior, risk perception and communication), management theory (systems, chaos, decision, etc. theories), integration (local, national, and international networking and collaboration, and preparedness and improvisation) (Shamim & Nasreen, 2016).

This contradiction might have arisen as result of the fact that, disaster management is on one end and economic development on the other hand but on a similar continuum, which forms important public policy in countries around the globe. For economic development, policymakers and public managers are obliged to design development policies and programs that can resuscitate business formation, industrial development, and solve the problems of economic cycles (Kapucu & Liou, 2014). When an emergency erupts, too much focus is given to response to disaster where the interplay between researchers and practitioners *"in disaster and/or development must contend just not with varying disciplinary and political perspectives but also with the tension between academic endeavor and practitioner-led interests"* (Fordham 2006, p. 341 cited in Kapucu & Liou, 2014, p. 2)". To support this view, effective disaster management requires well-coordinated and interdisciplinary response of various professionals such as paramedics, physicians, nurses, etc. with timely provision or delivery of essential resources that encompass but not limited to transportation, food, water, medical supplies, among others (Basyah et al., 2023). This translates the need to look at the disaster from a more complex perspective than ever imagined. This is among the reason why Burger et al., (2021) conducted a research and tried to link all the ingredients in an interwoven complexity. This is to signal policymakers on what to do to understand disasters in all ramifications. According to them, there is a paradigm shift of focus of decision-makers from economic growth to the sustainability and resilience of urban infrastructure and communities. Looking at this paradigm from complex adaptive system reveals that, understanding disaster requires knowledge of intersection between physical and social environments and individuals in somewhat reversible direction (Burger et al., 2021). Prior to this study, van Niekerk, (2011) has already captured most of the things raised in the above study. In a similar vein, the work of Kim and Sohn, (2018) on disaster theory has shone a light on providing a conclusive definition on cataclysmic events of disasters, interconnectivity of *societies, people, diseases, technology*, among others, leading to modern definition of disaster itself. Perhaps this is one of the reasons why Zulch, (2019) conducted a research contributing a segment of disaster management by delving into studying the psychological preparedness before, during, and after the disaster.

STATEMENT OF THE PROBLEM

When a disaster erupts, be it natural or manmade, one of the critical challenges faced by researchers in many situations; is framing questions that could best address the lingering hazards at that moment. In case of a pandemic, for example, these questions could be what is the organism responsible for the pandemic? What are the characteristics e.g., pathogenesis, pathogenicity, transmission route, developing vaccines, to mention but a few. That is why researchers navigate through a plethora of sources to quench their thirsts for urgency to treat research process to bring the causative agent under control. Fortunately, theories provide immensely rich information for formulating research questions and interpretation of findings (Musa, 2013). To universalize disasters as a unique or uniform phenomenon-taking place across the regions of the world is challenging but, historically, each country or region has an exceptional experience of such happenings time to time. Even though no country desires such disasters to occur, when it happens, it means that, each country is unique in terms of its populace, conditions, natural environment, among others, and ways put forward to respond to and recover from such disasters.

For long, there have been a series of debates on deductive reasoning and inductive reasoning, which are on the rise within the scholarly community. While the former forms the building blocks upon which theories are tested for predicting the future, the latter concerns with generating constructs that can be falsified in a series of experiments. Perhaps this demarcation arose due to the paradigmatic debates that have been in existence for time immemorial. This led us to the reality that, in the arena of disaster management, there are many cross-boundaries, inter-agency, inter-organization, etc. for trying to bring community back to its normalcy. This is to show how country's level of vulnerability implies the urgency to *"strengthen disaster risk reduction"* since *"disasters are the final effects of collision of natural hazards with vulnerabilities and exposures"* and thus, are *"failure[s] of human development"* (Kapucu & Liou, 2014, p. 3). It is not possible to call upon researchers to stop expanding the frontiers of testing other theories that have not been applied in mitigating disasters. Rather the need arises to ponder and look back and think to understand why, where, how, etc. the previous scholars applied those theories to understand the phenomenon at hand. That is, instead of focusing on using other theories to guide practice, policies should direct agencies involved in managing the disasters to ponder on the previous research outputs and build therefrom because only then can the possibility of reversing or managing such hazards is attainable.

Many local, national, and international organizations are continually developing strategies, plans, etc. on how to mitigate such disasters. However, such initiatives are hanging. This is because, disaster theories are expanding as the human actions evolve that cause alterations in the world and consequences that follow. Instead, studying the pattern of using constructs/variables of theories by researchers, and why they cite other scholarly outputs is critical for nesting these initiatives within a theoretically tested framework to guide practice. This research is an attempt to add to the body of knowledge.

OBJECTIVE OF THE STUDY

The current study tries to:

1. Understand the roles of researchers in the field of disaster-resilience infrastructure
2. Understand the theories used during COVID-19 pandemic

CURRENT STATE OF RESEARCH IN DISASTER RESILIENCE

One of the important aspects this chapter attempts to highlight is scanty of studies on "current state of research in disaster resilience" as it escapes the sights of many renowned scholars. This challenge has been annealed by attempting to bring to the limelight of audience small fragments of this issue to supply its shortage with something significant. To begin with, there is a correlation between migrating to risk areas in mostly lower-middle-income countries exacerbated by economic opportunity and disaster, where, for example, 90% of large floods occur in flood-exposed areas people live (Keating et al., 2017). One of the key areas that usher in large number of people to places is tourism. Some researchers observed that, *"tourism industry is not immune to crises"* (Ritchie & Walters, 2017, p. 2 cited in Nair & Dileep, 2020, p. 1496) such as *"natural disasters, pandemics, climate change, geopolitical issues, economic downturns, and terrorism"* (Paraskevas et al., 2013; Dahles & Susilowati, 2015 cited in Nair & Dileep, 2020, p. 1496). The call by researchers that governments failed to put in place measures that could avert disasters has been buried under carpet. This is to the extent that, risk reduction and vulnerability were not considered important until disaster occurs. Researchers such as de la Llera et al., (2018) were able to draw a framework of measurement for comparative assessments of disaster at local and community levels. This stemmed from the statement made by the Subcommittee on Disaster Reduction, (2005, p. 10) cited in de la Llera et al., (2018, p. 598) that, *with consistent factors and regularly updated metrics, communities will be able to maintain report cards that accurately assess the community's level of disaster resilience. This, in turn, will support comparability among communities and provide a context for action to further reduce vulnerability.*

From this promulgation, it can be argued that, R&D in disaster resilience is continually amassing giving rise to multitude approaches to mitigate ever-increasing disasters globally. For instance, US government spends about $57 billion annually and increases *ad infinitum* yearly (Gilbert, 2010). On the other hand, it is available in the literature that, despite advanced research in science and technology in disaster management and ever-increasing huge sums of money dedicated to mitigating man-made and natural disasters, it challenges commerce and financial networks, dares stakeholders to mobilize personnel, equipment, among others (Gilbert, 2010). In addition, developing frameworks for community resilience that best assist service providers, policymakers, and practitioners is challenging. Fortunately, Price-Robertson and Knight, (2012) have developed such a framework that best explains the strategy to enhance community resilience. According to them, the first step is to understand *"community's strengths and weaknesses as well as its physical characteristics (e.g., local infrastructure), procedural characteristics (e.g., disaster policies and plans) and social characteristics (e.g., level of community cohesion)"* (Price-Robertson & Knight, 2012, p. 1). It is of relevance to understand the current state of research in disaster resilience. From the perspective of global economy, when the COVID-19 pandemic erupted, as S&P500 indicated the highest possible record reached on 19th February, the decline in the following week attested the resemblance of ransacking profits observed since financial crisis of 2008. This necessitated different organizations and institutions such as Harvard Business School to finance several research outputs (Cheema-Fox et al., 2020).

Some researchers have observed that, there is a relationship between great financial crisis (GFC) and COVID-19 because; they are both exogenous shocks rather than business-cycle fluctuations and that, they affected companies through reduced liquidity. That is, scholars found that, innovation fails following a disaster at the estimated proportion of 40% and 90% (Rhaiem and Amara 2019 cited in Roper & Turner, 2020, p. 504-505). With reference to great financial crisis (GFC) of 2008-2010, Roper and Turner, (2020) have compared the willingness of small and medium enterprises (SMEs) in their quest to invest in R&D in the post-COVID-19 pandemic. From the study of Roper and Turner, (2020), it follows that, GFC had great impact on the global economy considering inability of smaller firms to access finance internationally, which drastically reduced innovations globally. From theoretical and practical perspectives, it is available in the literature that, economic crisis *"creates conditions for new innovations by lowering factor prices and creating a stock of idle resources"*. That is, at the time of crisis, *"there is reallocation of resources towards new entrants"* where *"if access to credit in order to finance innovative activities becomes limited during a recession, firms may become cash constrained, and R&D investment becomes procyclical"* (Aghion et al., 2012 cited Roper & Turner, 2020, p. 505). Interestingly, another study by Battist et al., (2023, p. 3) showed that,

Patent-intensive firms and firms with larger patent portfolios showed a reduced probability of exit. [In this way], firm size is associated with a higher probability of survival across the pandemic. [That is], patents have played a less relevant role in sectors that faced increasing demand due to health emergency and sectors with high asset specificity. [In this perspective], firm productivity has an important transmission channel. [Averagely], innovative firms show productivity levels of 40% higher than non-innovative firms [to the extent that] firms holding patents raised their productivity by 3.7% than [can their counterparts]

This corroborates with the submission of McKinsey and Company (2020, p. 2) that indicated that, COVID-19 pandemic has set a stage for many research firms. As at April 2020, 90% of the 34 companies studied by McKinsey and Company showed that, they *"had implemented the emergency procedures that were delineated under their business continuity plan"*. This is to the extent that, R&D leaders spend 40-50% of their time of crisis management. This follows from the recognition that, COVID-19 has tempered with Pharma and MedTech R&D where leaders must prepare to conserve research from total clampdown and protect it from not contributing to the wellbeing of people. To perform such roles, R&D leaders have to invest in *"—safeguarding patients and employees, adapting operations for a recovery, and building for the next normal"* (McKinsey & Company, 2020).

THEORETICAL FRAMEWORKS IN DISASTER MANAGEMENT

In many domain-specific disciplines, theories provide many purposes as an idealistic template for scholars to propagate in society or as a template for giving complete knowledge available about a phenomenon. That is, theories provide meaningful definitions to research, debate that can formulate concepts. In short, models, classification, and typology in theories assist in defining and linking variables to give meanings to scholarly constructs (Isser, 2022). In addition, there is possibility to change from one predominantly usable theories to another as the time passes by thereby incorporating some important variables to augment the already predominant theory or change them completely in a given scenario. For instance, Isser, (2022) observed that, the theory of comprehensive emergency management had narrowed or centralized issues on disaster to focus mainly on reaction. However, as the newer ideas such as emergency, chaos theory, theories based on community resilience, etc. developed, newer and insights that are more meaningful emerged (Isser, 2022). That is why researchers differ markedly in choosing a theory, its constructs, type of disaster, perspectives, etc. For instance, in the field of public health emergency management, Hu et al., (2006) used capacity assessment theory and emergency management theory. For the capacity assessment theory, it has three (3) components: the broader system, entity (organization), and individual. Capacity assessment theory posits that, *"for determining a complete set of emergency management function is the precondition to make a scientific capacity assessment for public health emergency response"* (UNDP cited in Hu et al., 2006, p. 44).

Some researchers such as Mojtahedi and Lan-Oo (2013) have faulted the use of emergency management theories (i.e. they described them as misnomer and oxymoron) and socio-economic conditions as these perspectives failed to guide stakeholders on the development of disaster response indices. In response to the lack of such a framework directing stakeholders to take a decisive action, these authors developed a framework using the stakeholder and decision-making theories. Stakeholder theory *"determines power and legitimacy of stakeholders whether they have tendency towards proactive or reactive approaches"*. On the other hand, decision-making theory gives *"optimized decisions for stakeholders with considering expectations, asset integration, and risk aversion"* (Mojtahedi & Lan-Oo, 2013, p. 2). In order to develop a framework for stakeholders in built environment, Mojtahedi and Lan-Oo, (2013) posit that, stakeholder response index should encompass stakeholder direct comparison, high level disaster planning, and developing preparedness procedure. To follow the same suit, Aid and Rassoul, (2017) have cautioned that, it is the *information* that is critical for understanding and engaging appropriate response actions on the disaster at hand. That is, situation-awareness, which has three (3) layers— perception, comprehension, and projection—requires accurate, timely, and reliable information to support quality decision-making process. Contemporarily, there is a shift from rooting theory based on response and preparedness to a more holistic disaster reduction approach (Isser, 2022). To support this view, Burger et al., (2021) observed that, decision-makers have now shifted their focus from economic growth to sustainability and resilience of urban infrastructure and communities. In order to comprehend the complexity of human behavior in relation to built environments, the need for interdisciplinary perspective is compulsory, which requires complex adaptive system to unravel the mysteries inherent therein (Burger et al., 2021). This may include *"operative and practical reflections on the typology of social relations*

established among diverse ethnical social groups at a time of disaster management"
(Lucini, 2014, p. 154)

SOCIO-ECONOMIC IMPACTS OF DISASTERS ON MARGINALIZED AND UNDERREPRESENTED COMMUNITY

This is an important transitioning, as it indicates the factors that reveal why a single phenomenon can affect community variably, indifferently, proportionately. This requires many inputs to make a claim more presentable to convince the audience. To begin with, researchers observed that, *"disasters tend to reveal existing national, regional, and global power structures, as well as power relations within intimate relations"* (Enarson & Morrow, 1998, p. 2 cited in Bradshaw, 2004, p. 7). In the 21st century, natural disasters have been recognized as one of the major catastrophes to humanity where people belonging to different categories of culture and nations experience climate change differently (Southard, 2017). This is true as there are marked differences between those who have and have-nots before, during, and after a disaster happens. That is, when disaster occurs, it does so in variance with respect to individual or community in poverty, considering their income, and socio-economic status (SES) in terms of perception, preparedness, response, and recovery. To be precise, *"when a disaster strikes, it affects not only households' physical assets but also their income levels and ability to contribute to the local economy"* (World Bank, 2021, p. 4).

For instance, in a Supplemental Research Bulletin released by Substance Abuse and Mental Health Services Administration (SAMHSA), (2017, p.3; 2014), a US-based organization, defined SES as *"related to many factors, including occupational prestige and education, yet . . . primarily associated with income level"*. This group of people is mostly vulnerable and at greater risk of negative experiences, effects, and reactions as disaster occurs in their community. They might be living in fragile housing, experience difficulty in accessing resources or trauma during or after disaster; or at a disadvantage of receiving timely, reliable, and accurate information (e.g. warning to evacuate). Similarly, perception about disaster risk is relative, malleable, and individually or community-centric. In a research conducted by Fothergill and Peek (2004) cited in SAMHSA, (2017), even among SES community, there is a difference on what constitutes perception about disaster risk. Some feel more concerned thereby perceiving more risk regarding both natural and artificial disasters while others especially working class whose work exposes them to risk are less concerned with risks associated with their work. Furthermore, there are mixed findings as regard preparedness, response, and recovery among SES individuals or community (SAMHSA, (2017). One of the ways to mitigate these happenings is by the ways these people are able to express their needs and opinions to government and this can be weakened or strengthened by social, political, or economic factors not merely natural environment (Southard, 2017).

In the same vein, gender has also been associated with disaster especially considering *"individual and strategies adopted, response of government and coordinated bodies of civil society, and reconstruction initiatives carried out by national and international organizations"* (Bradshaw, 2004, p. 5). Despite allocation of 6% to support pre-disaster hazard mitigation through the Building Resilience Infrastructure and Communities (BRIC) initiative, Coastal Resilience Centre (2021) has found disproportionate effects on socially marginalized groups and resource-constraint neighborhoods. This brought the research team constituted by the Coastal Resilience Centre at the University of North Carolina at Chapel Hill to use critical race theory (CRT) as an equity lens to address the flaring gaps in disaster management (Coastal Resilience Centre, 2021).

ROLE OF TECHNOLOGIES IN DISASTER-RESILIENCE

Technological advances can reduce death toll following increased occurrence of such events worldwide (UNDP, 2023). Approximately, three in five cities globally are vulnerable to natural hazards (Samarakkody et al., 2023). Disasters cause fatalities, destroy urban areas, challenge planning, and adaptation to climate change, which necessitates building a resilient city (Berawi, 2018). Countries such as India who have been experiencing a plethora of disasters in form of floods, droughts, cyclones, earthquakes, etc. are employing scientific and technic knowledge to disaster risk reduction (Rathore, 2016). That is, technologies used include database management system (DBMS), management information system (MIS), decision support system (DSS), geographical information system (GIS), remote sensing, among others. DBMS functions in providing large amount of data, data sharing, concurrency and locking, data security, data integrity, data recovery, support for languages. MIS operates in decision support systems and disaster information network, and DSS built on Hebert Simon model of decision-making uses intelligence, design, and choice for mitigating, preparing, responding, and recovering from disaster (Rathore, 2016). In their study, Samarakkody et al., (2023) were able to outline 20 citywide geo-data for improving disaster resilience in Smart Cities. These include cloud computing, internet of things (IoT), big data, geo-visualization and geographical information system (GIS), sensor networks, grid technologies, wireless wide communication and wireless local area network, location-based services (LBS), geographical positioning technique, blockchain, data warehouse, digital twins, unmanned aerial vehicle (UAV), cyber-physical system (CPS), building information model (BIM), smart disaster response

Table 1. Year-wise publications

Year of publication	No. of Publications	N% (N=10053)
2019	250	2.48682
2020	1528	15.19944
2021	2277	22.64996
2022	1629	16.20412
2023	4369	43.45966
	10053	100

system, early warning systems, virtual reality (VA), augmented reality (AU), and mixed reality (MR), AI/ML, crowdsourcing platforms, volunteered geographical information (VGI), web-based participatory tools, social media, living labs.

METHODOLOGY

The study used data from Scopus between 16-23rd January 2024 employing the search strategy "COVID-19 and theories OR ("Community resilience OR Disaster-resilience Infrastructure") where it yielded about 5,266,065 documents, which shows how scholars attached importance to disaster-resilience research. This result is too large and filtered by restricting a range of 2019-203 and excluding certain keywords thereby giving rise to 16,000 documents. Through data pruning, 10,053 documents were used for scientometric analysis. For content analysis, MAXQDA Analytics Pro (24.1.0) software was used for extracting relevant information. Theories, theorists, constructs/variables, etc. were identified.

FINDINGS

Scientometric Analysis Results

Table 1 shows the year-wise publications of scholars. It can be seen that, there is a pattern of increment from 2019 (2.48682%) to 2023 (43.45966%). This might be attributed to the fact that, scholars are now attaching increased importance to the studies on disaster. From the graph shown below, it can be seen that, this pattern continues at logarithmic phase. Perhaps the incidence of COVID-19 pandemic has added advantage to the recognition on the part of scholars to delve into studying

Figure 1. Year-wise publications

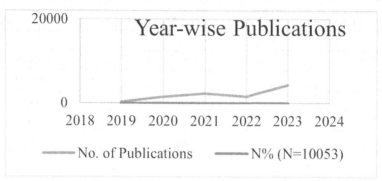

disaster. Even though there is a decline from 2021 to 2022, there was a rapid increment in 2023, which might be due to sudden alteration in the COVID-19 variant.

Table 2 shows the top 20 most cited authors in the field of disaster studies. In this sense, Griffith has the highest number of citations 2532 followed by Lin 2501, among others. This means that, there is a need to dig into seeking deeper understanding of why other researchers cited these authors. This will give an impression of the probable use of theories, their constructs, or developing models that could best give directions to understanding the phenomenon under investigation. The implication of this finding is that, based on the superficial analysis done on most literature consulted, the highly cited works are commonly scanty in templates designed by countries stroked disasters. This might indicate that, the strategies developed in those documents may not be in tandem with the current trends as regard physical, social environments or individuals. If this happens, then there is always a gap between what it is and what it ought to be in bridging the flaring gap or bringing the disaster under control.

It is obvious from the above figure that, a number of authors have contributed immensely on the disaster research. Despite this important co-authorship pattern, there is evidence suggesting that, not all collaboration led to the large number of publications. This implies that, there is some peculiarity as regard collaboration, citation, and number of publications among researchers.

Table 3 shows the most productive authors in the field of disaster. From the table, it is apparent that, Shah Kamal had 20 publications, Prestyo Yogi Tri had 18 publications, Van de Lindt (17), Yuen Kim Fai (17), Persada Satira Fadil (15), among others. Conversely, critical look at tables 2&3 reveals that, not all members in table 2 are present in table 3, which means that there is a marked difference between the most cited works and number of publications by authors. This shows an important indicator into investigating those studies that generated high citations and comparing

Table 2. Top 20 highly cited authors in the field of disaster

S/No.	Author	Citations
1	Griffiths, Mark D.	2532
2	Lin, Chung-Ying	2501
3	Pakpour, Amir H.	2472
4	Ivanov, Dmitry	2022
5	Dwivedi, Yogesh K.	954
6	Moktadir, MD. Abdul	952
7	Laato, Samuli	942
8	Shah, Kamal	773
9	Rana, Nripendra P.	673
10	Paul, Sanjoy Kumar	667
11	Van de Lindt, John W.	588
12	Gardoni, Paolo	586
13	Prasetyo, Yogi Tri	543
14	Dhir, Amandeep	537
15	Douglas, Karen M.	519
16	Dabija, Dan-Cristian	488
17	Ali, Syed Mithun	487
18	Morrison, Alastair M.	477
19	Naeem, Muhammad	467
20	Alvarez-Risco, Aldo	457

Figure 2. The co–authorship network map of authors on disaster-resilience infrastructure

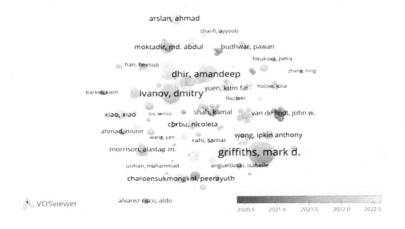

Table 3. Most productive authors in the field of disaster

S/No.	Author	No. of Publications	N% (10053)
1	Shah, Kamal	20	0.198946
2	Prasetyo, Yogi Tri	18	0.179051
3	Van de Lindt, John W.	17	0.169104
4	Yuen, Kum Fai	17	0.169104
5	Persada, Satria Fadil	15	0.149209
6	Wang, Xueqin	15	0.149209
7	Chatterjee, Sheshadri	15	0.149209
8	Chaudhuri, Ranjan	15	0.149209
9	Dhir, Amandeep	14	0.139262
10	Mostafavi, Ali	14	0.139262
11	Nadlifatin, Reny	12	0.119367
12	Ong, Ardvin Kester S.	11	0.10942
13	Vrontis, Demetris	11	0.10942
14	Griffiths, Mark D.	10	0.099473
15	Dwivedi, Yogesh K.	10	0.099473
16	Gardoni, Paolo	10	0.099473
17	Dabija, Dan-Cristian	10	0.099473
18	Cimellaro, Gian Paolo	10	0.099473
19	Gupta, Shivam	10	0.099473
20	Lin, Chung-Ying	9	0.089526

them with the authors that have produced a large number of research outputs. The implication of this finding is that, there is absence of highly prolific authors in most of the decision-making processes concerning prevention, preparedness, recovery, etc. This might create two opposing views about a single phenomenon between policymakers and researchers. Bringing expert researchers to intermingle with policymakers will pave the way for the development of appropriate preparedness, preventive, responsive, and recovery measures to the nagging disasters holistically.

From this figure, it is clear that, authors have played important roles in theoretical aspects of disaster, critically analyze practical cases, explore events in relation to local, national, international, inter-agency, inter-organizational proposals for mitigating disasters round the globe. To be precise, researchers are more interested in COVID-19, human, females, decision-making, resilience, non-human, vaccines, questionnaire, mental health, herd immunity, among others.

Figure 3. Co-occurrence of keywords and their associations

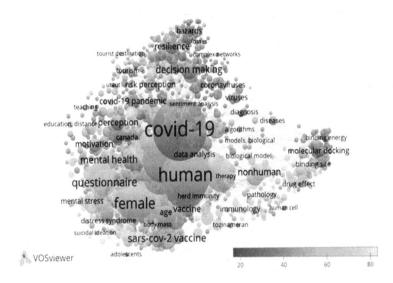

Table 4 shows the core journals that demonstrated higher participation in publishing disaster-related research. From this table, it follows that, there are specialized journals in the field of disaster, sustainability, and environment. This implies that, these journals might provide in-depth coverage of relevant research topics. Notably among these topics include: multilevel intervention to support school/education communities during COVID-19 pandemic. Examination of marketable skills gained by university students during COVID-19, Dysmorphia in e-teaching, Zero queue maintenance system using smart medicare application for COVID-19 pandemic, Youth without COVID-19 immunization, Young university students, COVID-19, and differential vulnerabilities, the fear of COVID-19 scale, Risk and resilience in family well-being during COVID-19, to mention but a few.

Content Analysis Results

At the onset of this discussion, it is important to make a remark that, the theory on disaster is an ever-expanding phenomenon since it contains both humans (subjective world) and natural world (objective world) alike. A study by Kapucu and Liou, (2014) highlighted a challenge that, there was artificial dissimilarity in literature and research between disaster and economic development researchers. From the perspective of content analysis, there is a seeming dissimilarity among them. For instance, to the best of their ability, in most of the documents consulted by the authors of this paper, there was no single document that used literature to compliment the work of

Table 4. Core journals (top 20)

S/No.	Journal	No. of Publications
1.	Sustainability (Switzerland)	181
2.	Frontiers in Psychology	162
3.	Heliyon	72
4.	PLOS one	63
5.	International Journal of Disaster Risk Reduction	61
6.	International Journal of Environmental Research and Public Health	60
7.	Current Psychology	60
8.	AIP Conference Proceedings	50
9.	ACM International Conference Proceeding Series	43
10.	International Journal of Hospitality Management	39
11.	Vaccines	38
12.	Technological Forecasting and Social Change	37
13.	Chaos, Solitons and Fractals	36
14.	Journal of Business Research	36
15.	lecture notes in computer science (including subseries lecture notes in artificial intelligence and lecture notes in bioinformatics)	36
16.	BMC Public Health	35
17.	E3S Web of Conferences	35
18.	International Journal of Contemporary Hospitality Management	34
19.	Personality and Individual Differences	32
20.	Computers in Human Behavior	31

most highly cited or prolific authors. either researchers. To begin with, the study by Griffith et al., (2022) that centered on "the fear of COVID-19 scale: Development and initial validation". The study ventured into developing instruments for measuring fear during a pandemic. The study did not touch aspects of economic development at all, which indicates that, there is a seeming approach between the researchers. Testing mental state with economic consequences during a pandemic would provide a significant avenue for balancing perceived vulnerability, perceived infectability in relation to economic challenges. This duality will enhance policymaking process developing a tool that can take care of the investigated phenomena squarely.

Similarly, Begum et al., conducted a research on "Adaptation of the Fear of COVID-19 Scale: Its association with psychological distress and life satisfaction in Turkey". The study used Item response theory (IRT), in conjunction with test theory, and psychometric theory. The study employed online survey method for collecting data. The population comprised 1304 participants where 917(70.3%) were females and 387(29.7%) were males between the ages of 18-64. The study administered Fear of COVID-19 Scale, Depression Anxiety Stress Scale-Short Form and Satisfaction with Life Scale. The data analysis involved confirmatory factor analysis (CFA) for validating the factor structure of the Fear of COVID-19 after which the discrimination, difficulty, and informativeness of the scale were examined using IRT. The study tried to link between Fear of COVID-19 Scale with Satisfaction with Life through the use of latent variables such as depression, anxiety, and stress. From CFA, it was found that, all the indices were within the acceptable limit and the association between them is significant.

Equally, to support the above views, a study by Ivanov Dimitry employed viable system model (VSM) in conjunction with dynamic system theory, control theory, game theory, ecological modeling, general system theory, accident theory, among others. Within the study, the author used VSM with the sole aim of understanding how the interconnected operations communicate with changing market environment and meta-systems such as markets, policy, and society. In addition, the author developed a supply chain (SC) that can adapt to both positive changes (i.e. agility angle) and absorb negative disturbances thereby recover and survive during short-, long-term global shocks (i.e. resilience and sustainability angles). It is obvious from the findings of this study that, the author focused mainly on economic development initiatives to assist industries or companies deal with economic shocks and devised a means through which they can recover and sustain their economic growth.

In addition, Michael P.A. Murphy used securitization theory in such a way that the state has the right to respond to the disaster as utterance emanates from the primary reality (i.e. populace). The theory tries to form a strong relationship between agency and structure as developed by Anthony Giddens, (1984) in structuration theory. When the securitization is activated, like a need for social distancing, there is a need for de-securitization programs to calm the nerves of the populace. The study captures the classic Copenhagen School Securitization theory. The study raises the awareness for de-securitization measure as to reverse securitization that can bring the heightened discourse of security into space of political discourse. De-securitization has certain difficulties of acceptance as policy decisions would be within the realm of political discourse rather than neutral applications of scientific principles. From securitization theory, the construction of a threat is a social process not a deterministic response to objective conditions. Furthermore, this study can be compared with Barnes's (2021) social identity theory that, understanding the roles of each member of a community is critical before, during, and after a disaster.

On the other hand, Islam et al., used conspiracy theory in their research. Based on infordemic, they referred to it as rumors, stigma, and conspiracy theory. The study comprised a team of social scientists, medical doctors, and epidemiologists. They reviewed a wide range of sources such as fact-checking agency websites, Facebook, Twitter, websites for television networks and newspapers (both national and international). Misinformation fueled by rumors, stigma, and conspiracy theories can have potentially severe implications on public health if prioritized over scientific guidelines. Governments and health agencies must understand the pattern of rumors, stigma, and conspiracy theories circulating globally for developing appropriate risk communication messages. Publishing correct and context-appropriate information supported by evidence-based practices should be available on the health agencies websites and that, national and international agencies including fact-checking agencies, should engage social media companies to spread correct information. Prioritizing rumor, stigma, and conspiracy theories over scientific evidence has negative effects on public health. The fact that, conspiracy theories are based on beliefs (i.e. psychological, political or social factors) communicated to individuals through traditional and social media platforms associated with societal risks implies the necessity for stakeholders to mitigate its spread by all means (Douglas et al., 2019).

Shima et al., investigated a research on "Does density aggravate the COVID-19 pandemic? Early findings and lessons for planners" using structural equation modeling. This research attempted to show that, density leads to close contact and more interaction among residents, which makes them potential hotspots for the rapid spread of emerging infectious diseases. This is especially the case in urban areas and other relatively densely populated areas.

CONCLUSION

Disaster is a manifestation of reversibly interacting objective and subjective worlds where the actions, inactions, interactions of humans have considerable impacts on disaster-causing incidences around the globe. This makes interdisciplinary research a viable option for preparing, preventing, responding, and recovering from such occurrences. Looking back at the theories used in previous studies remains one of the key elements that is badly needed but sadly neglected in mitigating disasters. Even though theories predict what might the cause be of or solution to the possible disasters, most disasters remain unpredictable a consequence of putting more pressure on policymakers and researchers in the field. One of the most fundamental concerns raised by this study is that, there is disconnect between disaster and development researchers to the extent most documents on disaster management do not contain current studies on mitigating disasters. To be precise, this study showed that, theories about disasters are continuous, expansive, and dynamic. Further attempts are needed to augment the already existing theories to provide directions to bringing disasters to barest minimum. Even though theories are deterministically used to direct policy formulations and implementation in an indeterministic disaster conditions, they quickly provide a means of understanding and enumerating possible variables for tackling such hazards.

REFERENCES

Adakawa, M. I., Balachandran, C., Kumar, P. K., & Harinarayana, N. S. (2023). History of pandemics—A critical pathway to challenge scholarly communications? National Conference on *Exploring the Past, Present, and Future of Library and Information Science*. IEEE.

Aid, A., & Rassoul, I. (2017). Context-aware framework to support situation-awareness for disaster management. *International Journal of Ad Hoc and Ubiquitous Computing, 25*(3), 120–132. doi:10.1504/IJAHUC.2017.083597

Ashraf, M. A., & Azad, M. A. K. (2015). Gender issues in disaster: Understanding the relationships of vulnerability, preparedness and capacity. *Environment and Ecology Research, 3*(5), 136–142. doi:10.13189/eer.2015.030504

Barnes, J.M. (2021). *Theoretical foundations of community disaster resilience.* [Submitted in partial fulfillment of the requirements for the degree of Master of Arts in Security Studies (Homeland Security and Defense) from the Naval Postgraduate School].

Basyah, N. A., Syukri, M., Fahmi, I., Ali, I., Rusli, Z., & Putri, E. S. (2023). Disaster prevention and management: A critical review of the literature. *Jurnal Penelitian Pendidikan IPA, 9*(11), 1045–1051. doi:10.29303/jppipa.v9i11.4486

Battisti, M., Belloc, F., & Del Gatto, M. (2023). *COVID-19, innovative firms and resilience.* (Economic Research Working Paper No. 73). https://www.wipo.int/edocs/pubdocs/en/wipo-pub-econstat-wp-73-en-covid-19-innovative-firms-and-resilience.pdf

Beekharry, D., & Baroudi, B. (2015). *Emergency risk management: Decision making factors and challenges when planning for disaster events.* Research Gate. https://www.researchgate.net/publication/280644351

Berawi, M. A. (2018). The role of technology in building a resilient city: Managing natural disasters. *International Journal of Technology, 5*(5), 862–865. doi:10.14716/ijtech.v9i5.2530

Bradshaw, S. (2004). *Socio-economic impacts of natural disasters: A gender analysis.*

Burger, A., Kennedy, W. G., & Crooks, A. (2021). Organizing theories for disasters into a complex adaptive system framework. *Urban Science (Basel, Switzerland), 5*(61), 61. Advance online publication. doi:10.3390/urbansci5030061

Cheema-Fox, A., LaPerla, B. R., Serafeim, G., & Wang, H. S. (2020). *Corporate resilience and response during COVID-19.* (Working Paper 20-108).

Coastal Resilience Centre. (2021). *Support strategies for socially marginalized neighborhoods likely impacted by natural hazards.* The University of North Carolina at Chapel Hill. https://coastalresiliencecenter.unc.edu/wp-content/uploads/sites/845/2021/07/Support-Strategies-for-Socially-Marginalized-Neighborhoods.pdf

Costine, K. J. (2015). *The four fundamental theories of disasters.* Research Gate. https://www.researchgate.net/publication/273694762_the_four_fundamental_theories_of_disasters

Cuthbertson, J., & Penney, G. (2023). Ethical decision making in disaster and emergency management: A systematic review of the literature. *Prehospital and Disaster Medicine, 00*(00), 1–6. doi:10.1017/S1049023X23006325 PMID:37675490

de la Llera, J. C., Rivera, F., Gil, M., & Schwarzhaupt, Ú. (2018). Mitigating risk through R&D+ innovation: Chile's national strategy for disaster resilience. *16th European Conference on Earthquake Engineering.* Research Gate.

Emergency Database (EM-DAT). (2024). *Inventorying hazards and disasters worldwide since 1988.* EM-DAT. https://www.emdat.be

Gilbert, S. W. (2010). *Disaster resilience: A guide to the literature.* National Institute of Standards and Technology Building and Fire Research Laboratory. https://www.govinfo.gov/content/pkg/GOVPUB-C13-724c6a83323886beabf2a2910a8fb81e/pdf/GOVPUB-C13-724c6a83323886beabf2a2910a8fb81e.pdf

Hu, G., Rao, K., & Sun, Z. (2006). A preliminary framework to measure public health emergency response capacity. *Zeitschrift für Gesundheitswissenschaften, 14*(1), 43–47. doi:10.1007/s10389-005-0008-2

Inan, D. I., Beydoun, G., & Othman, S. H. (2023). Risk assessment and sustainable disaster management. *Sustainability (Basel), 15*(6), 1–5. doi:10.3390/su15065254

Independent Commission Multilateralism (ICM) & International Peace Institute (IPI) (2016). *Discussion paper: Humanitarian engagements-Independent Commission on Multilateralism.* https://reliefweb.int/report/world/discussion-paper-humanitarian-engagements-independent-commission-multilateralism

Isser, R. (2022). *Disaster management: Theory, planning and preparedness.* CAPS. https://capsindia.org/wp-content/uploads/2022/08/Rajesh-Isser-1.pdf

Kapucu, N., & Liou, K. T. (2014). Disaster and development: Examining global issues and cases. In N. Kapucu & K. T. Liou (Eds.), *Disaster and development, environmental hazards.* Springer International Publishing Switzerland. doi:10.1007/978-3-319-04468-2

Keating, A., Campbell, K., Mechler, R., Magnuszewski, P., Mochizuki, J., Liu, W., Szoenyi, M., & McQuistan, C. (2017). Disaster resilience: What it is and how it can engender a meaningful change in development policy. *Development Policy Review, 35*(1), 65–91. doi:10.1111/dpr.12201

Khan, H., Vasilescu, L. G., & Khan, A. (2008). Disaster management cycle – A theoretical approach. *Management and Marketing Journal, 6*(1), 43–50.

Kim, Y.-k., & Sohn, H.-G. (2018). Disaster theory. In Y.-k. Kim & H.-G. Sohn (Eds.), *Disaster risk management in the Republic of Korea.* Disaster Risk Reduction., doi:10.1007/978-981-10-4789-3_2

Lord, F. M. (1951). *A theory of test scores and their relation to the trait measured (Research Bulletin No. RB-51-13).* Princeton: Educational Testing Service. doi:10.1002/j.2333-8504.1951.tb00922.x

Lord, F. M. (1952). *A theory of test scores (Psychometric Monograph No. 7).* Psychometric Corporation.

Lord, F. M. (1953). The relation of test score to the trait underlying the test. *Educational and Psychological Measurement, 13*(4), 517–549. doi:10.1177/001316445301300401

McKinsey& Company. (2020). *COVID-19 implications for life sciences R&D: Recovery and the next normal.* Pharmaceuticals & Medical Products Practice. https://www.mckinsey.de/~/media/McKinsey/Industries/Pharmaceuticals%20and%20Medical%20Products/Our%20Insights/COVID%2019%20implications%20for%20life%20sciences%20R%20and%20D%20Recovery%20and%20the%20next%20normal/COVID19implicationsforlifesciencesRDRecoveryandthenextnormalvF.pdf

Mojtahedi, S. M. H., & Lan-Oo, B. (2013). *Theoretical framework for stakeholders' disaster response index in the built environment.* IRBnet. https://www.irbnet.de/daten/iconda/CIB_DC27260.pdf

Musa, A. I. (2013). Understanding the intersections of paradigm, meta-theory, and theory in library and information science research: A social constructionist perspective. *Samaru Journal of Information Studies*, *13*(1 & 2). https://www.researchgate.net/publication/309479829

Nair, B. B., & Dileep, M. R. (2020). A study on the role of tourism in destination's disaster and resilience management. *Journal of Environmental Management and Tourism*, *11*(6), 1496–1507. doi:10.14505/jemt.11.6(46).20

Pepinsky, T. B. (2022). *Theory, comparison, and design: A review essay. Book review essay*. Cambridge University Press.

Perry, R. W. (2018). Defining disaster: An evolving concept. In H. Rodriguez, W. Donner, & J. Trainor (Eds.), *Handbook of disaster research*. Springer International Publishing. doi:10.1007/978-3-319-63254-4_1

Roper, S., & Turner, J. (2020). R&D and innovation after COVID-19: What can we expect? A review of prior research and data trends after the great financial crisis. *International Small Business Journal*, *38*(6), 504–514. doi:10.1177/0266242620947946

Rust, J., Stillwell, D., Loe, A., & Sun, L. (2017). Introduction to item response theory. https://www.psychometrics.cam.ac.uk/system/files/documents/SSRMCIRT2017.pdf

Salamati, S. P. N., & Kulatunga, U. (2017). *The importance of disaster management & impact of natural disasters on hospitals*. The 6th World Construction Symposium 2017: What's New and What's Next in the Built Environment Sustainability Agenda? Colombo, Sri Lanka.

Salamati, S. P. N., & Kulatunga, U. (2017). The challenges of hospital disaster managers in natural disaster events. *5th International Conference on Disaster Management and Human Health: Reducing Risk, Improving Outcomes*, Seville, Spain.

Samarakkody, A., Amaratunga, D., & Haigh, R. (2023). Technological innovations for enhancing disaster resilience in Smart Cities: A comprehensive urban scholar's analysis. *Sustainability (Basel)*, *15*(15), 12036. doi:10.3390/su151512036

SAMHSA. (2017). *Greater impact: How disasters affect people of low socioeconomic status*. Disaster Technical Assistance Centre, Supplemental Research Bulletin. https://www.samhsa.gov/sites/default/files/dtac/srb-low-ses_2.pdf

ShamimM. A. B.NasreenM. (2016). Theoretical approach to disaster management. doi:10.13140/RG.2.2.25708.97927

Singh, Z. (2020). Disasters: Implications, mitigation, and preparedness. *Indian Journal of Public Health, 64*(1), 1–3. doi:10.4103/ijph.IJPH_40_20 PMID:32189674

Southard, N. (2017). *The socio-political and economic causes of natural disasters.* [Thesis, CMC]. https://scholarship.claremont.edu/cmc_theses/1720

Suleyman, M., & Bhaskar, M. (2023). *The coming wave: Technology, power, and the 21ˢᵗ century greatest dilemma.* New York: Crown Publishing Group. https://pdfget.com/please-wait-for-few-moments-6/

Thurstone, L. L. (1927). A law of comparative judgment. *Psychological Review, 34*(4), 273–286. doi:10.1037/h0070288

UNDP. (2023). *Innovation in disaster management: Leveraging technology to save more lives.* One United Nations Plaza, New York. https://www.undp.org/sites/g/files/zskgke326/files/2024-03/innovation_in_disaster_management_web_final_compressed.pdf

van Niekerk, D. (2011). *USAID disaster risk reduction training course for Southern Africa.* Prevention Web. https://www.preventionweb.net/files/26081_kp1concepdisasterrisk1.pdf

Weston, R., & Gore, P. A. Jr. (2006). Brief guide to structural equation modeling. *The Counseling Psychologist, 34*(5), 719–751. doi:10.1177/0011000006286345

World Bank. (2021). *Overlooked: Examining the impact of disasters and climate shocks on poverty in the Europe and Central Asia region.* World Bank. https://documents1.worldbank.org/curated/en/493181607687673440/pdf/Overlooked-Examining-the-Impact-of-Disasters-and-Climate-Shocks-on-Poverty-in-the-Europe-and-Central-Asia-Region.pdf

World Health Organization. (2019). *Health emergency and disaster risk management framework.* WHO.

Zulch, H. (2019). *Psychological preparedness for natural hazards – improving disaster preparedness policy and practice.* Contributing Paper to GAR 2019. UNDRR. https://www.undrr.org/publication/psychological-preparedness-natural-hazards-improving-disaster-preparedness-policy

Chapter 9
Strategic Deployment of AI and Drones Enhancing Disaster Management in Natural Disasters

Ravinder Singh

 https://orcid.org/0009-0007-7161-6709
Indian Army, India

Geetha Manoharan

 https://orcid.org/0000-0002-8644-8871
SR University, India

Samrath Singh

 https://orcid.org/0009-0009-3720-1432
Vellore Institute of Technology, Chennai, India

ABSTRACT

The emergence of artificial intelligence (AI) and unmanned aerial vehicles (UAVs), commonly known as drones, has revolutionized disaster management practices, particularly in the context of natural disasters. In recent years, there has been a notable surge in the integration of innovative smart connected devices and platforms, including drones and UAVs, into the extensive network of the internet of things (IoT). The integration of AI and drones into the IoT network presents numerous challenges, opportunities, and implications for leveraging them in disaster management. This study offers an in-depth overview of the several applications of AI and drones in various stages of disaster management. In this review, the authors have explored the evolution of disaster management paradigms, the applications of AI and drones in disaster management, and the challenges and disruptive technologies shaping this field. The study reveals that the application of AI and drones has great potential in disaster management and can enhance the resilience of the community.

DOI: 10.4018/979-8-3693-3896-4.ch009

INTRODUCTION

Evolution of Disaster Management Paradigms. The turbulent landscape of disaster management underwent a profound evolution in the 20th century that was characterized by reorienting paradigms and strategic initiatives meant to protect lives and lessen the devastating effects of natural calamities. During this era, there emerged a discernible shift towards centralized measures for citizen protection, catalyzing the genesis of civil defense mechanisms (Coppola, 2015; Hecker & Domres, 2018). Governments worldwide began to recognize the imperative for coordinated responses to potential crises, prompting the establishment of frameworks dedicated to disaster mitigation and preparedness.

However, it was not until the 1970s that the spotlight intensified on disaster response, signaling a pivotal juncture in the trajectory of disaster management. Governments and international bodies increasingly prioritize strategies geared towards swift and effective interventions in the aftermath of catastrophic events, underscoring the paramount importance of rapid mobilization and resource allocation.

Global Policy Frameworks for Disaster Risk Reduction. A seminal moment in the annals of disaster management occurred on December 11, 1987, when the United Nations General Assembly decreed the 1990s "The International Decade for Natural Disaster Reduction (IDNDR)" (United Nations, 1999). This epochal declaration aimed to galvanize global efforts towards heightened awareness and concerted action in mitigating the risks posed by natural disasters. In May 1994, against the backdrop of escalating environmental vulnerabilities and escalating human tolls, the United Nations convened in Yokohama to formulate a strategy and action plan (United Nations, 1994). This landmark initiative, known as the Yokohama Strategy, heralded a paradigm shift towards proactive risk management, acknowledging the imperative of fortifying global resilience in the face of mounting threats.

The Great Hanshin Awaji earthquake, which ravaged the Kobe area of Japan on January 17, 1995, served as a poignant reminder of the urgent need for enhanced disaster preparedness and response mechanisms (Britannica, 2021). In its wake, the international community convened in Kobe in 2005 to promulgate the Hyogo Framework for Action (HFA), a comprehensive blueprint aimed at bolstering global resilience over the ensuing decade. Three important goals were spelled out in the HFA plan: disaster risk reduction (DRR) programs should be built into plans for sustainable development; institutions should be strengthened to make them more resilient; and risk reduction methods should be systematically added to plans for emergency preparedness and response (UNISDR, 2005). This concerted endeavor brought to the fore a collective commitment to forging a more secure and resilient world in the face of escalating environmental exigencies.

Integration of Disaster Risk Management into Sustainable Development Goals. The strength of Japan's buildings and infrastructure in the wake of the devastating tsunami that resulted from a 9-magnitude earthquake in Sendai on March 11, 2011, highlighted the country's strict enforcement of earthquake-related technology standards (Zaré, 2011). This resilience, coupled with effective law enforcement measures, played a pivotal role in mitigating the impact of the disaster. The Sendai Framework for Disaster Risk Reduction 2015–2030 (SFDRR) stands as a landmark global policy framework within the United Nations' post-2015 agenda. It emphasizes mental and physical health, resilience, and well-being in disaster risk reduction (DRR) strategies, elevating priorities like safe schools and hospitals. The third world conference on disaster reduction, convened in 2015 to assess the outcomes of preparedness efforts initiated under the Hyogo Framework for Action 2005–2015, marked a critical milestone in global disaster management (Aitsi-Selmi et al., 2015). Additionally, the 2030 Agenda for Sustainable Development, which all United Nations Member States ratified in 2015, lists 17 Sustainable Development Goals (SDGs). These goals, encompassing poverty eradication, improved health and education, reduced inequality, economic growth, climate action, and environmental conservation, reflect a collective commitment to global partnership and sustainable progress (THE 17 GOALS | Sustainable Development, n.d.).

Transition to Disaster Risk Management (DRM). In recent times, there has been a notable shift in the approach towards disasters, moving away from merely responding to them to actively mitigating risks. This shift has given rise to the concept of Disaster Risk Management (DRM), where measures aimed at reducing risks are viewed as investments rather than mere expenses (Wisner et al., 2004). DRM encompasses various strategies focused on risk reduction, acknowledging that effective disaster management is part of a broader, more holistic system. It emphasizes managing the underlying risks rather than solely focusing on responding to disasters as they occur. While aspects such as risk identification, monitoring, and assessment were outlined in the Hyogo Framework, the Sendai document significantly emphasizes understanding and addressing disaster risk (Aitsi-Selmi et al., 2016).

The Role of AI and Drones in Harnessing Advanced Technologies for Disaster Management. The emergence of AI and unmanned aerial vehicles (UAVs), commonly known as drones, has revolutionized disaster management practices, particularly in the context of natural disasters. The strategic deployment and utilization of AI and drones to enhance disaster management in the face of natural calamities is going to be a force multiplier for the nations affected by disasters. With the increasing frequency and intensity of natural disasters worldwide, there is a growing imperative to leverage advanced technologies for DRM activities. AI algorithms, powered by vast datasets and machine learning techniques, enable predictive analytics, early warning systems, and decision support tools that enhance situational awareness and response coordination. Meanwhile, drones equipped with sophisticated sensors and imaging capabilities allow people to run activities that are not accessible to humans during disasters (Vermiglio et al., 2022).

EVOLUTION OF AI AND DRONES

Evolution of Drones: From Military Origins to Civilian Integration. Drones, also known as Unmanned Aerial Vehicles (UAVs) or Unmanned Aerial Systems (UAS), are pilotless flying machines controlled remotely via radio waves or autonomously following preset routes. They come in various sizes and configurations and are commonly equipped with optoelectronic devices for surveillance and monitoring purposes. Originally developed for military and police use, drones have evolved over time. The earliest documented use of an unmanned flying vehicle dates back to August 1849, when Austrians employed balloons loaded with explosives for military purposes. Civilian drones, which emerged much later, primarily in the 1980s, differ from military drones in size and propulsion, often utilizing electric motors for power. They are commonly used for photography and videography purposes (Varalakshmi, 2019).

In recent years, there has been a notable surge in the integration of innovative smart connected devices and platforms, including drones and UAVs, into the extensive network of the Internet of Things (IoT). UAVs not only provide novel avenues for delivering value-added IoT services across various applications such as monitoring, surveillance, on-demand last-mile delivery, and transportation of people, but they also offer a practical solution to the constraints of fixed terrestrial IoT infrastructure. Due to their significant potential, they are anticipated to become an essential component of urban environments, exerting dominance over shared low-altitude airspace (Labib et al., 2021).

Expanding Roles for Drones. Drones, once predominantly military assets, are now diversifying into various civilian roles, prompting consideration. While limited by evolving technology, they offer vast versatility, potentially revolutionizing logistics and services. COVID-19 has showcased its potential for safe, efficient delivery, hinting at broader transformative impacts. Commercially, drones show promise for cost-effective operations across industries like mining and agriculture. Urban parcel delivery is an emerging frontier, promising efficiency but raising regulatory and safety concerns. Despite challenges, drones hold promise for reshaping various sectors and warrant proactive management for their integration into daily life. Drone usage is being exploited in the areas of monitoring, inspection, and data collection; photography/image collection; recreation; and logistics. Industries are grappling with managing the drone revolution, with operators aiming to develop intricate management systems tailored to their needs, such as for search and rescue (Mohsin et al., 2016), complex distribution networks (Shavarani, 2019), and ad hoc routing (Suteris et al., 2018). While industry and government are considering overarching coordinating networks, some advocate for integrating drones into existing air transport management systems for enhanced safety and oversight (Merkert & Bushell, 2020).

The future of drones heralds an era of unprecedented innovation and utility. With the ability to autonomously navigate both natural landscapes and man-made environments, drones are poised to revolutionize a multitude of civilian tasks. Beyond their traditional role in defense, drones hold promise for transforming transportation, communication, agriculture, disaster mitigation, and environmental preservation efforts. However, this transition to autonomous flight within confined spaces poses significant scientific and technical hurdles. Overcoming the energetic demands of sustained flight and developing the requisite perceptual intelligence to navigate complex environments are paramount challenges that must be addressed. Nevertheless, the potential benefits of integrating drones into various facets of daily life are vast, promising a future where these aerial robots play an integral role in advancing human endeavors (Floreano, 2015). The utilization of drones raises significant social and economic expectations owing to their diverse capabilities and wide-ranging business applications. However, the realization of their full potential hinges crucially on social acceptance. Experts assert that achieving social acceptance is contingent upon striking a delicate balance between the benefits they offer and the inconveniences associated with their use.

Additionally, strict safety regulations and policies governing airspace, as well as the requirements of current airspace users, have an impact on this equilibrium within the aeronautical sector. Finding harmony among these factors is essential for the comprehensive development of drone technology (Macias et al., 2019).

The Evolution of AI. For a significant amount of time, drone development has been progressing steadily. However, it is the surge in the progression of AI in recent years that has truly captured attention and accelerated innovation. 'AI refers to the development of computer systems that can perform tasks that would typically require human intelligence, such as visual perception, speech recognition, decision-making, and language translation' (Vashishth et al., 2023).

Alan Turing, a pioneering figure in artificial intelligence, introduced the concept of a "universal machine" capable of emulating any human intellectual task, yet early AI development faced setbacks due to limited computing power and data scarcity. During the 1950s and 1960s, AI researchers concentrated on developing rule-based systems to emulate human reasoning, but these systems faced challenges in adapting and learning from new information until the exploration of machine learning algorithms in the 1970s and 1980s allowed computers to enhance their performance through data-driven learning. The 2010s witnessed a resurgence in AI advancements, particularly with the emergence of deep learning techniques. These methods leverage neural networks to process and analyze vast datasets, revolutionizing various industries such as healthcare, finance, and transportation with AI-powered applications (Vashishth et al., 2023). AI research has primarily concentrated on exploring the following aspects of intelligence: learning, logical reasoning, problem-solving, perception, and comprehension of language (Copeland, 2020).

Applications of AI Technologies. Since those initial advancements, artificial intelligence has made significant strides, with a plethora of technologies now available, encompassing natural language generation (NLG), natural language understanding (NLU), speech recognition (SR), machine learning (ML), virtual agents (VA), expert systems (ES), decision management (DM), deep learning (DL), robotic process automation (RPA), text analytics (TA), natural language processing (NLP), biometrics, cyber defense (CD), emotion recognition (ER), and image recognition (IR) (Moloi & Marwala, 2021; Edureka, 2020).

We currently reside in the era of "big data," where we possess the capability to gather vast amounts of information that surpass human processing capacities. The integration of artificial intelligence has proven highly beneficial across various sectors like technology, banking, marketing, and entertainment. Despite potential stagnation in algorithmic advancement, the sheer volume of data and computational power enable AI to learn effectively through sheer force. While there may be indications of a slowdown in Moore's Law, the exponential growth of data shows no signs of diminishing momentum (Anyoha, 2017).

The proliferation of AI opportunities extends across all sectors of society, with industries and enterprises increasingly leveraging their potential. This trend is particularly noticeable due to the recent surge in data collection and analysis, facilitated by reliable IoT connectivity and rapid computer processing. While certain sectors have already reaped the benefits, others are just beginning to grasp the possibilities AI offers. These sectors encompass manufacturing, healthcare, energy, environment, transportation, education, media, customer service, space, e-commerce, navigation, lifestyle, sociology, robotics, agriculture, gaming, marketing, social media, chatbots, finance, and economics (Groumpos, 2023).

Challenges and Risks Associated with AI. AI signifies a transformative shift in scientific methodology, which traditionally leans on theories and models developed through rigorous scientific inquiry. While humans historically crafted mathematical models, methods, and algorithms, recent years have witnessed machines assuming these tasks independently, prompting concerns about the dependability and oversight of such models. Furthermore, numerous AI failures have occurred, often with far-reaching consequences and insufficient scrutiny. Noteworthy failures include compromised Face ID, Google AI misidentifying objects, and fatal collisions due to system malfunctions, exemplifying the risks associated with AI. Addressing these failures poses a significant challenge, necessitating a thorough understanding of AI limitations and the implementation of robust safeguards to ensure the reliability, trustworthiness, and control of AI models and algorithms. It is crucial to prioritize addressing AI failures seriously to mitigate potential harm and foster responsible AI development (Groumpos, 2023).

AI threats persist and, rather than diminishing, have multiplied and intensified, presenting a more tangible and perilous risk (Kieslich et al., 2021). AI presents a myriad of potential dangers, demanding vigilant risk management strategies. The lack of transparency and explainability in AI models hinders understanding and may result in biased or unsafe decisions. Job losses loom large as AI automation penetrates various industries, threatening up to 30% of current U.S. jobs by 2030. Social manipulation via AI algorithms, exemplified by political exploitation on platforms like TikTok, underscores the urgent need for safeguards against misuse. Social surveillance utilizing AI technology, as seen in China's widespread facial recognition systems, poses grave privacy concerns. Additionally, AI tools often collect personal data, raising questions about data privacy and usage transparency. Biases inherent in AI systems, including algorithmic and human biases, perpetuate societal inequalities and discriminatory practices. The rise of AI-powered autonomous weapons amplifies risks, particularly in the hands of malicious actors capable of unleashing catastrophic consequences. Financial crises fueled by AI algorithms in trading processes further accentuate the need for prudent regulation. The pervasive integration of AI may lead to a loss of human influence and functioning in crucial sectors like healthcare and creativity, while the prospect of uncontrollable, self-aware AI poses existential threats. Effectively managing these risks necessitates robust ethical frameworks, stringent regulations, and ongoing research to ensure the responsible and beneficial deployment of AI technologies (Thomas, 2023).

Navigating the Future: Managing Risks and Maximizing Benefits. Drones, initially developed for military and police use, have evolved to serve various civilian roles, notably in photography, videography, and logistics. Their integration into the Internet of Things (IoT) network presents novel opportunities, especially in monitoring, surveillance, and delivery services. Despite their potential benefits, drones pose challenges, such as regulatory concerns and privacy issues. Furthermore, AI advancements have expanded AI applications across sectors, raising concerns about transparency, job displacement, social manipulation, surveillance, biases, and autonomous weapons. Effective management strategies, including ethical frameworks and stringent regulations, are crucial to harnessing the benefits of AI while mitigating its risks.

UNDERSTANDING DISASTER MANAGEMENT CHALLENGES

Foundations of Disaster Management and Disaster Risk Management. According to the Disaster Management Act of India of 2005, "disaster management" refers to an ongoing and coordinated process aimed at planning, organizing, coordinating, and executing measures essential for preventing, reducing, capacity-building, preparedness, rapid response, impact assessment, and facilitating recovery efforts in handling any disaster situation. Disaster management encompasses coordinating resources and responsibilities to address humanitarian emergencies, emphasizing preparedness, response, and recovery through administrative decisions and operational activities covering prevention, mitigation, preparedness, response, recovery, and rehabilitation (National Institute of Disaster Management, n.d.). However, Disaster Risk Management can be stated as 'The application of disaster risk reduction policies and strategies to prevent new disaster risk, reduce existing disaster risk, and manage residual risk, contributing to the strengthening of resilience and reduction of disaster losses' (UNDRR, n.d.).

Disaster management encompasses a multifaceted approach aimed at mitigating the impact of disasters on communities and infrastructure. This involves proactive measures such as preparedness, which includes creating emergency plans, conducting drills, and stockpiling essential supplies, as well as mitigation efforts like land-use planning and infrastructure improvements to reduce vulnerability (Tabish, 2015). During a disaster, response activities such as search and rescue operations, medical assistance, and evacuation are crucial for saving lives and protecting property (Born et al., 2007). Following a disaster, recovery efforts focus on restoring communities and addressing long-term needs (Finucane, 2020). Additionally, risk reduction strategies aim to identify and address underlying risk factors, while community engagement ensures that local voices are heard and resilience is built from within (Wamsler & Johannessen, 2020). Leveraging technology, international cooperation, and innovation further enhances disaster management efforts, ultimately fostering more resilient communities globally.

Consequences of Natural Disasters. Among the prevalent consequences of disasters are various mental health issues, including peritraumatic stress reactions, Post Traumatic Stress Disorder (PTSD), complicated grief symptoms, depression, anxiety disorder, substance abuse disorders, distorted perceptions, pessimism, and suicidal ideation and attempts. These impacts are particularly profound for children. To address these challenges, the deployment of psychological interventions like psychological first aid and grief counseling could prove highly effective in mitigating the psychological toll of disasters (Sandhu, 2013). In recent decades, economic losses stemming from natural disasters have seen a steady rise, largely attributed to the burgeoning population and economic activity in disaster-prone regions.

Projections suggest that future losses will continue to escalate due to sustained growth in economic exposure and the effects of climate change (Singh & Manoharan, 2024). This underscores the critical need for policy measures aimed at reducing the adverse effects of these disasters on both the economy and society at large. The direct consequences of disasters can be property damage as well as indirect impacts, including effects on gross domestic product growth and trade (Botzen et al., 2019). A recent analysis of the impacts of cyclones, floods, and droughts on agriculture, food security, and the natural environment revealed significant declines in food production after a disaster. Moreover, they also substantially jeopardized food security for households in affected regions. Various coping mechanisms were observed among households, encompassing both consumption and non-consumption strategies. The environmental impacts based on available evidence suggest significant repercussions on natural resources and ecosystems (Israel & Briones, 2012). Natural disasters affect children in many ways, such as getting sick due to problems such as malnutrition, psychological disorders due to various negligence, traumas, and interruptions of educational opportunities (Kousky, 2016). Natural disasters pose significant threats to global supply chains due to the centralization of production, supplier networks, and distribution channels, which concentrates risks and limits alternatives. Reliance on specific transportation routes for cross-border production further exacerbates supply disruptions during infrastructure failures. Moreover, strategies aimed at enhancing business efficiency within supply chains may inadvertently intensify the adverse effects of natural disasters (Ye, 2012).

The Evolving Nature of Disaster Risks. Natural disasters unleash a myriad of consequences that ripple through societies and ecosystems. Loss of life is perhaps the most immediate and heart-wrenching outcome, as communities mourn the lives lost and families are forever changed. Displacement of populations exacerbates the crisis, forcing individuals to flee their homes and seek refuge elsewhere, often straining resources and exacerbating social tensions. The destruction of infrastructure compounds the chaos, disrupting essential services and impeding rescue and recovery efforts. Economic upheaval follows suit, as businesses falter, jobs are lost, and entire economies grapple with the aftermath. Environmental degradation adds another layer of complexity as ecosystems suffer irreparable harm from the forces of nature, further exacerbating long-term vulnerabilities. Psychological issues emerge as survivors grapple with trauma, grief, and uncertainty, often struggling to cope with the emotional toll of their experiences. Furthermore, disruptions to food production exacerbate food insecurity, leaving vulnerable populations at risk of hunger and malnutrition. Finally, the threat to global supply chains looms large, as disruptions in one region can have far-reaching implications, affecting industries and consumers worldwide. The consequences of natural disasters are far-reaching and multifaceted, underscoring the urgent need for robust preparedness, response, and recovery efforts.

The Importance of AI and Drones in Disaster Management. The evolving nature of disaster risks presents a multifaceted challenge in today's world. Climate change exacerbates extreme weather events like cyclones, floods, and wildfires, affecting both urban and rural areas (Ebi et al., 2021). Rapid urbanization further compounds these risks as growing populations concentrate in vulnerable regions (Singh, 2023). Additionally, infrastructure failures and pandemics introduce new perils, amplifying the vulnerabilities of communities worldwide. Socio-economic disparities deepen these risks, disproportionately impacting marginalized populations (Qiang, 2019). To effectively address these challenges, comprehensive risk assessment, adaptive planning, and resilient infrastructure are essential, along with proactive measures to foster community engagement, strengthen social cohesion, and promote sustainable development practices. The evolving landscape of disaster risks presents a formidable challenge for disaster management efforts worldwide. Addressing these evolving risks requires a holistic approach that integrates robust risk assessment, proactive mitigation measures, and community engagement initiatives. It also necessitates agile and adaptive disaster management frameworks capable of responding effectively to dynamic and interconnected hazards while prioritizing the protection and resilience of vulnerable populations. Ultimately, managing evolving disaster risks demands a collaborative and interdisciplinary approach that transcends traditional boundaries and embraces innovation and resilience-building across all levels of society. In this endeavor, the integration of AI and drones plays a crucial role. AI can analyze vast amounts of data in real-time to enhance early warning systems, predict disaster impacts, and optimize resource allocation for response efforts. Drones equipped with advanced sensors and imaging technologies offer rapid aerial reconnaissance, enabling emergency responders to assess damage, identify survivors, and deliver supplies to remote or inaccessible areas more efficiently. By harnessing the power of AI and drones alongside other innovative technologies, disaster management agencies can improve situational awareness, enhance decision-making processes, and ultimately save lives and reduce the impact of disasters on communities worldwide.

Enhancing Coping Capacity through AI and Drones. In addition to enhancing disaster management efforts, the integration of AI and drones can significantly improve the coping capacity of communities, response organizations, and administrations. AI-powered predictive analytics can help communities better understand their vulnerabilities and develop targeted preparedness plans tailored to their specific needs. By leveraging AI algorithms to analyze historical data and simulate disaster scenarios, communities can identify potential risks and prioritize mitigation measures to bolster their resilience. Drones equipped with AI capabilities can provide real-time situational awareness during disaster events, allowing response organizations to quickly assess the extent of damage and allocate resources more effectively. Furthermore, AI-driven communication systems can facilitate coordination between different response agencies and streamline information sharing, enabling a more efficient and coordinated response effort. By empowering communities, response organizations, and administrations with AI and drone technologies, coping capacity can be significantly enhanced, ultimately leading to more effective disaster preparedness, response, and recovery outcomes.

APPLICATIONS OF AI IN DISASTER MANAGEMENT

AI's Revolution in Computing: Transforming Data Processing, Automation, and Decision-making. The transformative impact of AI on computing cannot be overstated, as AI technologies have revolutionized how computers process data, automate tasks, and make decisions. Previously unthinkable feats, such as image recognition, natural language understanding, and complex decision-making, are now within the realm of possibility for machines thanks to AI. By enhancing data analysis and decision-making capabilities, AI empowers computers to discern patterns and trends, leading to the creation of intelligent systems applicable across diverse fields like healthcare, finance, transportation, and manufacturing (Manoharan et al., 2024). Research findings underscore AI's potential to redefine computing and reshape industries, offering advantages such as heightened efficiency, accuracy, and adaptability compared to traditional computing methods (Vashishth et al., 2023).

Role of AI in Disaster Management. The application of AI in disaster management represents a transformative shift in how we approach mitigation, preparedness, response, and recovery efforts. AI and machine learning algorithms can analyze vast amounts of data from various sources, including satellite imagery, social media feeds, and sensor networks, to provide real-time insights into disaster situations (Balaji et al., 2021; Shi et al., 2020). This enables emergency responders to make more informed decisions regarding resource allocation, evacuation routes, and the prioritization of critical infrastructure. Additionally, AI-powered predictive analytics can forecast the trajectory of natural disasters, allowing authorities to proactively implement mitigation measures and alert affected populations well in advance (Tan et al., 2021). Furthermore, AI-driven technologies like machine learning and natural language processing facilitate the automation of administrative tasks, freeing up valuable time and resources for responders to focus on frontline operations and community outreach (Abid et al., 2021).

Future Directions: Expanding AI Applications in Disaster Management. As AI continues to mature, its applications in disaster management are expected to expand further, encompassing areas such as risk modeling, decision support systems,

Figure 1. AI methods and application areas in disaster management (Sun, Wenjuan, Bocchini, Paolo, and Davison, Brian, (2020)

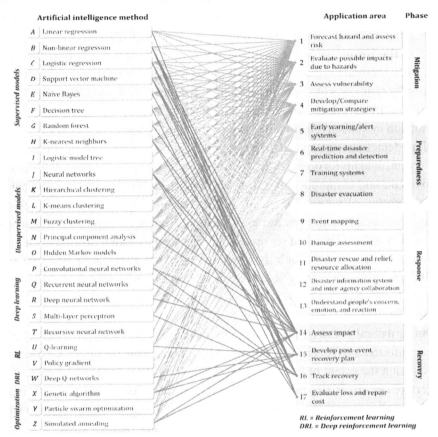

and humanitarian assistance. Through ongoing research and collaboration between academia, industry, and government agencies, AI has the potential to revolutionize how we prepare for and respond to disasters, ultimately saving lives and minimizing the impact on communities worldwide.

A study by Sun W et al. in 2020 on AI applications in disaster management identified a total of 26 AI methods and 17 application areas as representative examples (Figure 1).

AI in the Preparedness Stage. Artificial intelligence has the potential to predict upcoming events, like cyclone trajectories and storms, earthquakes, floods, volcano eruptions, and fires, thus enabling early warnings to be issued. Response agencies can utilize AI-based tools to predict probable damage locations and estimate the duration of service outages before a disaster occurs, allowing them to prepare in advance. In readiness for evacuations, it's essential to thoroughly assess potential challenges and devise strategies specifically tailored for cyclone evacuations in coastal regions. Given the diverse routes taken by large crowds during evacuations, AI can facilitate anticipating crowd dynamics, pinpointing optimal evacuation routes, and establishing supportive evacuation systems (Sun et al., 2020).

AI in the Response Stage. Swift responses to disasters are critical for saving lives. Decision-makers must exert maximum effort to comprehend the situation and enhance the efficiency of response endeavors. This necessitates situational awareness to facilitate effective decision-making, as outlined in, along with user-friendly disaster information systems to ensure effective coordination (Sun et al., 2020).

Maps depicting events and damage, produced through various AI methodologies, offer crucial insights for organizing search and rescue missions, coordinating resource deployment, and comprehending short-term housing requirements in response to natural disasters (Vieweg, 2012; Lin, 2015). Satellites, unmanned aerial vehicles, robots, and social media platforms consistently produce large amounts of data related to disasters. This data serves as the basis for generating disaster event maps (Sun et al., 2020), and through the comparison of pre-event and post-event maps and images, disparities in features can be identified and utilized to evaluate the damage to structures and infrastructure. This process aids in prioritizing response efforts effectively.

During natural disasters, millions of individuals are swiftly resorting to social media to share information (Aisha et al., 2015). AI has proven valuable in various tasks such as evaluating the extent of building damage resulting from earthquakes using aerial imagery (Ci et al., 2019), distinguishing between informative and non-informative tweets during a disaster (Madichetty & Sridevi, 2019), and detecting disaster victims amidst highly cluttered backgrounds (Sulistijono & Risnumawan, 2016).

In disaster rescue and relief efforts, the integration of social media and mobile phone data often facilitates timely and effective decision-making. Social media platforms serve as potent communication channels, enabling individuals and local communities to seek assistance, while governments and organizations can efficiently disseminate disaster relief information (Li & Rao, 2010; Tatsubori et al., 2012).

Social media data contains valuable time, geo-location, and disaster-related information, making it a valuable resource for constructing disaster information systems (Goodchild & Glennon, 2010; Srivastava et al., 2012). Ultimately, this supports decision-making processes concerning disaster relief efforts and resource allocations.

During a disaster, there is a notable increase in telecommunications activity (Bagrow et al., 2011). Disaster management agencies must promptly sort through the information from these calls and relay urgent public needs to the appropriate authorities. Machine listening technology can assist in automatically recognizing voices and prioritizing key words, facilitating the swift processing of voice data from diverse regions (Ramchurn et al., 2016).

A disaster causes not only physical damage to buildings and infrastructure but also induces psychological stress for the victims, resulting in differing decisions and responses (Greifeneder et al., 2011). Recognizing the emotions and psychological needs of victims is essential for effective disaster relief. Social media data includes emotional text and images, as well as information about time and place. This information is very useful for figuring out how people behave, how people move, and their mental and medical needs change over time and space (Bengtsson et al., 2011; Caragea et al., 2014).

AI in the Recovery Stage. During the stage of disaster recovery, the primary objective is to expedite the recovery process, which entails implementing effective measures such as devising recovery and reconstruction plans, allocating both internal and external resources, analyzing post-disaster data, restoring infrastructure, and providing disaster assistance (Sahebjamnia et al., 2015). Achieving this necessitates comprehensive decision-making to swiftly grasp the situation's complexity, recognize operational requirements and recovery strategies, and execute rehabilitation and reconstruction efforts.

For the recovery process to proceed as quickly and accurately as possible, disaster impact evaluation is essential. While visual inspection serves as a primary method for assessing physical damage, it is often laborious and time-consuming. AI methods offer a solution by automating this process using data from aerial images, social media imagery, and sensor measurements (Khaloo et al., 2017). AI models can assess the severity of disasters and generate damage maps (Rodríguez et al., 2013).

In addressing psychosocial stress issues post-disaster, surveys are commonly employed. Both supervised and unsupervised AI models, including regression methods, dimension reduction methods, and neural networks, are frequently utilized to analyze survey results. These methods help identify risk factors and evaluate the effectiveness of preventive interventions (Gao et al., 2006).

Additionally, AI-powered website applications aid in extracting public sentiment and loss data from social media posts, contributing to a comprehensive understanding of community impact (Yang et al., 2019). AI methods are also effective in estimating the economic impacts of hazards and identifying potential stimuli for economic growth (Zhang & Peacock, 2009).

In the realm of post-disaster reconstruction and assistance, (Singh, 2007) developed a prototype Expert System for monitoring and evaluating food aid provided by international disaster relief organizations. Furthermore, during the recovery process, assessing how a community recovers from a disaster over time is essential for building resilience. Comparing nighttime light data at different times can offer valuable insights into how the regional economy recovers quantitatively (Wang et al., 2018). Additionally, Google Street View can assist in remotely tracking disaster recovery efforts (Curtis et al., 2010).

DRONE TECHNOLOGY IN DISASTER MANAGEMENT

The term drone refers to any unpiloted aircraft ranging from large unmanned air vehicles (UAVs) covering extensive distances to compact models navigating confined spaces. These aerial vehicles, devoid of human operators, operate remotely or autonomously and are capable of carrying payloads that can be lethal or nonlethal (Gupta et al., 2013). Their versatility spans civil and military domains, executing missions both outdoors and indoors, even in the most demanding conditions (Rodríguez et al., 2013). Equipped with a variety of sensors and cameras, drones excel in intelligence, surveillance, and reconnaissance tasks. Their applications can be classified based on mission types (military/civil), flight zones (outdoor/indoor), and environmental settings (underwater/on water/ground/air/space) (Hassanalian & Abdelkefi, 2017). Drones have diverse applications, including search and rescue missions, environmental protection, goods delivery, and studying marine environments (Wang et al., 2023). They aid in locating missing individuals during emergencies, monitor ecosystems for conservation efforts, transport items efficiently, and conduct research on marine organisms and oil spill detection (Jiménez López & Mulero-Pázmány, 2019; Yang et al., 2022). Additionally, drones serve recreational purposes, such as hobbies, and contribute to agricultural practices by monitoring crops and assessing fields. Moreover, industries leverage drones for various tasks like infrastructure inspection and aerial photography. With their versatility, drones play integral roles across sectors, offering innovative solutions to complex challenges in a compact and agile form.

The challenge of accessing vast terrain post-disaster despite the urgent need for support is a significant obstacle in the management process. However, with the advancement, affordability, and adaptability of drones, innovative communication, delivery, and sensing solutions can swiftly be deployed in high-risk areas. Drones offer numerous benefits to first responders and disaster managers, particularly as they become more cost-effective and technologically advanced. Examples of drone applications in disaster scenarios encompass various situations like fires in buildings or forests, cyclones, floods, structural collapses, crowd surveillance, search and rescue operations, and post-disaster monitoring of critical infrastructure. Drones present a valuable tool for addressing the complexities of crisis and disaster management.

Technology advancements and the increasing complexity of disaster scenarios are driving constant change in the field of AI in disaster management. One emerging area of focus is the integration of AI with drones and robotics to enhance situational awareness and response capabilities in disaster-affected areas.

Drone Mapping. Drones play a crucial role in disaster management by facilitating monitoring, mapping, and damage assessment. Their utility spans various disasters, including floods, landslides, forest fires, hurricanes, tsunamis, volcanic eruptions, earthquakes, and radiological incidents like Chernobyl. Compared to traditional first responder methods, drones offer greater efficiency, delivering high-quality images in less time than satellite imagery. They excel at assessing erosion and flooding, providing detailed and continuous data with minimal on-site presence. Drones demonstrate adaptability in terms of resources and timing, enabling remote revisits to study sites. Additionally, they enhance information sharing during the initial disaster stages and effectively assess building damage in inaccessible areas. Multi-rotor drones offer flexibility for surveying small or isolated locations, while fixed-wing drones are optimal for mapping extensive areas such as floods and wildfires (Daud et al., 2022). Drones can collect crucial information, establish communication infrastructure, and enhance situational awareness in affected areas. Examples of valuable first responder information include assessing road conditions, identifying surface types of roads and paths, determining potential collection points, analyzing contamination results, and gathering other necessary data for first responders (Glantz et al., 2020). Drones equipped with AI algorithms can autonomously navigate through hazardous environments, gather aerial imagery, and assess damage to infrastructure with unprecedented speed and accuracy (Kim et al., 2022). Images captured by drones serve multiple purposes in disaster response efforts. They can be utilized to generate 3D models, topographical maps, and various visualizations, aiding emergency responders in planning operations and prioritizing efforts effectively. Additionally, drones are commonly deployed to observe ongoing disasters, providing real-time updates to response teams and enabling informed decision-making in dynamic situations (Nihad & Nadjat, 2023).

Search and Rescue. Drones significantly expedite rescue operations while ensuring the safety of responders and covering larger areas in shorter timeframes, particularly in challenging terrains (Nihad & Nadjat, 2023). A pioneering approach to detecting victims involves analyzing drone data to identify cardiopulmonary motion, manifested through periodic chest movements. Additionally, employing a swarm of drones rather than a single unit has proven effective in searching for missing individuals across expansive territories. These unmanned aerial vehicles have been deployed at high altitudes and in remote locations, diminishing response and evacuation durations while safeguarding responders. Moreover, drones have demonstrated efficacy in locating buried subjects, further enhancing their utility in critical situations (Daud et al., 2022). Moreover, AI-powered robotic systems can assist in search and rescue operations by detecting signs of life in rubble or debris, thereby increasing the chances of survival for trapped individuals (Ramadan et al., 2023). A suggested method to locate buried victims involves detecting the electromagnetic emissions from their mobile devices using drones. The aim is to pinpoint areas within the rubble where rescue teams are more likely to locate victims (Tanzi et al. 2016). In countries where flood response infrastructure is slow to develop, citizens are often at risk of becoming trapped in affected areas. Drones offer a demonstrably effective solution to this pressing issue, particularly through their capacity for imagery collection. In the event of floods where floodwaters can rise rapidly and access to affected regions may be limited, drones provide a crucial advantage in surveying and assessing the extent of flooding. Equipped with high-resolution cameras and sensors, drones can capture detailed imagery of flooded areas, allowing emergency responders to identify the locations of stranded individuals or communities in need of assistance (Restas, 2017). Integration of Wireless Sensor Networks (WSN) with drones for disaster management in search and rescue operations enhances response capabilities. WSNs consist of dispersed sensors collecting environmental data, while drones equipped with sensors autonomously navigate disaster areas, gathering real-time information. Research efforts focus on efficient communication protocols, deployment strategies, and autonomy enhancement to optimize response effectiveness. This integration offers a promising approach to improve situational awareness and resilience in disaster scenarios (Glantz et al., 2020). Drones can be equipped with speakers or microphones to facilitate communication with survivors or broadcast messages over a larger area. This capability aids in locating and rescuing survivors more quickly by providing essential communication channels in disaster situations (Nihad & Nadjat, 2023).

Delivery of Goods. For numerous years, researchers have focused on utilizing drones for transporting medical or emergency supplies to regions impacted by disasters or during emergency scenarios. Additionally, drones offer the capability to evaluate a patient's condition before ambulance arrival and facilitate the delivery of medical supplies and blood samples during crises. Various studies have documented the successful transportation of goods to their designated destinations using drones in disaster situations. Furthermore, drones possess the ability to reach locations faster than emergency medical services under favorable conditions (Daud et al., 2022). Specialized drones, despite payload weight limitations, have the capability to autonomously deliver critical supplies necessary for sustaining human life, even in scenarios where transport infrastructure is destroyed or roads are inaccessible (Bushnaq et al., 2022). Drones hold the potential to transport critical aid and provisions, including food, clean water, and medical resources, to disaster-affected regions that are difficult to access using traditional methods (Nihad & Nadjat, 2023).

Establishing Communication. Drones can serve a critical role in establishing temporary communication infrastructure, especially in disaster-stricken or remote areas where traditional communication networks may be compromised or non-existent (Tanzi et al., 2016; Bushnaq et al., 2022). In disaster scenarios like earthquakes, hurricanes, or floods, where terrestrial communication infrastructure may be severely damaged, drones can quickly deploy and bridge the communication gap.

Drones are utilized throughout forest fire management processes: they detect hot spots before fires, provide real-time information during interventions, and conduct post-fire monitoring. They offer critical data for decision-making, enhancing firefighting efficiency and safety (Restas, 2017).

Decision Support. Drones can be a highly effective tool in the hands of disaster managers. After deployment, the drone can continuously provide real-time data, allowing for immediate decision-making support for commanders within the initial period. One key aspect of this decision support is that even before the drone returns, it can establish the extent of the affected area and request assistance from additional units. This capability saves a significant amount of time in disaster response efforts (Restas, 2017). Through a variety of methods, such as structural health monitoring and video inspection by UAVs, UAVs play a crucial role in determining the extent of disaster-related damage (Erdelj et al., 2017).

CHALLENGES AND LIMITATIONS FOR AI AND DRONES IN DISASTER MANAGEMENT.

Challenges in Disaster Management Using AI and Drones

The application of AI in disaster management presents substantial challenges, encompassing various aspects such as the requirement for comprehensive and varied datasets, alignment with pre-existing systems and technologies, ethical and societal considerations, and continual research and advancement (Velev & Zlateva, 2023). The integration of AI and drones in natural disaster management introduces complex ethical considerations. While AI algorithms enhance drones' capabilities for efficient search, rescue, and damage assessment, concerns arise regarding data privacy, algorithmic bias, and autonomy. Drones equipped with AI may inadvertently infringe on individuals' privacy rights through indiscriminate data collection. Moreover, biases embedded in AI algorithms could lead to unequal treatment or resource allocation, exacerbating social disparities in disaster response.

Ethical frameworks must address these issues to ensure responsible AI deployment in disaster management. Transparent data collection protocols, informed consent mechanisms, and strict privacy safeguards are essential to protect individuals' rights. Additionally, continuous monitoring and auditing of AI algorithms are crucial to detect and mitigate biases.

Furthermore, ethical guidelines should prioritize the preservation of human autonomy and dignity. While AI can enhance efficiency, it should supplement human decision-making rather than replace it entirely. Collaborative approaches that involve local communities and prioritize their needs and perspectives are vital for ethical AI implementation in disaster response.

Computation Challenges. There are three challenging issues related to computation. Firstly, the availability of human-labeled training data may be insufficient, especially considering the escalating volume of data and the limited workforce in the aftermath of a disaster (Pouyanfar et al., 2018). Secondly, the computational complexity escalates significantly with the magnitude, diversity, and frequency of data updates, posing a challenge to processing, managing, and learning data within a reasonable response time during a disaster scenario. The efficient handling, storage, and processing of large datasets are crucial for disaster management, particularly in the context of disaster response (Tanaka et al., 2019). Leveraging crowdsourcing alongside real-time AI analyses can aid in completing essential computations within time constraints and reduce the burden of laborious tasks traditionally performed on-site (Bevington et al., 2015). Thirdly, the development of user-friendly tools for disaster management is paramount for practitioners. This entails designing AI-based tools with interfaces that demand minimal technical expertise for practical utilization.

Limitations of Data Collection and Integration. The data collected by smart sensors primarily originates from remote sensing, monitoring devices, and social media. This data can be utilized to create and update disaster-sensitive maps regularly, thereby enhancing the information available about disaster situations. However, the integration and analysis of multi-source data pose challenges due to the absence of theoretical frameworks and algorithmic strategies to represent heterogeneous data (Nweke et al., 2019).

Challenges in Disaster Response. In the context of disaster response, the focal point is on making appropriate emergency decisions in the disaster environment. During the process of disaster emergency decision-making, it is crucial to explore and implement methods such as hierarchical reinforcement learning and the analytic hierarchy process to elucidate the potential relationship between objectives and various path choices. This aids decision-makers in identifying the optimal task sequence (Tissera et al., 2012).

Challenges in Disaster Recovery. During the disaster recovery phase, remote sensing mapping may exhibit different characteristics regarding sensor fields, image proportions, and scene complexity. Information extraction from these sources may be challenging, thereby limiting the accuracy of loss assessment to some extent. Additionally, evaluating the effectiveness of emergency response through post-disaster data analysis and identifying and addressing issues such as suboptimal resource allocation is vital to enhancing the efficiency of disaster management processes and upgrading the efficacy of disaster emergency management. This remains a significant challenge that requires resolution in the disaster recovery stage (Tan et al., 2021).

Challenges of Using Drones in Disaster Management

Using drones for Search and Rescue (SAR) operations in disaster management confronts several key challenges. Legal constraints impose regulatory hurdles, requiring meticulous adherence to airspace regulations and permissions, which can impede swift deployment. Moreover, adverse weather conditions, such as strong winds or heavy rain, often hamper drone operations, limiting their effectiveness in critical situations. Additionally, the success of SAR missions heavily relies on the capacity and cooperation of local communities, necessitating proactive engagement to ensure acceptance and collaboration. Overcoming these obstacles demands a comprehensive approach that addresses legal frameworks, weather forecasting, and community outreach, enabling drones to fulfill their potential as valuable assets in disaster response efforts (Clark et al., 2018).

Regulatory Constraints and Privacy Concerns. The regulatory framework governing the use of drones varies significantly from one country to another. These diverse sets of rules and regulations can pose significant barriers to the widespread adoption and acceptance of drones. The low altitude and autonomous navigation capabilities of a drone raise concerns about the potential for injuries to nearby victims or rescuers in the event of a crash. The security of data sensed and stored on-board UAVs can pose significant privacy concerns, particularly regarding the privacy of victims. There have been instances in the past where images of identifiable victims have been circulated in the media without their consent, highlighting the potential risks associated with data handling by drones (Tanz et al,. 2016). Drones have the capacity to gather extensive data via images, videos, and mapping coordinates, typically disseminated to users via a network. However, the transmission and storage of this data can raise privacy concerns and challenges. Potential risks include breaches of location privacy, linking attacks, man-in-the-middle attacks, eavesdropping attacks, and other forms of privacy infringements (Erdelj et al., 2017).

Training and Certification Requirements. Training and certifications are crucial, especially for advanced operations like remote piloting. They ensure that operators possess the necessary skills and knowledge to operate drones safely and effectively. Comprehensive training programs and recognized certifications are essential for maintaining high standards in drone operations (Glantz et al., 2020).

Technological Challenges. Drones require a system known as Sense and Avoid, considered a significant advancement in modern drone development. This system enables drones to autonomously avoid collisions with unforeseen obstacles, people, and vehicles by utilizing onboard sensors and real-time decision-making capabilities. Although substantial progress has been made in this field in recent years, it will take time before it is fully operational for real-world applications. Additionally, delivery drones must demonstrate reliability in all weather conditions, including rain, strong winds, and nighttime operations, a capability that remains limited in current drone technology (Pathak et al., 2019).

Cyber security Risks. Cyberattacks targeting UAV control systems pose significant risks. An attack on a UAV's control system could lead to malfunctions or crashes, potentially resulting in harm to individuals or damage to property (Erdelj et al., 2017).

Physical Challenges. In disaster management, drones face a myriad of physical challenges that impede their efficacy. These challenges include limitations in battery capacity, necessitating frequent recharging and careful deployment planning; constraints in processing power and technology, leading to rapid battery drainage; difficulties in network optimization and design, hindering communication reliability and data sharing; restrictions on payload capacity, limiting the amount of equipment they can carry; and challenges in maneuverability, particularly in harsh weather conditions like storms and heavy rainfall. Addressing these physical limitations is crucial for optimizing the use of drones in disaster response and recovery efforts (Bushnaq et al., 2022).

DISCUSSION

The integration of AI and drones in disaster management represents a paradigm shift in how we approach mitigation, preparedness, response, and recovery efforts in the face of natural disasters. AI and drones present numerous challenges, opportunities, and implications for leveraging them in disaster management.

Emerging Trends in Disaster Management Paradigms. The evolution of disaster management paradigms from traditional approaches to more comprehensive DRM frameworks underscores the need for innovative solutions to address evolving threats. The transition to DRM necessitates a holistic approach that integrates robust risk assessment, proactive mitigation measures, and community engagement initiatives. By harnessing AI and drone technologies, stakeholders can enhance situational awareness, optimize resource allocation, and expedite decision-making processes.

Applications of AI and Drones in Disaster Management. AI plays a pivotal role across all stages of disaster management, from preparedness to recovery. By leveraging AI-driven technologies such as predictive analytics, natural language processing, and machine learning, responders can gain valuable insights into disaster situations, enabling more effective response efforts. Drones offer unique capabilities for disaster management, including aerial mapping, search and rescue operations, goods delivery, and establishing communication networks. Drones have been found to be useful in search and rescue operations in mountains (Karaka et al., 2018). Studies have also shown the utility of drones in inspecting damage caused to bridges by drones (Mandirola et al., 2022). Despite their potential benefits, challenges such as regulatory constraints, training requirements, and technological limitations must be addressed to maximize their effectiveness in disaster scenarios.

Challenges and Disruptive Technologies. The application of AI and drones in disaster management presents substantial challenges, ranging from data collection and integration to regulatory constraints and cybersecurity risks. Addressing these challenges requires a comprehensive approach that encompasses technological innovation, regulatory reform, and capacity building. By investing in research and development, fostering international collaboration, and enhancing interdisciplinary partnerships, stakeholders can overcome these challenges and unlock the transformative potential of AI and drones in disaster management.

Future Directions. Looking ahead, the integration of AI and drones holds immense promise for enhancing disaster management strategies and improving outcomes in natural disaster scenarios. By addressing existing challenges and harnessing opportunities for innovation, stakeholders can build more resilient and adaptive disaster management systems. However, realizing this vision requires concerted efforts from governments, academia, industry, and civil society to prioritize investment in AI and drone technologies, strengthen regulatory frameworks, and promote knowledge sharing and capacity building initiatives. Ultimately, by embracing innovation and collaboration, we can build more resilient communities and better prepare for the challenges of an increasingly uncertain future.

CONCLUSION

The strategic deployment of AI and drones in disaster management represents a significant advancement in our ability to mitigate, respond to, and recover from natural disasters. Throughout this chapter, we have explored the evolution of disaster management paradigms, the applications of AI and drones in disaster management, and the challenges and disruptive technologies shaping this field. As natural disasters continue to increase in frequency and severity, it is imperative that we leverage cutting-edge technologies to enhance our preparedness and response capabilities. AI-driven technologies offer unprecedented insights into disaster situations, enabling more effective decision-making and resource allocation. Drones, equipped with advanced sensors and imaging technologies, provide rapid aerial reconnaissance and logistical support in disaster-affected areas. However, the integration of AI and drones in disaster management is not without its challenges. Regulatory constraints, privacy concerns, and technological limitations pose significant barriers to the widespread adoption and effectiveness of these technologies. Addressing these challenges requires a multi-faceted approach that encompasses technological innovation, regulatory reform, and capacity building.

Looking ahead, the future of disaster management lies in embracing innovation, collaboration, and resilience-building strategies. By investing in research and development, fostering international cooperation, and strengthening community engagement, we can harness the full potential of AI and drones to build more resilient and adaptive disaster management systems.

In conclusion, the strategic deployment of AI and drones offers a transformative opportunity to enhance our ability to prepare for, respond to, and recover from natural disasters. By leveraging these technologies alongside traditional disaster management approaches, we can build more resilient communities and better protect lives and livelihoods in the face of increasingly complex and dynamic disaster scenarios.

REFERENCES

Abid, S. K., Sulaiman, N., Chan, S. W., Nazir, U., Abid, M., Han, H., Ariza-Montes, A., & Vega-Muñoz, A. (2021). Toward an integrated disaster management approach: How artificial intelligence can boost disaster management. *Sustainability (Basel)*, *13*(22), 12560. doi:10.3390/su132212560

Aisha, T. S., Wok, S., Manaf, A. M. A., & Ismail, R. (2015). Exploring the Use of Social Media during the 2014 Flood in Malaysia. *Procedia: Social and Behavioral Sciences*, *211*, 931–937. doi:10.1016/j.sbspro.2015.11.123

Aitsi-Selmi, A., Egawa, S., Sasaki, H., Wannous, C., & Murray, V. (2015). The Sendai framework for disaster risk reduction: Renewing the global commitment to people's resilience, health, and well-being. *International Journal of Disaster Risk Science*, *6*(2), 164–176. doi:10.1007/s13753-015-0050-9

Aitsi-Selmi, A., Murray, V., Wannous, C., Dickinson, C., Johnston, D., Kawasaki, A., Stevance, A.-S., & Yeung, T. (2016). Reflections on a science and technology agenda for 21st century disaster risk reduction: Based on the scientific content of the 2016 UNISDR science and technology conference on the implementation of the Sendai framework for disaster risk reduction 2015–2030. *International Journal of Disaster Risk Science*, *7*(1), 1–29. doi:10.1007/s13753-016-0081-x

Anyoha, R. (2017). The history of artificial intelligence: Can machine think? HMS. https://sitn.hms.harvard.edu/flash/2017/history-artificial-intelligence/

Bagrow, J. P., Wang, D., & Barabási, A. L. (2011). Collective response of human populations to large-scale emergencies. *PLoS One*, *6*(3), e17680. doi:10.1371/journal.pone.0017680 PMID:21479206

Balaji, T. K., Annavarapu, C. S. R., & Bablani, A. (2021). Machine learning algorithms for social media analysis: A survey. *Computer Science Review*, *40*, 100395. doi:10.1016/j.cosrev.2021.100395

Bengtsson, L., Lu, X., Thorson, A., Garfeld, R., & von Schreeb, J. (2011). Improved response to disasters and outbreaks by tracking population movements with mobile phone network data: A post-earthquake geospatial study in Haiti. *PLoS Medicine*, *8*(8), e1001083. doi:10.1371/journal.pmed.1001083 PMID:21918643

Bevington, J. S., Eguchi, R. T., Gill, S., Ghosh, S., & Huyck, C. K. (2015). A comprehensive analysis of building damage in the 2010 Haiti earthquake using high-resolution imagery and crowdsourcing. *Time-sensitive remote sensing*, 131-145.

Born, C. T., Briggs, S. M., Ciraulo, D. L., Frykberg, E. R., Hammond, J. S., Hirshberg, A., Lhowe, D. W., & O'Neill, P. A. (2007). Disasters and Mass Casualties: I. General Principles of Response and Management. *The Journal of the American Academy of Orthopaedic Surgeons*, *15*(7), 388–396. doi:10.5435/00124635-200707000-00004 PMID:17602028

Botzen, W. W., Deschenes, O., & Sanders, M. (2019). The economic impacts of natural disasters: A review of models and empirical studies. *Review of Environmental Economics and Policy*, *13*(2), 167–188. doi:10.1093/reep/rez004

Britannica. (2021). *Kōbe earthquake of 1995*. Encyclopedia Britannica. https://www.britannica.com/event/Kobe-earthquake-of-1995

Bushnaq, O. M., Mishra, D., Natalizio, E., & Akyildiz, I. F. (2022). Unmanned aerial vehicles (UAVs) for disaster management. In *Nanotechnology-Based Smart Remote Sensing Networks for Disaster Prevention* (pp. 159–188). Elsevier. doi:10.1016/B978-0-323-91166-5.00013-6

Caragea, C., Squicciarini, A., Stehle, S., Neppalli, K., & Tapia, A. (2014). Mapping moods: geo-mapped sentiment analysis during Hurricane Sandy. In: *Proceedings of the 11th international conference on information systems for crisis response and management (ISCRAM 2014)*. ISCRAM.

Ci, T., Liu, Z., & Wang, Y. (2019). Assessment of the degree of building damage caused by disaster using convolutional neural networks in combination with ordinal regression. *Remote Sensing (Basel)*, *11*(23), 2858. doi:10.3390/rs11232858

Clark, D. G., Ford, J. D., & Tabish, T. (2018). What role can unmanned aerial vehicles play in emergency response in the Arctic: A case study from Canada. *PLoS One*, *13*(12), e0205299. doi:10.1371/journal.pone.0205299 PMID:30562340

Copeland, B. J. (2020). *What is artificial intelligence?* Alan Turning. https://www.alanturing.net/turing_archive/pages/Reference%20Articles/What%20is%20AI.html

Coppola, D. (2015). Introduction to International Disaster Management. Butterworth-Heinemann. Elsevier.

Curtis, A., Duval-Diop, D., & Novak, J. (2010). Identifying spatial patterns of recovery and abandonment in the post-Katrina Holy Cross neighborhood of New Orleans. *Cartography and Geographic Information Science*, *37*(1), 45–56. doi:10.1559/152304010790588043

Daud, S. M. S. M., Yusof, M. Y. P. M., Heo, C. C., Khoo, L. S., Singh, M. K. C., Mahmood, M. S., & Nawawi, H. (2022). Applications of drone in disaster management: A scoping review. *Science & Justice*, *62*(1), 30–42. doi:10.1016/j.scijus.2021.11.002 PMID:35033326

Ebi, K. L., Vanos, J., Baldwin, J. W., Bell, J. E., Hondula, D. M., Errett, N. A., Hayes, K., Reid, C. E., Saha, S., Spector, J., & Berry, P. (2021). Extreme weather and climate change: Population health and health system implications. *Annual Review of Public Health*, *42*(1), 293–315. doi:10.1146/annurev-publhealth-012420-105026 PMID:33406378

Edureka, (2020). *Top 15 hot artificial intelligence technologies*. Edureka. https://www.edureka.co/blog/top-15-hot-artificial-intelligence-technologies/

Erdelj, M., Natalizio, E., Chowdhury, K. R., & Akyildiz, I. F. (2017). Help from the sky: Leveraging UAVs for disaster management. *IEEE Pervasive Computing*, *16*(1), 24–32. doi:10.1109/MPRV.2017.11

Finucane, M. L., Acosta, J., Wicker, A., & Whipkey, K. (2020). Short-Term Solutions to a Long-Term Challenge: Rethinking Disaster Recovery Planning to Reduce Vulnerabilities and Inequities. *International Journal of Environmental Research and Public Health*, *2020*(17), 482. doi:10.3390/ijerph17020482 PMID:31940859

Floreano, D., & Wood, R. (2015). Science, technology and the future of small autonomous drones. *Nature*, *521*(7553), 460–466. doi:10.1038/nature14542 PMID:26017445

Gao, Y., Chen, Y. X., Ding, Y. S., & Tang, B. Y. (2006). Immune genetic algorithm based on network model for food disaster evaluation. *J Nat Disaster*, *15*, 110–114.

Glantz, E. J., Ritter, F. E., Gilbreath, D., Stager, S. J., Anton, A., & Emani, R. (2020). UAV use in disaster management. In *Proceedings of the 17th ISCRAM Conference* (pp. 914-921). IEEE.

Goodchild, M. F., & Glennon, J. A. (2010). Crowdsourcing geographic information for disaster response: A research frontier. *International Journal of Digital Earth*, *3*(3), 231–241. doi:10.1080/17538941003759255

Greifeneder, R., Bless, H., & Pham, M. (2011). When do people rely on afective and cognitive feelings in judgment? A review. *Personality and Social Psychology Review*, *15*(2), 107–141. doi:10.1177/1088868310367640 PMID:20495111

Groumpos, P. P. (2023). A Critical Historic Overview of Artificial Intelligence: Issues, Challenges, Opportunities, and Threats. In *Artificial Intelligence and Applications*, *1*(4). https://ojs.bonviewpress.com/index.php/AIA/article/view/689/580

Gupta, S. G., Ghonge, D. M., & Jawandhiya, P. M. (2013). Review of unmanned aircraft system (UAS). [IJARCET]. *International Journal of Advanced Research in Computer Engineering and Technology*, *2*.

Hassanalian, M., & Abdelkefi, A. (2017). Classifications, applications, and design challenges of drones: A review. *Progress in Aerospace Sciences*, *91*, 99–131. doi:10.1016/j.paerosci.2017.04.003

Hecker, N., & Domres, B. D. (2018). The German emergency and disaster medicine and management system—history and present. *Chinese Journal of Traumatology*, *21*(2), 64–72. doi:10.1016/j.cjtee.2017.09.003 PMID:29622286

Israel, D. C., & Briones, R. M. (2012). Impacts of Natural Disasters on Agriculture, Food Security, and Natural Resources and Environment in the Philippines. In Y. Sawada & S. Oum (Eds.), *Economic and Welfare Impacts of Disasters in East Asia and Policy Responses. ERIA Research Project Report 2011-8* (pp. 553–599). ERIA.

Jiménez López, J., & Mulero-Pázmány, M. (2019). Drones for conservation in protected areas: Present and future. *Drones (Basel)*, *3*(1), 10. doi:10.3390/drones3010010

Karaca, Y., Cicek, M., Tatli, O., Sahin, A., Pasli, S., Beser, M. F., & Turedi, S. (2018). The potential use of unmanned aircraft systems (drones) in mountain search and rescue operations. *The American Journal of Emergency Medicine*, *36*(4), 583–588. doi:10.1016/j.ajem.2017.09.025 PMID:28928001

Khaloo, A., Lattanzi, D., Cunningham, K., Dell'Andream, R., & Riley, M. (2017). Unmanned aerial vehicle inspection of the Placer River trail bridge through image-based 3D modelling. *Structure and Infrastructure Engineering*, *14*(1), 124–136. do i:10.1080/15732479.2017.1330891

Kieslich, K., Lünich, M., & Marcinkowski, F. (2021). The threats of artificial intelligence scale (TAI) development, measurement and test over three application domains. *International Journal of Social Robotics*, *13*(7), 1563–1577. doi:10.1007/s12369-020-00734-w

Kim, S. S., Shin, D. Y., Lim, E. T., Jung, Y. H., & Cho, S. B. (2022). Disaster Damage Investigation using Artificial Intelligence and Drone Mapping. *The International Archives of the Photogrammetry, Remote Sensing and Spatial Information Sciences*, *43*, 1109–1114. doi:10.5194/isprs-archives-XLIII-B3-2022-1109-2022

Kousky, C. (2016). Impacts of natural disasters on children. *The Future of Children*, *26*(1), 73–92. doi:10.1353/foc.2016.0004

Labib, N. S., Brust, M. R., Danoy, G., & Bouvry, P. (2021). The rise of drones in internet of things: A survey on the evolution, prospects and challenges of unmanned aerial vehicles. *IEEE Access : Practical Innovations, Open Solutions*, *9*, 115466–115487. doi:10.1109/ACCESS.2021.3104963

Li, J., & Rao, H. (2010). Twitter as a rapid response news service: An exploration in the context of the 2008 China earthquake. *The Electronic Journal on Information Systems in Developing Countries*, *42*(1), 1–22. doi:10.1002/j.1681-4835.2010.tb00300.x

Lin, Y. R. (2015). Event-related crowd activities on social media. In B. Gonçalves & N. Perra (Eds.), *Social phenomena* (pp. 235–250). Springer. doi:10.1007/978-3-319-14011-7_12

Macias, M., Barrado, C., Pastor, E., & Royo, P. (2019). The future of drones and their public acceptance. In 2019 IEEE/AIAA 38th Digital Avionics Systems Conference (DASC) (pp. 1-8). IEEE. 10.1109/DASC43569.2019.9081623

Madichetty, S., & Sridevi, M. (2019). Detecting informative tweets during disaster using deep neural networks. In *2019 11th international conference on communication systems & networks (COMSNETS)* (pp. 709-713). IEEE. 10.1109/COMSNETS.2019.8711095

Mandirola, M., Casarotti, C., Peloso, S., Lanese, I., Brunesi, E., & Senaldi, I. (2022). Use of UAS for damage inspection and assessment of bridge infrastructures. *International Journal of Disaster Risk Reduction, 72*, 102824. doi:10.1016/j. ijdrr.2022.102824

Manoharan, G., Razak, A., Rao, B. S., Singh, R., Ashtikar, S. P., & Nivedha, M. (2024). Navigating the Crescendo of Challenges in Harnessing Artificial Intelligence for Disaster Management. In D. Satishkumar & M. Sivaraja (Eds.), *Predicting Natural Disasters With AI and Machine Learning* (pp. 64–94). IGI Global. doi:10.4018/979-8-3693-2280-2.ch003

Merkert, R., & Bushell, J. (2020). Managing the drone revolution: A systematic literature review into the current use of airborne drones and future strategic directions for their effective control. *Journal of Air Transport Management, 89*, 101929. doi:10.1016/j.jairtraman.2020.101929 PMID:32952321

Mohsin, B., Steinhäusler, F., Madl, P., & Kiefel, M. (2016). An innovative system to enhance situational awareness in disaster response. *Journal of Homeland Security and Emergency Management, 13*(3), 301–327. doi:10.1515/jhsem-2015-0079

Moloi, T., & Marwala, T. (2021). A High-Level Overview of Artificial Intelligence: Historical Overview and Emerging Developments. In *Artificial Intelligence and the Changing Nature of Corporations. Future of Business and Finance.* Springer., doi:10.1007/978-3-030-76313-8_2

National Institute of Disaster Management. (n.d.). *Understanding Disasters.* https://nidm.gov.in/PDF/Disaster_about.pdf

Nihad, B., & Nadjat, A. (2023). *Drones Use in Disaster Management* [Doctoral dissertation, university center of abdalhafid boussouf-MILA].

Nweke, H. F., Teh, Y. W., Mujtaba, G., & Al-Garadi, M. A. (2019). Data fusion and multiple classifier systems for human activity detection and health monitoring: Review and open research directions. *Information Fusion, 46*, 147–170. doi:10.1016/j. inffus.2018.06.002

Pathak, P., Damle, M., Pal, P. R., & Yadav, V. (2019). Humanitarian impact of drones in healthcare and disaster management. *Int. J. Recent Technol. Eng, 7*(5), 201–205.

Pouyanfar, S., Sadiq, S., Yan, Y., Tian, H., Tao, Y., Reyes, M. P., Shyu, M.-L., Chen, S.-C., & Iyengar, S. S. (2018). A survey on deep learning: Algorithms, techniques, and applications. *ACM Computing Surveys, 51*(5), 1–36. doi:10.1145/3234150

Qiang, Y. (2019). Disparities of population exposed to flood hazards in the United States. *Journal of Environmental Management, 232*, 295–304. doi:10.1016/j.jenvman.2018.11.039 PMID:30481643

Ramadan, M., Hilles, S. M., & Alkhedher, M. (2023). Design and Study of an AI-Powered Autonomous Stair Climbing Robot. *El-Cezeri, 10*(3), 571–585. doi:10.31202/ecjse.1272769

Ramchurn, S. D., Huynh, T. D., Ikuno, Y., Flann, J., Wu, F., Moreau, L., Jennings, N. R., Fischer, J., Jiang, W., Rodden, T., Simpson, E., Reece, S., & Roberts, S. J. (2015). HAC-ER: a disaster response system based on humanagent collectives. In: *Proceedings of the 2015 international conference on autonomous agents and multiagent systems (AAMAS 2015)*, (pp. 533–541). IEEE.

Restas, A. (2017). Disaster management supported by unmanned aerial systems (UAS) focusing especially on natural disasters. *Zeszyty Naukowe SGSP/Szkoła Główna Służby Pożarniczej, 61*.

Rodríguez, R. M., Alarcón, F., Rubio, D. S., & Ollero, A. (2013). *Autonomous management of an UAV Airfield*. In proceedings of the 3rd international conference on application and theory of automation in command and control systems, Naples, Italy.

Sahebjamnia, N., Torabi, S. A., & Mansouri, S. A. (2015). Integrated business continuity and disaster recovery planning: Towards organizational resilience. *European Journal of Operational Research, 242*(1), 261–273. doi:10.1016/j.ejor.2014.09.055

Sandhu, D., & Kaur, S. (2013). Psychological impacts of natural disasters. *Indian Journal of Health and Wellbeing, 4*(6), 1317.

Shavarani, S. M. (2019). Multi-level facility location-allocation problem for post-disaster humanitarian relief distribution: A case study. *Journal of Humanitarian Logistics and Supply Chain Management, 9*(1), 70–81. doi:10.1108/JHLSCM-05-2018-0036

Shi, W., Zhang, M., Zhang, R., Chen, S., & Zhan, Z. (2020). Change detection based on artificial intelligence: State-of-the-art and challenges. *Remote Sensing (Basel), 12*(10), 1688. doi:10.3390/rs12101688

Singh, C. R., & Manoharan, G. (2024). Strengthening Resilience: AI and Machine Learning in Emergency Decision-Making for Natural Disasters. In D. Satishkumar & M. Sivaraja (Eds.), *Internet of Things and AI for Natural Disaster Management and Prediction* (pp. 249–278). IGI Global., doi:10.4018/979-8-3693-4284-8.ch012

Singh, N. (2007). Expert system prototype of food aid distribution. *Asia Pacific Journal of Clinical Nutrition, 16*, 116–121. PMID:17392088

Singh, R. (2023). *Turkey Earthquake 2023: Reducing Risks in the Indian Context.* MP-IDSA. https://idsa.in/idsacomments/turkey-earthquake-2023-rsingh-120423

Srivastava, M., Abdelzaher, T., & Szymanski, B. (2012). Human-centric sensing. Philos Trans R Soc A Math Phys. *Engineering and Science, 370*, 176–197.

Sulistijono, I. A., & Risnumawan, A. (2016). From concrete to abstract: Multilayer neural networks for disaster victims detection. In *2016 International electronics symposium (IES)* (pp. 93-98). IEEE.

Suteris, M. S., Rahman, F. A., & Ismail, A. (2018). Route schedule optimization method of unmanned aerial vehicle implementation for maritime surveillance in monitoring trawler activities in Kuala Kedah, Malaysia. *Int. J. Supply Chain Manag, 7*(5), 245–249.

Tabish, S. A., & Syed, N. (2015). Disaster Preparedness: Current Trends and Future Directions. [IJSR]. *International Journal of Scientific Research*, 438.

Tan, D., Qu, W., & Tu, J. (2010). The damage detection based on the fuzzy clustering and support vector machine. In *2010 International Conference on Intelligent System Design and Engineering Application* (Vol. 2, pp. 598-601). IEEE. 10.1109/ISDEA.2010.404

Tan, L., Guo, J., Mohanarajah, S., & Zhou, K. (2021). Can we detect trends in natural disaster management with artificial intelligence? A review of modeling practices. *Natural Hazards, 107*(3), 2389–2417. doi:10.1007/s11069-020-04429-3

Tanaka, G., Yamane, T., Héroux, J. B., Nakane, R., Kanazawa, N., Takeda, S., Numata, H., Nakano, D., & Hirose, A. (2019). Recent advances in physical reservoir computing: A review. *Neural Networks, 115*, 100–123. doi:10.1016/j.neunet.2019.03.005 PMID:30981085

Tanzi, T. J., Chandra, M., Isnard, J., Camara, D., Sebastien, O., & Harivelo, F. (2016). Towards" drone-borne" disaster management: future application scenarios. In *XXIII ISPRS Congress, Commission VIII* (Volume III-8) (Vol. 3, pp. 181-189). Copernicus GmbH.

Tatsubori, M., Watanabe, H., Shibayama, A., Sato, S., & Imamura, F. (2012). Social web in disaster archives. In: *The proceedings of the 21st international conference companion on world wide web—WWW'12 Companion.* ACM. 10.1145/2187980.2188190

Thomas, M. (2023). Dangerous risks of artificial intelligence. *Built In.* https://builtin.com/artificial-intelligence/risks-of-artificial-intelligence

Tissera, P. C., Printista, A. M., & Luque, E. (2012). A hybrid simulation model to test behaviour designs in an emergency evacuation. *Procedia Computer Science, 9,* 266–275. doi:10.1016/j.procs.2012.04.028

UNDRR. (n.d.). *Sendai Framework Terminology on Disaster Risk Reduction.* UNDRR. https://www.undrr.org/terminology/disaster-risk-management

United Nations. (1994). *Yokohoma Strategy and Plan of Action for a Safer World guidelines for Natural Disaster Prevention, Preparedness and Mitigation.* Prevention Web. https://www.preventionweb.net/files/8241_doc6841contenido1.pdf

United Nations. (1999). *Activities of the International Decade for Natural Disaster Reduction; Report of the General Secretary.* UN. https://www.un.org/esa/documents/ecosoc/docs/1999/e1999-80.htm#:~:text=Faced%20with%20an%20increasing%20frequency,United%20Nations%2C%20would%20pay%20special

United Nations Office for Disaster Risk Reduction (UNISDR). (2005). Hyogo Framework of Action for 2005-2015: Building the Resilience of Nations and Communities to Disasters. *World Conference on Disaster Reduction; Hyogo, Japan: International Strategy for Disaster Reduction (ISDR).* UN. https://www.unisdr.org/2005/wcdr/intergover/officialdoc/L-docs/Hyogo-framework-for-action-english.pdf

Varalakshmi, M. S. (2019). Evolution and Significance of Drones in Modern Technology. *8 IJRAR, 6*(1). https://ijrar.org/papers/IJRAR19J1225.pdf

Vashishth, T. K., Kumar, B., Sharma, V., Chaudhary, S., Kumar, S., & Sharma, K. K. (2023). The Evolution of AI and Its Transformative Effects on Computing: A Comparative Analysis. In *Intelligent Engineering Applications and Applied Sciences for Sustainability* (pp. 425–442). IGI Global. doi:10.4018/979-8-3693-0044-2.ch022

Velev, D., & Zlateva, P. (2023). Challenges of artificial intelligence application for disaster risk Management. *The International Archives of the Photogrammetry, Remote Sensing and Spatial Information Sciences, 48,* 387–394. doi:10.5194/isprs-archives-XLVIII-M-1-2023-387-2023

Vermiglio, C., Noto, G., Rodríguez Bolívar, M. P., & Zarone, V. (2022). Disaster management and emerging technologies: A performance-based perspective. *Meditari Accountancy Research, 30*(4), 1093–1117. doi:10.1108/MEDAR-02-2021-1206

Vieweg, S. (2012). *Situational awareness in mass emergency: a behavioral and linguistic analysis of microblogged communications*. [PhD dissertation, University of Colorado at Boulder].

Wamsler, C., & Johannessen, A. (2020). Meeting at the crossroads? Developing national strategies for disaster risk reduction and resilience: Relevance, scope for, and challenges to, integration. *International Journal of Disaster Risk Reduction, 45*, 101452. doi:10.1016/j.ijdrr.2019.101452

Wang, J., Zhang, J., Gong, L., Li, Q., & Zhou, D. (2018). Indirect seismic economic loss assessment and recovery evaluation using nighttime light images—Application for Wenchuan earthquake. *Natural Hazards and Earth System Sciences, 18*(12), 3253–3266. doi:10.5194/nhess-18-3253-2018

Wang, J., Zhou, K., Xing, W., Li, H., & Yang, Z. (2023). Applications, evolutions, and challenges of drones in maritime transport. *Journal of Marine Science and Engineering, 11*(11), 2056. doi:10.3390/jmse11112056

Wisner, B., Blaikie, P., Cannon, T., & Davis, I. (2004). *At risk: Natural hazards, people's vulnerability, and disasters* (2nd ed.). Routledge.

Yang, T., Xie, J., Li, G., Mou, N., Li, Z., Tian, C., & Zhao, J. (2019). Social media big data mining and spatio-temporal analysis on public emotions for disaster mitigation. *ISPRS International Journal of Geo-Information, 8*(1), 29. doi:10.3390/ijgi8010029

Yang, Z., Yu, X., Dedman, S., Rosso, M., Zhu, J., Yang, J., Xia, Y., Tian, Y., Zhang, G., & Wang, J. (2022). UAV remote sensing applications in marine monitoring: Knowledge visualization and review. *The Science of the Total Environment, 838*, 155939. doi:10.1016/j.scitotenv.2022.155939 PMID:35577092

Ye, L., & Abe, M. (2012). *The impacts of natural disasters on global supply chains (No. 115)*. ARTNeT working paper series.

Zaré M, & Afrouz SG. (2011). Crisis management of Tohoku; Japan earthquake and tsunami. *Iran J Public Health, 41*(6), 12-20

Zhang, Y., & Peacock, W. G. (2009). Planning for housing recovery? Lessons learned from Hurricane Andrew. *Journal of the American Planning Association, 76*(1), 5–24. doi:10.1080/01944360903294556

Chapter 10
Blockchain Technology With the Internet of Drones (IOD) to Address Privacy and Security Issues in Disaster Management

Naser Hussein
University of Tunis El Manar, Tunisia

Hella Kaffel Ben Ayed
University of Tunis El Manar, Tunisia

ABSTRACT

The portability and automation of the internet of drones (IOD) have drawn increasing attention in recent years, and it is being used in various fields (such as military, rescue and entertainment, and disaster management). Unmanned aerial vehicles (UAVs) have recently established their capacity to provide cost-effective and credible solutions for various real-world scenarios. since security and privacy are among the main concerns for the IOD, in this chapter, the authors perform a complete analysis of security issues and solutions for IOD security, analyzing IOD-related security in disaster requirements and identifying the newest improvement in IOD security research. This analysis explores many essential security technologies emphasizing authentication mechanisms and blockchain-powered schemes. Based on a rigorous review, the authors discuss the issues faced by current approaches in disaster management and offer future IOD security research areas.

DOI: 10.4018/979-8-3693-3896-4.ch010

INTRODUCTION

Natural disasters, such as earthquakes, floods, and storms, and so on, can have serious repercussions, including endangering the lives of citizens and destroying infrastructure. Furthermore, first responders face threats in the post-disaster environment in most circumstances, which makes their task much more difficult. Time is of the essence in this situation. Effective resource management and timely operational choices are critical to the success of a search and rescue mission. To effectively and promptly handle new issues and guarantee the safety of both staff and victims, decision-makers must therefore have a clear and comprehensive understanding of the situation. Blockchain and IOD together can offer integrated solutions that support decision makers with time- and life-critical decision support systems and valuable insights about the current condition.

The Internet of Drones (IOD) is a novel paradigm that uses the Internet to do various jobs in many areas by allowing a group of flying vehicles/devices to exchange information with a Ground Control Station (GCS) and one another, such as emergency response, smart agriculture, military, smart city management, surveillance, and healthcare, and, most recently, managing the COVID-19 epidemic (Ayamga, Akaba, & Nyaaba, 2021).

In the last ten years, the application of drones has expanded quickly, encompassing a range of industries such as agriculture, commerce, the military, and disaster relief and management. Unfortunately, the evidence of its application in large-scale catastrophes is currently sparse and uncertain. In the purpose of motivating and inspiring potential future work, this paper intends to assess the present drone feasibility projects and explore several problems linked to the implementation of internet of drones with blockchain in mass disasters. This study outlines the research trend of drone use in disaster relief by summarizing the findings of 31 research articles published over the past 10 years, from 2014 to 2024, using the Arksey and O'Malley framework, revised by the Joanna Briggs Institute Framework for Scoping Reviews methodology(Arksey & O'malley, 2005).

A search for literature was conducted using Google Scholars, Scopus, individual journals, grey literature, and Google search, with the content and significance of each search being evaluated. Drones have a wide range of potential uses in emergency situations. Drone applications in catastrophes are divided into four categories based on the publications that were found;

(1)   mapping or disaster relief, which has made the biggest impact,
(2)   rescue and search operations,
(3)   transportation,
(4)   training.

Despite the fact that the number of articles about drone use in disaster relief has significantly increased over the last five years as shown in the table 2, However, there hasn't been much consideration of the post-disaster healthcare scenario, particularly when it comes to identifying disaster victims. It is obvious that more research needs to be done on drone applications, with a particular emphasis on using drones to help people, particularly with victim identification. It is anticipated that as drone technology advances, drone applications on the internet will become more effective, particularly in disaster relief.

Disasters are unforeseen, abrupt occurrences that have the potential to seriously harm both human life and property. Therefore, it is extremely difficult for governments, scientists, and academics to figure out how to manage the situation and information flow when a disaster strikes given the unpredictability and devastation that ensue. Using one or more drones connected to the network, the Internet of Drones (IoD) with blockchain offers a solution that improves the effectiveness and efficiency of data and information transfer. The difficulty raised above is addressed in this study, along with the need for increased accuracy, mobility, and the effectiveness of the command center's communication with other parties involved in disaster operations (Barnabas, Citrawati, Santoso, & Surantha, 2023).

On the 6th of February, 2023, A string of powerful earthquakes rocked Syria and Turkey, millions of people being displaced, causing almost $119 billion in direct damages to private property and infrastructure, and more than 59 000 deaths. Turkey, considering the lengthy history of seismic activity, had considerable experience dealing with the cataclysm https://en.wikipedia.org/wiki/2023_Turkey%E2%80%93Syria_earthquake [Accessed: (08/24/23)].. In fact, over the last decades, Numerous nations have experienced significant loss of life and financial damages as a result of the frequent and extensive occurrence of various natural and man-made calamities. Given the growing urbanization (Tchappi et al., 2022) Considering population density, it's possible that in the ensuing decades, the losses from these kinds of disasters would rise significantly. Therefore, Increasing the ability for disaster assessment, response, and forecast is essential. Furthermore, after a tragedy strikes, the rescue teams are typically overworked and unable to save every casualty. Recently, a Red Cross report recommended using newly developed One of the most potent and promising new technologies for disaster response is unmanned aerial vehicles (UAVs)https://americanredcross.github.io/rcrc-drones/Aid_from_the_Air.pdf..

Data from UAVs is used to control natural disasters like bushfires and floods(Ejaz, Ahmed, Mushtaq, & Ibnkahla, 2020). Unmanned Aerial Vehicle (UAV) reconnaissance of disaster-affected regions provides valuable aerial imagery that facilitates emergency evacuation and rescue operations by identifying safe routes to deliver aid to the affected areas. The use of UAVs in a variety of fields is seen in (Aggarwal & Kumar, 2020).

Motivation

This study presented a number of novel approaches to overcome the obstacles found in relation in order to protect consensus methods on blockchain and heterogeneous network interoperability, threat protection, and flexible smart contracts for changing disaster management requirements. The main goal is to revolutionize disaster response tactics by utilizing blockchain integration to unleash UAV networks' full potential. The specific goal is to enable autonomous, effective, and distributed UAV-based operations, especially in challenging post-disaster scenarios. Therefore, in this Chapter holistically tackles these motivations by study that included disaster management by applying the notion of blockchain technology and its integration with the Internet of drones and some techniques that lead to improving results.

Contributions and Organization

This study offers a thorough analysis of the main concerns surrounding IOD and blockchain, including security and privacy, difficulties and solutions related to IOD security, and possible areas for future research on IOD security (such as the integration of IOD with developing technologies). This IOD-oriented study to disaster management and will benefit future research and development by offering insightful commentary on security. The following is a summary of this work's primary contributions:

1. This article analyzes several key technologies for IOD security and focuses on IOD security challenges and solutions. The majority of the current surveys in the table 1, Security solutions, such new blockchain-powered schemes and authentication methods, have not been examined as thoroughly as this article.
2. This review paper examines a number of significant technologies for IOD security and focuses on IOD security challenges and solutions. Most of the surveys or reviews that are currently in use (e.g., (Bine, Boukerche, Ruiz, & Loureiro, 2023),(Heidari, Jafari Navimipour, Unal, & Zhang, 2023),(Derhab et al., 2023), (Wandelt, Wang, Zheng, & Sun, 2023)) only covered broad IOD topics, like drone types, applications, architectures, and communications. There aren't many articles that have examined the security challenges of IODs thus far (e.g., (R. L. Kumar, Pham, Khan, Piran, & Dev, 2021) – (Abdelmaboud, 2021), (Michailidis, Maliatsos, Skoutas, Vouyioukas, & Skianis, 2022)), and the state-of-the-art security measures relevant to the IOD were not sufficiently covered in these research. Security options related to disaster management, like emerging blockchain-powered schemes and authentication mechanisms, have not been examined as thoroughly as they are in this study.

3. There are promising research directions provided that provide insight into how to integrate various technologies, such artificial intelligence and cloud computing, to improve the protection of the IOD. With a detailed presentation of IOD, UAV applications that are compatible with disaster management.
4. Based on a thorough investigation, the difficulties with blockchain-powered schemes and authentication are illustrated and talked about.

In order to highlight the most recent developments in the field, we conduct a literature review on the cutting edge cross blockchain frameworks.

We also offer a current assessment of blockchain-based IOD and UAV applications. We provide a range of scenarios linked to IOD and Disasters that may take advantage of the potentials of the cross-blockchain solutions that are now accessible, based on the results of our survey. Finally, in an effort to steer future research efforts, We've found unresolved issues and possible difficulties with applying a cross-blockchain technique to UAV networks.

This is how the remainder of the paper is structured. There is a research methodology in Section 2. The Background is presented in Section 3. The relevant work utilizing blockchain in IOD for disaster management is covered in depth in Section 4, and the applications of UAVs and IOD in disaster management are covered in Section 5. Open Research Issues are presented in Section 6. Section 7 wraps up the investigation and looks forward to future research. Figure 1. illustrates the article's structure.

METHODOLOGY

The methods used to select the research articles examined in this report were similar to those outlined by Syed et al. Mohd Daud et al. The following keywords were entered into the Google Scholar scientific database: "Blockchain", "Blockchain technology"," UAVs for disaster management"," UAV"," Internet of Drone"," Search and Rescue UAVs"," Security and privacy of Disaster" as well as various combinations of these. 309 papers were initially taken under consideration. A first filter based on the publication date was used because this study focuses on the most recent developments in the use of UAVs for disaster response. Additional requirements for acceptance included results and novelty of the work, as well as abstract and content analysis, machine-learning orientation, and level of technicality. Therefore, out of the 309 papers, 31 papers were selected, as presented in Table 1, 2.

Figure 1. Organization of chapter

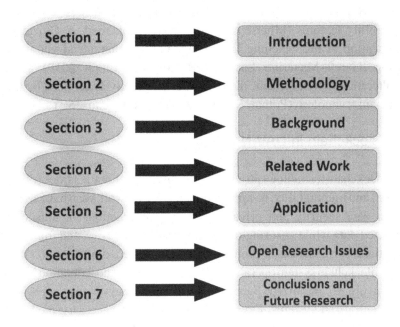

BACKGROUND

Blockchain Technology

Blockchain is a Collaborative,

tamper-proof ledger with block-based organization for tracking transactions NIST, Information technology, Blockchain. https://www.nist.gov/topics/blockchain. (accessed June 22,

2020). After a transaction is verified, a block in the blockchain is permanent(Jo, Heo, Jung, Kim, & Min, 2017). Each block uses a distinct identifier to establish a connection to the block before it. Every modification to the data block modifies a unique identification, which is communicated to all users. Every such altered block is rejected by the nodes. As a result, the blockchain network is difficult to eliminate or disrupt, making it a robust approach to cooperative record keeping NIST, Information technology, Blockchain. https://www.nist.gov/topics/blockchain. (accessed June 22, 2020). Its security is improved by its distributed, immutable, and uncentralized authorization(Jo et al., 2017). Blockchain uses public key infrastructure (PKI) to encrypt data(Rana, Shankar, Sultan, Patan, & Balusamy, 2019). Consensus techniques and smart contracts, such as the current work, can be used to automate dynamic UAV systems. It is driven by the blockchain's automated business logic execution and traceability features (Mhaisen, Fetais, Erbad, Mohamed, & Guizani, 2020). The verification of the matching UAV's signature is guaranteed via asymmetric encryption. (García-Magariño, Lacuesta, Rajarajan, & Lloret, 2019). The danger of data alteration by malevolent, illegal parties is effectively reduced by the usage of blockchain technology(Puthal, Malik, Mohanty, Kougianos, & Yang, 2018).

Role of Blockchain in UAV Networks

Because blockchain technology provides audibility, smart contracts, decentralization, encryption, and immutability, its application in UAV networks can have a substantial impact on data security and privacy. The system functions in unison to store decentralized data while ensuring immutability, data integrity, and transparency at the same time. Nevertheless, a central server is still needed on occasion to manage and keep a blockchain system operating efficiently. Private and federation blockchains are deceptively decentralized. Security on the blockchain is also superior to that on other centralized systems. Cryptography is another necessary method for the ledger to maintain sensitive information secure. Cryptography is a secure approach to encrypting data to deter malevolent cyber criminals (Hafeez et al., 2024).

A thorough analysis of blockchain-assisted UAV communication systems can address the following points.

Confidentiality: It comprises limiting data access for unauthorized individuals. Similar to other networks, Secrecy threats including data theft, sniffing, eavesdropping, and replay are possible on UAV networks. There are multiple situations in which a low-cost, impenetrable blockchain-based solution could protect UAV networks' privacy(Wu, Song, & Wang, 2020).

With the use of drone swarms, this distributed crowd-monitoring system seeks to maintain the confidentiality, security and up-to-dateness of the supervision data(Xiao et al., 2021).

Integrity: However, a blockchain comparison improved their efficiency and the precision of the data that drones exchanged in an environment known as the "internet of drones." IOD employs the DLT to safeguard its data; yet, one of the biggest challenges is making sure that UAV airspace services are available. Since blockchains are decentralized as well, a good network should be immune to hostile actors.

Non-repudiation: is an additional crucial factor for UAV network security. This expression describes the inability to use vital public infrastructure to dodge accountability for one's actions or to deny them, a UAV signing messages before transmitting them over the internet, for instance. For UAV networks, non-repudiation satisfies the prerequisite. Therefore, it implies that a UAV might decline to deliver images containing illicit material.

Authenticity: Lastly, UAV networks need to guarantee the veracity of both users and communications. Authentication, such as cloaking, is the capacity to identify real user identity authentication issues in UAV network attacks. Cryptographic data is also included in the UAV network for privacy and authentication, as is discussed in (M. S. Kumar, Vimal, Jhanjhi, Dhanabalan, & Alhumyani, 2021).

Internet of Drone (IOD)

It is anticipated that a significant number of drones operating in various form factors and carrying out various tasks would arise from the numerous autonomous flight capabilities of drones (as seen in Figure 2). For a large number of drones to coexist with the ongoing airspace for airplanes, which is regulated by organizations like the FAA, a clearly defined ecosystem is required for scalability and sustainability. Thus, the Internet of Drones (IOD) emerged as a promising domain.

Architecture

IOD must address a variety of complex issues, including drone navigation, shared coordination, varied connection, external variables, and immediate security concerns. As suggested in (Gharibi, Boutaba, & Waslander, 2016) For flexibility and scalability, the design must be modular, and a layered model is a good option for this. Furthermore, because the IOD operation region shares existing regulated airspace, drone movement needs to be managed differently depending on the area. As suggested, Common aviation airspace must be regulated according to a set flight route plan and is referred to as airways. The drones must travel independently in areas where free flights can be scheduled, using their surroundings—which are considered nodes—to guide them. As seen in Figure 2, the operational area is separated into zones, and drones fly between them via permanent locations called gateways. Ground stations are in charge of overseeing each zone.

Applications

Because of its autonomy and scalability, IOD is widely used in the applications that follow:

- Drone swarms are collections of drones that fly in unison to achieve a common goal, like conducting surveys. Since it is impractical to manually

Figure 2. Architecture IOD

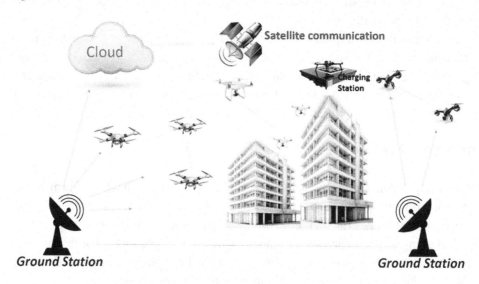

control a large number of drones flying together, autonomous operations are a potential solution.

- Drones' reach, low operating costs, and speedier delivery are seen as a logistics game-changer. Drone delivery is a popular idea for businesses like UPS and Amazon, but it requires a fully autonomous system to carry out the tasks.

- Provision of services Drone-based relaying is another field that is seeing enormous investment. If the model is to advance, it must transform into IOD for sustainability. Facebook's ambitious effort to utilize a solar-powered drone to bring Internet access to remote regions is a start in this direction.

Security Requirements

An IOD is a cyber-physical system that must guarantee control and information security. In our case, security, alteration, disclosure of information, or loss of assets might be defined as the IOD's capacity to accurately perform the intended activities without unauthorized access. IOD must adhere to many security standards, include those pertaining to privacy, nonrepudiation, integrity, availability, Confidentiality, and authentication.

- **Authentication**: It guarantees that access to IoD resources is limited to authorized devices (like drones and GCS) and authorized users (like operators and end users). A malevolent actor may pose as a drone, obtain private data, and insert malicious commands and data if authentication is not guaranteed.

- **Confidentiality**: It guarantees that data is not revealed to unapproved users. We may find several kinds of information in IOD: (i) information and instructions that are shared between the various IOD components, (ii) data transferred via the network; (iii) data captured by the drones (e.g., sensed values, captured images); and (iv) data shared between subsets of the components (e.g., system states transmitted via communication channels between the sensors, controllers, actuators, and transmission system).
- **Availability**: It guarantees that each and every IOD component carries out the necessary tasks accurately (i.e., surveillance, processing, command, action, transfer, and archiving) and ought should be accessible upon request.
- **Integrity:** Data integrity and system integrity are the two integrity attributes we take into account. Data integrity guarantees that there is no malicious or unintentional data alteration or removal, such as orders and telemetry data, is carried out. A flight control system that receives inaccurate GPS data, for example, could undermine availability and have a detrimental impact on the drone's mission planning if data integrity is not met.
- **Non-repudiation**: This condition relates to legal matters. Given the variety of applications that drones are used for, in order to establish liability in the event of a drone accident or incident, user activity must be tracked. Nonrepudiation in the context of communication guarantees that neither the sender nor the drone can refuse to generate messages or send photos.
- **Privacy**: User-specific data, such as places visited and information gathered, may be revealed by the drone mission. Consequently, in order to safeguard user data privacy, drone operations need to be secured against traffic analysis and data collection. Moreover, In order to protect user privacy when processing and outsourcing drone-collected data (such in a cloud system), appropriate cryptographic procedures (like homomorphic encryption) must be put in place(Lin et al., 2018).

RELATED WORK

Integration Blockchain with IOD

In (Nguyen, Katila, & Gia, 2023) The system under consideration integrates blockchain technology with security methodologies to satisfy security prerequisites, including integrity, authorization, access control, transparency, traceability, trust, and tolerance. It applied and contrasted two different blockchain types: public Ethereum and private

Hyperledger. In order to promote automaticity, smart contracts were also created and put into use. The outcomes demonstrated that the solution being described might provide cutting-edge edge services to enhance drone energy efficiency and Search and Rescue SAR mission quality of service. Additionally, the plan hopes to meet SAR's security standards by utilizing Hyperledger Fabric.

Drone usage is increasing more than it has in the past few years. They have been employed in military services on numerous occasions. Because drones offer validity and secrecy in data transmission, it is imperative that their intelligence, privacy, and security be increased. This research(Maurya, Rauthan, Verma, & Ahmad, 2022) A blockchain-based secure solution for the internet of drones has been proposed and put into operation by study. It offers smooth drone registration and verification along with improved data storage security. Dummy drones are used to test the suggested approach and ensure system security and secrecy.

In (Oláh, Molnár, & Huszti, 2023) It has become increasingly important to handle and regulate the risks associated with security incidents and physical damage as our environment adopts increasingly sophisticated surroundings. To guarantee proper drone registration in the IOD systems, they combined a private blockchain with the PUF function. During the design, they took into consideration the problem of scalability and successfully kept the keys on the ARP table. They discovered that LoRa is a suitable physical layer for our application because of its extended range, low power consumption, immunity to interference, and reasonable data rate.

With individuals stranded in their homes and governments attempting to impose restrictions on public transportation, coronavirus disease 2019 COVID-19 brought the world to a halt. But in order to do this, a major issue that surfaced and outweighed all others was providing for people's daily needs without involving humans. In (Singh et al., 2020)this regard, suggest using drones that are accessible from commercial retail suppliers to deliver goods in COVID-19-like scenarios over a blockchain-enabled secure communication infrastructure. Since payments are performed through the execution of smart contracts, the blockchain system is used to develop smart contracts that help buyers and sellers gain confidence in the framework. The order processing using blockchain technology guarantees the accuracy and legitimacy of the data.

Nonetheless, UAVs can be utilized to compromise the efficacy of military operations due to the open nature of wireless communication. In (Pancholi, Jadav, Tanwar, Garg, & Zanjani, 2023), They created a framework based on blockchain technology to reduce security risks posed by unmanned aerial vehicles (UAVs) used in combat. UAVs, sensors, adversaries, and friendly nodes are all included in the simulation of a war scenario that takes place in the MATLAB virtual environment. An adversary that persistently manipulates the UAVs to jeopardize battlefield operations is also present in the UAV scenario. To identify malevolent UAVs, the proposed framework makes use of machine learning algorithms. Additionally, the suggested system makes use of blockchain technology to improve security and reduce attempts at data tampering. To validate data, a smart contract is implemented on a public blockchain based on IPFS. In addition, utilizing 5G technology, effective communication between UAVs and the control station is made feasible with reduced packet error rates and delay.

In (Islam & Shin, 2019) suggested a safe outdoor health monitoring system integrating MEC, UAV, and blockchain technology. Under the suggested approach, users' carried sensors gather Health Data (HD), which are then transmitted by a UAV to the subsequent MEC server. The health data is encrypted before being sent to MEC in order to guard against cyberattacks.

In (Islam & Shin, 2019) suggested a blockchain-enabled data acquisition plan that uses blockchain technology with unmanned aerial vehicle swarming to offer security and data integrity. Specifically, before beginning data collection, the unmanned aerial vehicle swarm and the Internet of Things devices exchange a common key for communication. The hash bloom filter and digital signature algorithm prevent and resist two forms of man-in-the-middle attacks: manipulation and eavesdropping.

In(Alsumayt, El-Haggar, Amouri, Alfawaer, & Aljameel, 2023) The proposed FDSS (Flood Detection Secure System) allowed us to measure the hazard posed by flooding by estimating the inundated regions and monitoring the quick changes in dam water levels. The suggested approach is simple, easily adjustable, and provides suggestions for local administration and Saudi Arabian decision-makers to handle the increasing risk of flooding. This study's consideration of the suggested approach and its difficulties in utilizing blockchain technology and artificial intelligence to manage floods in remote areas comes to a close.

Li et al. (K. Li, Lau, Au, Ho, & Wang, 2020) indicated that 5G drones are essential for a variety of uses, particularly in the military and private sectors. During the epidemic, this IoD is able to track people and enforce social separation. But it has problems with privacy and security, for example. The aforementioned problem is resolved by blockchain technology, which has also demonstrated that it works best in operational situations.

Bera et al.(Bera, Das, & Sutrala, 2021), developed a concept for an access control method that might be applied in an Internet of drones (IoD) environment to detect and lessen the impact of unapproved UAVs. They saved the transactional data over a private blockchain, which was completely genuine and valid. It comprised the regular secure data, which was transferred between the drone and the ground station server and the anomalous data, which was also known as suspected data. They used these data to identify unauthorized UAVs.

In (Allouch et al., 2021) propose UTM-Chain, a low-weight, blockchain-based security solution that matches the constraints of UAVs' computing and storage capacity by utilizing Hyperledger Fabric for UTM of low-altitude UAVs. Additionally,

UTM-Chain gives the UAVs' ground control stations safe, unchangeable traffic data. cAdvisor is used to analyze the suggested system's performance in terms of transaction latency and resource usage. Finally, the suggested UTM-Chain strategy for the safe sharing of UAV data is workable and expandable, as shown by the examination of security considerations

In (Xu, Wei, Chen, Chen, & Pham, 2022), This paper presents LightMAN, a lightweight microchained fabric for data assurance and resilience-oriented UAM networks . LightMAN is designed for smaller-scale permissioned UAV networks, wherein a microchain – a lightweight distributed ledger – provides 'enough security assurance' . This allows users to authenticate drones and verify the data received to/ from drones without requiring a third party. Furthermore, as smaller UAV network actively relies on such mechanisms, the computational cost is significantly lower in comparison to the PoW. In addition, In UAM networks, a hybrid on-chain and off-chain storage method is used to preserve sensitive data privacy while simultaneously increasing performance metrics like latency and throughput.

An innovative deep learning-based blockchain system for access control and accurate DoA estimation for unapproved UAV localization in the Internet of Medical Things. The (Akter, Golam, Doan, Lee, & Kim, 2022) A peer-to-peer authentication method based on blockchain has been used in the proposed scheme to authenticate users. In order to prevent unwanted access, the suggested approach also includes a CNN-based object position detection technique.

In (Bera, Wazid, Das, & Rodrigues, 2021) When the drones are launched, they can sense and collect information from their environment for security considerations. They can then safely transmit that information to the ground station computer. After securely transforming the acquired data into encrypted transactions, the ground station server sends it to the peer-to-peer cloud server (P2PCS) network. Ultimately, the P2PCS network's consensus methods are used to construct blocks from the encrypted transactions and add them to a blockchain.

In (Z. Li et al., 2024) a unique architecture for UAV network security services that combines deep learning, machine learning, and blockchain is proposed. has been provide a trusted UAV network service architecture that is built on blockchain technology. It addresses trusted identity verification for UAV clusters and mitigates security threats by lowering network security risks and using a data-trustworthy interconnection platform.

To enable secure communication in the Internet of Things context, this study (Subramani, Maria, Rajasekaran, & Lloret, 2024) presents a lightweight, anonymous mutual authentication and handover authentication mechanism based on blockchain technology. Drones may quickly re_authenticate by sending drone authentication codes to the following UAV operators using the blockchain network.

In order to defend against several types of assaults, such as botnet attacks, false node attacks, sinkhole attacks, and black hole attacks, this work (Gilani et al., 2024) aims to provide an affordable, scalable, and secure communication system. an startup and registration procedure for various entities, including drones and ground stations, is also being implemented as part of this research.

Review IOD-Related Survey Papers

We review survey articles linked to IOD-UAV in this section. An overview of IOD with blockchain in terms of communications, applications, and security is given in these survey articles. Table 2 provides an overview of the current IOD-UAV related surveys and this work.

APPLICATION

Application of IOD in Disaster Management

The use of Internet of Drones (IOD) in disaster management can bring significant advantages and improvements to response efforts. Here are some applications of IOD in disaster management as in the Figure 3:

1. **Aerial Surveillance**: Aerial observation of disaster-affected areas can be done in real time using drones fitted with cameras and sensors. They can offer high-definition photos and videos to help determine the extent of the devastation, locate survivors, and organize rescue efforts.
2. **Search and Rescue Operations**: In disaster areas, drones can be used to look for people who have gone missing. They can swiftly and effectively cover wide regions, cutting down on the time and materials required for search and rescue operations.
3. **Delivery of Aid and Supplies**: Drones can transport vital supplies like food, water, medication, and communication devices to places that are hard to reach with conventional transportation. This can help in reaching affected populations faster and more effectively.
4. **Mapping and Damage Assessment**: Drones can create 3D maps of disaster-affected areas, allowing responders to assess the damage, plan reconstruction efforts, and prioritize areas for assistance.
5. **Communication and Connectivity**: Drones equipped with communication devices can establish temporary communication networks in areas where

Table 1. A summary of IOD- blockchain related papers

Ref.	Year	Type	Platform	Security consideration	Storage	Application	limitations
(Nguyen et al., 2023)	2023	Private, public blockchain	Ethereum and Hyperledger	integrity, authorization, authentication, transparency, traceability, trust, access control, and tolerance.	Used Smart contracts	SAR missions	need for humans control, energy waste and its effects on quality of service (QoS)
(Maurya et al., 2022)	2022	NA	NA	Confidentiality, authentication, data storage, Privacy	NA	Military application	enhancing drone intelligence, security, and privacy
(Oláh et al., 2023)	2023	Private blockchain + PUF	NA	authentication and confidentiality	NA	UAV networks	limited resources, scalability issues, insecure communication
(Singh et al., 2020)	2020	blockchain	NA	the integrity and authenticity	Smart contract	COVID-19	need without human involvement.
(Pancholi et al., 2023)	2023	Blockchain + ML+5G	NA	Confidentiality,	Smart contract, IPFS	Defence and Military application	latency, and transaction and execution costs
(Islam & Shin, 2019a)	2019	Blockchain + MEC	NA	data is encrypted to protect	Data stored in blockchain	health monitoring system	cyber threats, Integrity
(Islam & Shin, 2019b)	2019	Digital signature algorithm	NA	security and data integrity	Data stored in blockchain	IOT Devices	energy usage and offers protection from external hazards.
(Alsumayt et al., 2023)	2023	Blockchain + AI	NA	Confidentiality, Privacy	IPFS	flooded areas	minimize communication costs, maximize accuracy.
(K. Li et al., 2020)	2020	Blockchain + 5G	NA	Authentications Privacy	NA	military and civilian	transparency, efficiency,
(Bera, Das, et al., 2021)	2021	Private blockchain	NA	Access control, Authorization	NA	IOD environment.	Access Control unauthorized, malicious packages
(Allouch et al., 2021)	2021	Blockchain UTM-Chain +Cloud server	Hyperledger	Data Integrity, Availability, Authentication, Identity management, Privacy	IPFS	search and rescue	computational and storage resources
(Xu et al., 2022)	2022	Blockchain LightMAN	Hyperledger	Authentication, Integrity, Privacy	On_chain and Off_chain storage	UAV networks	latency and throughput

continued on following page

Table 1. Continued

Ref.	Year	Type	Platform	Security consideration	Storage	Application	limitations
(Akter et al., 2022)	2022	Blockchain + DL+ IOMT	NA	Authorization, Access control, authentication	CNN	military	Data manipulation, unauthorized access
(Bera, Wazid, et al., 2021)	2021	Blockchain+ AI+	NA	confidently	cloud server +Smart contract	environmental monitoring	lack of security and trust levels
(Z. Li et al., 2024)	2024	Blockchain + ML+DL	NA	Authorization, authentication	NA	All Application	Security risk of UAV
(Subramani et al., 2024)	2024	Blockchain + Pufs	NA	Authorization, Privacy	NA	All Application	Security communication in the IOD
(Gilani et al., 2024)	2024	Blockchain	NA	Authorization	NA	Military application	Security and privacy, Storage, communication

infrastructure has been damaged. This can help in coordinating response efforts and providing vital information to affected populations.

6. **Environmental Monitoring**: Drones can be used in disaster areas to monitor environmental threats such chemical spills, fires, and floods. By using this information, you may reduce future risks and make well-informed decisions.

Overall, the use of drones in disaster management through IOD can speed up reaction times, increase situational awareness, and perhaps save lives in emergency situations.

Applications of UAVs

Applications of UAVs in disaster management can be generally categorized as indicated in Figure 4 and vary depending on their involvement throughout different phases of disaster management.

Monitoring

Unmanned Aerial Vehicle (UAV) surveillance of the environment can aid in catastrophe prediction, preparedness, and prevention. In (Yim, Park, Kwon, Kim, & Lee, 2018), UAVs are used to gather sensor data and aerial photography in disaster-prone areas. It is suggested to use image stitching technology to identify disasters by examining the mapped sensor data on the combined image.

Table 2. A summary of IOD- blockchain related surveys or reviews

Review	Year	Highlight	Focus of security, Domain
(R. L. Kumar, Pham, Khan, Piran, & Dev, 2021)	2021	An overview of the integration of blockchain technology with aerial communications (BAC) is presented in this study.	Yes, UAV and Blockchain
(Harbi et al., 2023)	2023	Analyzing the present state of research on the security of IoD settings using the upcoming blockchain through a systematic analysis of the literature	Yes, IOD and Blokchain
(Yang, Wang, Yin, Wang, & Hu, 2022)	2022	significant security technologies, with a focus on blockchain-powered schemes and authentication methods	Yes, UAV and blockchain
(Mehta, Gupta, & Tanwar, 2020)	2020	study provides a thorough analysis of security concerns in UAV networks enabled by 5G.	Yes,5G and UAV
(Abdelmaboud, 2021)	2021	Has been offer an IoD taxonomy and discuss the needs for communication, security, and privacy.	Yes, IOD and blockchain
(Bine, Boukerche, Ruiz, & Loureiro, 2023)	2023	link between urban computing (UC) and the Internet of Drones (IOD)	No, IOD and UC
(Heidari, Jafari Navimipour, Unal, & Zhang, 2023)	2023	categorize ML in the context of IOD-UAVs based on its uses.	No, ML,DL and IOD
(Hafeez et al., 2023)	2023	thoroughly describes the integration of privacy and security in blockchain-assisted UAV communication	Yes, UAV and blokhain
(Derhab et al., 2023)	2023	survey pertaining to IOD physical and cyber security	No, Cyber and IOD
(Jameii, Zamirnaddafi, & Rezabakhsh, 2022)	2022	examine the body of knowledge already available in the IOFT and IOD security domains.	No, IOT, IOFT and IOD
(Michailidis, Maliatsos, Skoutas, Vouyioukas, & Skianis, 2022)	2022	A thorough explanation of the use cases and application scenarios for the Internet of Things enabled by Mobile Edge Computing (MEC) with assistance from UAVs	Yes, UAV and (MEC)
(Wandelt, Wang, Zheng, & Sun, 2023)	2023	literature reviews covering a range of topics related to UAV mobility, logistics, and monitoring activities	No, UAV mobility and Monitoring
(Tubis, Poturaj, Dereń, & Żurek, 2024)	2024	bibliometric analysis of pertinent works and a discussion of the primary lines of inquiry on the subject of drone application dangers	risks in drone applications
(Husman et al., 2021)	2021	scholarly sources on the application of unmanned aerial vehicles to crowd surveillance and analysis	No, UAV for crowd monitoring
(Kedys, Tchappi, & Najjar, 2024)	2024	focuses on the most recent developments in UAV use for disaster relief	Yes, UAV for Disaster management
This Work	2024	In this review, the security aspect was highlighted in a focused manner for its benefit in disaster management using the Internet of Drones and based on the important characteristics of blockchain technology	Yes, blockchain and UAV and IOD, Disaster Management

Search and Rescue

In times of disaster, unmanned aerial vehicles (UAVs) are utilized for search and rescue operations in order to identify and locate survivors within the affected area. In (Liu & Ansari, 2018), In the absence of conventional network infrastructure, human wearing Internet of things devices establish communication through UAV mounted base stations to send messages for disaster rescue. It is being discussed

Figure 3. Applications of IOD in disaster management

Figure 4. Classification of applications of UAVs

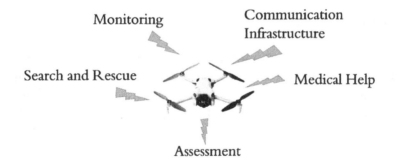

to use a multi-UAV system to search for and rescue victims of natural disasters (Miyano, Shinkuma, Mandayam, Sato, & Oki, 2019).

Assessment

Accurate understanding of the extent and gravity of damage resulting from a disaster is crucial for efficient disaster management. UAVs gather footage, pictures, and sensor data from the impacted region, which is then processed and examined to determine the extent of the damage. In (Lim, Kim, Cho, Gong, & Khodaei, 2016), UAVs are used to assess how much an extreme weather event has damaged a power network. The satellite image data can assist in carrying out accurate risk assessments of the impacted area in order to launch an appropriate disaster response(Munawar, Qayyum, Ullah, & Sepasgozar, 2020). UAVs have been investigated for post-disaster management in the field of disaster management employing IOT, energy-efficient task scheduling, and physiological evaluation (Ejaz et al., 2020).

Medical Help

In disaster-affected areas, unmanned aerial vehicles (UAVs) can be utilized to transport medical supplies and aid during emergency medical response. UAVs can be utilized to swiftly reach individuals in need before more medical assistance arrives, as time is of the essence in medical emergencies and survival prospects rapidly diminish with the passage of time. UAVs are being evaluated for the delivery of vital medical supplies in disaster-affected areas where the transportation infrastructure has been damaged or destroyed(Ejaz, Azam, Saadat, Iqbal, & Hanan, 2019). Following major catastrophes, the deployment of emergency response workers and the delivery of medical supplies are frequently fraught with logistical and time-sensitive issues, as well as the possibility of human injury(Hughes, 2023).

Communication Infrastructure

The traditional communication infrastructure might be damaged in a disaster. UAVs can be used in this situation to close the gaps in coverage. To cover the gaps in a cellular network's coverage after a disaster, drone-based small cell deployment is being researched in (Hayajneh, Zaidi, McLernon, Di Renzo, & Ghogho, 2018).

OPEN RESEARCH ISSUES

Integrating blockchain technology with the Internet of Drones (IOD) in disaster management scenarios presents unique research challenges and issues that need to be addressed. Some of the key research issues faced by this integration in disaster response include:

1. **Real-time Data Processing**: Ensuring that the blockchain network can handle real-time data processing and analysis from drones deployed in disaster areas to provide timely and accurate information to responders.

2.   **Decentralized Coordination**: Designing decentralized coordination mechanisms to enable efficient collaboration and communication between drones, emergency responders, and other stakeholders in disaster situations.

3.   **Data Integrity and Trust**: Ensuring the integrity and trustworthiness of data collected and shared by drones during disaster response operations to prevent tampering or manipulation of critical information.

4.   **Interoperability and Standardization**: Establishing interoperability standards and protocols to facilitate seamless communication and data exchange between different drones, blockchain networks, and disaster management systems.

5.   **Resource Optimization**: Developing algorithms and strategies to optimize the deployment of drones and allocate resources effectively in dynamic and unpredictable disaster environments.

6.   **Privacy and Security**: Implementing robust privacy and security measures to protect sensitive data collected by drones during disaster response operations and ensure compliance with data protection regulations.

7.   **Scalability and Performance**: Ensuring that the blockchain network can scale to accommodate the increased data volume and transaction throughput during large-scale disaster events without compromising performance or reliability.

8.   **Resilience and Redundancy**: Designing resilient and redundant communication and data storage mechanisms to ensure continuous operation of the blockchain network and data availability in the event of network disruptions or failures.

In order to fully utilize the potential of combining blockchain technology with the Internet of Drones for efficient disaster response and management, as well as to enhance coordination, communication, and data integrity in urgent situations, it will be imperative to address these research difficulties.

CONCLUSION AND FUTURE RESEARCH

In order to promote communication during a disaster, a flexible and trustworthy information transmission platform is needed. In priority order, a Blockchain-enabled IOD network greatly improves security (authentication, integrity, privacy, and access control), as listed in table 1. performance and reduce the cost for disaster management scenarios. In this paper, we describe an examination of a UAV-enabled IOD network using blockchain for disaster management, highlighting the main obstacles in the field such as data analytics, ground IoT network deployment, experimental validation, and wireless connectivity.

Unmanned aerial networks have shown to be extraordinarily successful in vital operations, reaching remote locations to deliver vital support. Even though this sector has seen a lot of research, there are still issues with communication, task scheduling, data processing, and trajectory optimization because there are no standardized methods in place. Because of their wide range of uses, UAVs necessitate targeted research in particular fields to handle the complex problems particular to each of their many applications. The article's writers have noted the current research trends and placed special emphasis on the need to prioritize studies pertaining to data processing capabilities and network security for both devices and data. A significant amount of research is presented in this article, which looks at publications on a range of network-related topics, including as deployment, routing, security, and trajectory optimization. It also lists the study topics that will be thoroughly investigated in the upcoming years.

REFERENCES

Abdelmaboud, A. (2021). The internet of drones: Requirements, taxonomy, recent advances, and challenges of research trends. *Sensors (Basel)*, *21*(17), 5718. doi:10.3390/s21175718 PMID:34502608

Aggarwal, S., & Kumar, N. (2020). Path planning techniques for unmanned aerial vehicles: A review, solutions, and challenges. *Computer Communications*, *149*, 270–299. doi:10.1016/j.comcom.2019.10.014

Akter, R., Golam, M., Doan, V.-S., Lee, J.-M., & Kim, D.-S. (2022). Iomt-net: Blockchain-integrated unauthorized uav localization using lightweight convolution neural network for internet of military things. *IEEE Internet of Things Journal*, *10*(8), 6634–6651. doi:10.1109/JIOT.2022.3176310

Allouch, A., Cheikhrouhou, O., Koubâa, A., Toumi, K., Khalgui, M., & Nguyen Gia, T. (2021). Utm-chain: Blockchain-based secure unmanned traffic management for internet of drones. *Sensors (Basel)*, *21*(9), 3049. doi:10.3390/s21093049 PMID:33925489

Alsumayt, A., El-Haggar, N., Amouri, L., Alfawaer, Z. M., & Aljameel, S. S. (2023). Smart flood detection with AI and blockchain integration in Saudi Arabia using drones. *Sensors (Basel)*, *23*(11), 5148. doi:10.3390/s23115148 PMID:37299876

Arksey, H., & O'malley, L. (2005). Scoping studies: Towards a methodological framework. *International Journal of Social Research Methodology*, *8*(1), 19–32. doi:10.1080/1364557032000119616

Ayamga, M., Akaba, S., & Nyaaba, A. A. (2021). Multifaceted applicability of drones: A review. *Technological Forecasting and Social Change*, *167*, 120677. doi:10.1016/j.techfore.2021.120677

Barnabas, G., Citrawati, H., Santoso, T., & Surantha, N. (2023). Internet of Things in disaster management systems: A systematic review. *Paper presented at the AIP Conference Proceedings*. AIP. 10.1063/5.0109493

Bera, B., Das, A. K., & Sutrala, A. K. (2021). Private blockchain-based access control mechanism for unauthorized UAV detection and mitigation in Internet of Drones environment. *Computer Communications*, *166*, 91–109. doi:10.1016/j.comcom.2020.12.005

Bera, B., Wazid, M., Das, A. K., & Rodrigues, J. J. (2021). Securing internet of drones networks using ai-envisioned smart-contract-based blockchain. *IEEE Internet of Things Magazine*, *4*(4), 68–73. doi:10.1109/IOTM.001.2100044

Bine, L. M., Boukerche, A., Ruiz, L. B., & Loureiro, A. A. (2023). Connecting Internet of Drones and Urban Computing: Methods, protocols and applications. *Computer Networks*, 110136.

Derhab, A., Cheikhrouhou, O., Allouch, A., Koubaa, A., Qureshi, B., Ferrag, M. A., Maglaras, L., & Khan, F. A. (2023). Internet of drones security: Taxonomies, open issues, and future directions. *Vehicular Communications*, *39*, 100552. doi:10.1016/j.vehcom.2022.100552

Ejaz, W., Ahmed, A., Mushtaq, A., & Ibnkahla, M. (2020). Energy-efficient task scheduling and physiological assessment in disaster management using UAV-assisted networks. *Computer Communications*, *155*, 150–157. doi:10.1016/j.comcom.2020.03.019

Ejaz, W., Azam, M. A., Saadat, S., Iqbal, F., & Hanan, A. (2019). Unmanned aerial vehicles enabled IoT platform for disaster management. *Energies*, *12*(14), 2706. doi:10.3390/en12142706

García-Magariño, I., Lacuesta, R., Rajarajan, M., & Lloret, J. (2019). Security in networks of unmanned aerial vehicles for surveillance with an agent-based approach inspired by the principles of blockchain. *Ad Hoc Networks*, *86*, 72–82. doi:10.1016/j.adhoc.2018.11.010

Gharibi, M., Boutaba, R., & Waslander, S. L. (2016). Internet of drones. *IEEE Access : Practical Innovations, Open Solutions*, *4*, 1148–1162. doi:10.1109/ACCESS.2016.2537208

Gilani, S. M., Anjum, A., Khan, A., Syed, M. H., Moqurrab, S. A., & Srivastava, G. (2024). A robust Internet of Drones security surveillance communication network based on IOTA. *Internet of Things : Engineering Cyber Physical Human Systems*, 25, 101066. doi:10.1016/j.iot.2024.101066

Hafeez, S., Khan, A. R., Al-Quraan, M., Mohjazi, L., Zoha, A., Imran, M. A., & Sun, Y. (2023). Blockchain-assisted UAV communication systems: A comprehensive survey. *IEEE Open Journal of Vehicular Technology*, 4, 558–580. doi:10.1109/OJVT.2023.3295208

Hafeez, S., Manzoor, H. U., Mohjazi, L., Zoha, A., Imran, M. A., & Sun, Y. (2024). Blockchain-Empowered Immutable and Reliable Delivery Service (BIRDS) Using UAV Networks. *arXiv preprint arXiv:2403.12060.*

Harbi, Y., Medani, K., Gherbi, C., Senouci, O., Aliouat, Z., & Harous, S. (2023). A systematic literature review of blockchain technology for internet of drones security. *Arabian Journal for Science and Engineering*, 48(2), 1053–1074. doi:10.1007/s13369-022-07380-6 PMID:36337772

Hayajneh, A. M., Zaidi, S. A. R., McLernon, D. C., Di Renzo, M., & Ghogho, M. (2018). Performance analysis of UAV enabled disaster recovery networks: A stochastic geometric framework based on cluster processes. *IEEE Access : Practical Innovations, Open Solutions*, 6, 26215–26230. doi:10.1109/ACCESS.2018.2835638

Heidari, A., Jafari Navimipour, N., Unal, M., & Zhang, G. (2023). Machine learning applications in internet-of-drones: Systematic review, recent deployments, and open issues. *ACM Computing Surveys*, 55(12), 1–45. doi:10.1145/3571728

Hughes, A. A. (2023). Unmanned aerial vehicle use in humanitarian activities. *University of Lynchburg DMSc Doctoral Project Assignment Repository*, 5(1), 137.

Husman, M. A., Albattah, W., Abidin, Z. Z., Mustafah, Y. M., Kadir, K., Habib, S., Islam, M., & Khan, S. (2021). Unmanned aerial vehicles for crowd monitoring and analysis. *Electronics (Basel)*, 10(23), 2974. doi:10.3390/electronics10232974

Islam, A., & Shin, S. Y. (2019). BHMUS: Blockchain based secure outdoor health monitoring scheme using UAV in smart city. *Paper presented at the 2019 7th international conference on information and communication technology (ICoICT)*. IEEE. 10.1109/ICoICT.2019.8835373

Islam, A., & Shin, S. Y. (2019). Bus: A blockchain-enabled data acquisition scheme with the assistance of uav swarm in internet of things. *IEEE Access : Practical Innovations, Open Solutions*, 7, 103231–103249. doi:10.1109/ACCESS.2019.2930774

Jameii, S. M., Zamirnaddafi, R. S., & Rezabakhsh, R. (2022). Internet of Flying Things security: A systematic review. *Concurrency and Computation, 34*(24), e7213. doi:10.1002/cpe.7213

Jo, K., Heo, J., Jung, J., Kim, B., & Min, H. (2017). A rendezvous point estimation considering drone speed and data collection delay. *Paper presented at the 2017 4th International Conference on Computer Applications and Information Processing Technology (CAIPT)*. IEEE. 10.1109/CAIPT.2017.8320706

Kedys, J., Tchappi, I., & Najjar, A. (2024). UAVs for Disaster Management-An Exploratory Review. *Procedia Computer Science, 231*, 129–136. doi:10.1016/j.procs.2023.12.184

Kumar, M. S., Vimal, S., Jhanjhi, N., Dhanabalan, S. S., & Alhumyani, H. A. (2021). Blockchain based peer to peer communication in autonomous drone operation. *Energy Reports, 7*, 7925–7939. doi:10.1016/j.egyr.2021.08.073

Kumar, R. L., Pham, Q.-V., Khan, F., Piran, M. J., & Dev, K. (2021). Blockchain for securing aerial communications: Potentials, solutions, and research directions. *Physical Communication, 47*, 101390. doi:10.1016/j.phycom.2021.101390

Li, K., Lau, W. F., Au, M. H., Ho, I. W.-H., & Wang, Y. (2020). Efficient message authentication with revocation transparency using blockchain for vehicular networks. *Computers & Electrical Engineering, 86*, 106721. doi:10.1016/j.compeleceng.2020.106721

Li, Z., Chen, Q., Li, J., Huang, J., Mo, W., Wong, D. S., & Jiang, H. (2024). A secure and efficient UAV network defense strategy: Convergence of blockchain and deep learning. *Computer Standards & Interfaces, 90*, 103844. doi:10.1016/j.csi.2024.103844

Lim, G. J., Kim, S., Cho, J., Gong, Y., & Khodaei, A. (2016). Multi-UAV pre-positioning and routing for power network damage assessment. *IEEE Transactions on Smart Grid, 9*(4), 3643–3651. doi:10.1109/TSG.2016.2637408

Lin, C., He, D., Kumar, N., Choo, K.-K. R., Vinel, A., & Huang, X. (2018). Security and privacy for the internet of drones: Challenges and solutions. *IEEE Communications Magazine, 56*(1), 64–69. doi:10.1109/MCOM.2017.1700390

Liu, X., & Ansari, N. (2018). Resource allocation in UAV-assisted M2M communications for disaster rescue. *IEEE Wireless Communications Letters, 8*(2), 580–583. doi:10.1109/LWC.2018.2880467

Maurya, S., Rauthan, M., Verma, R., & Ahmad, R. B. (2022). Blockchain based Secure System for the Internet of Drones (IoD). *Paper presented at the 2022 6th International Conference on Electronics, Communication and Aerospace Technology.*

Mehta, P., Gupta, R., & Tanwar, S. (2020). Blockchain envisioned UAV networks: Challenges, solutions, and comparisons. *Computer Communications*, *151*, 518–538. doi:10.1016/j.comcom.2020.01.023

Mhaisen, N., Fetais, N., Erbad, A., Mohamed, A., & Guizani, M. (2020). To chain or not to chain: A reinforcement learning approach for blockchain-enabled IoT monitoring applications. *Future Generation Computer Systems*, *111*, 39–51. doi:10.1016/j.future.2020.04.035

Michailidis, E. T., Maliatsos, K., Skoutas, D. N., Vouyioukas, D., & Skianis, C. (2022). Secure UAV-aided mobile edge computing for IoT: A review. *IEEE Access : Practical Innovations, Open Solutions*, *10*, 86353–86383. doi:10.1109/ACCESS.2022.3199408

Miyano, K., Shinkuma, R., Mandayam, N. B., Sato, T., & Oki, E. (2019). Utility based scheduling for multi-UAV search systems in disaster-hit areas. *IEEE Access : Practical Innovations, Open Solutions*, *7*, 26810–26820. doi:10.1109/ACCESS.2019.2900865

Munawar, H. S., Qayyum, S., Ullah, F., & Sepasgozar, S. (2020). Big data and its applications in smart real estate and the disaster management life cycle: A systematic analysis. *Big Data and Cognitive Computing*, *4*(2), 4. doi:10.3390/bdcc4020004

Nguyen, T., Katila, R., & Gia, T. N. (2023). An advanced Internet-of-Drones System with Blockchain for improving quality of service of Search and Rescue: A feasibility study. *Future Generation Computer Systems*, *140*, 36–52. doi:10.1016/j.future.2022.10.002

Oláh, N., Molnár, B., & Huszti, A. (2023). Secure Registration Protocol for the Internet of Drones Using Blockchain and Physical Unclonable Function Technology. *Symmetry*, *15*(10), 1886. doi:10.3390/sym15101886

Pancholi, D., Jadav, N. K., Tanwar, S., Garg, D., & Zanjani, S. M. (2023). *Blockchain-based Secure UAV-assisted Battlefield Operation underlying 5G*. Paper presented at the 2023 14th International Conference on Information and Knowledge Technology (IKT).

Puthal, D., Malik, N., Mohanty, S. P., Kougianos, E., & Yang, C. (2018). The blockchain as a decentralized security framework [future directions]. *IEEE Consumer Electronics Magazine*, *7*(2), 18–21. doi:10.1109/MCE.2017.2776459

Rana, T., Shankar, A., Sultan, M. K., Patan, R., & Balusamy, B. (2019). *An intelligent approach for UAV and drone privacy security using blockchain methodology*. Paper presented at the 2019 9th International Conference on Cloud Computing, Data Science & Engineering (Confluence). 10.1109/CONFLUENCE.2019.8776613

Singh, M., Aujla, G. S., Bali, R. S., Vashisht, S., Singh, A., & Jindal, A. (2020). Blockchain-enabled secure communication for drone delivery: A case study in COVID-like scenarios. *Proceedings of the 2nd ACM MobiCom Workshop on Drone Assisted Wireless Communications for 5G and beyond*. ACM.

Subramani, J., Maria, A., Rajasekaran, A. S., & Lloret, J. (2024). Physically secure and privacy-preserving blockchain enabled authentication scheme for internet of drones. *Security and Privacy*, *7*(3), 364. doi:10.1002/spy2.364

Tchappi, I., Mualla, Y., Galland, S., Bottaro, A., Kamla, V. C., & Kamgang, J. C. (2022). Multilevel and holonic model for dynamic holarchy management: Application to large-scale road traffic. *Engineering Applications of Artificial Intelligence*, *109*, 104622. doi:10.1016/j.engappai.2021.104622

Tubis, A. A., Poturaj, H., Dereń, K., & Żurek, A. (2024). Risks of Drone Use in Light of Literature Studies. *Sensors (Basel)*, *24*(4), 1205. doi:10.3390/s24041205 PMID:38400363

Wandelt, S., Wang, S., Zheng, C., & Sun, X. (2023). AERIAL: A Meta Review and Discussion of Challenges Toward Unmanned Aerial Vehicle Operations in Logistics, Mobility, and Monitoring. *IEEE Transactions on Intelligent Transportation Systems*, 1–14. doi:10.1109/TITS.2023.3343713

Wu, Y., Song, P., & Wang, F. (2020). Hybrid consensus algorithm optimization: A mathematical method based on POS and PBFT and its application in blockchain. *Mathematical Problems in Engineering*, *2020*, 2020. doi:10.1155/2020/7270624

Xiao, W., Li, M., Alzahrani, B., Alotaibi, R., Barnawi, A., & Ai, Q. (2021). A blockchain-based secure crowd monitoring system using UAV swarm. *IEEE Network*, *35*(1), 108–115. doi:10.1109/MNET.011.2000210

Xu, R., Wei, S., Chen, Y., Chen, G., & Pham, K. (2022). Lightman: A lightweight microchained fabric for assurance-and resilience-oriented urban air mobility networks. *Drones (Basel)*, *6*(12), 421. doi:10.3390/drones6120421

Yang, W., Wang, S., Yin, X., Wang, X., & Hu, J. (2022). A review on security issues and solutions of the internet of drones. *IEEE Open Journal of the Computer Society*, *3*, 96–110. doi:10.1109/OJCS.2022.3183003

Yim, J., Park, H., Kwon, E., Kim, S., & Lee, Y.-T. (2018). Low-power image stitching management for reducing power consumption of UAVs for disaster management system. *Paper presented at the 2018 IEEE International Conference on Consumer Electronics (ICCE)*. IEEE. 10.1109/ICCE.2018.8326248

Compilation of References

Abdalzaher, M. S., Elsayed, H. A., Fouda, M. M., & Salim, M. M. (2023). Employing Machine Learning and IoT for Earthquake Early Warning System in Smart Cities. *Energies*, *16*(1), 495. doi:10.3390/en16010495

Abdelmaboud, A. (2021). The internet of drones: Requirements, taxonomy, recent advances, and challenges of research trends. *Sensors (Basel)*, *21*(17), 5718. doi:10.3390/s21175718 PMID:34502608

Abid, F. (2021). A Survey of Machine Learning Algorithms Based Forest Fires Prediction and Detection Systems. In Fire Technology, 57(2). doi:10.1007/s10694-020-01056-z

Abid, S. K., Sulaiman, N., Chan, S. W., Nazir, U., Abid, M., Han, H., Ariza-Montes, A., & Vega-Muñoz, A. (2021). Toward an integrated disaster management approach: How artificial intelligence can boost disaster management. *Sustainability (Basel)*, *13*(22), 12560. doi:10.3390/su132212560

Abram, N. J., Abram, N. J., Henley, B. J., Gupta, A. S., Lippmann, T. J. R., Clarke, H., Dowdy, A. J., Sharples, J. J., Nolan, R. H., Zhang, T., Wooster, M. J., Wurtzel, J. B., Meissner, K. J., Pitman, A. J., Ukkola, A. M., Murphy, B. P., Tapper, N. J., & Boer, M. M. (2021). Connections of climate change and variability to large and extreme forest fires in southeast Australia. *Communications Earth & Environment*, *2*(8), 8. doi:10.1038/s43247-020-00065-8

Adakawa, M. I., Balachandran, C., Kumar, P. K., & Harinarayana, N. S. (2023). History of pandemics—A critical pathway to challenge scholarly communications? National Conference on *Exploring the Past, Present, and Future of Library and Information Science*. IEEE.

Adeel, A., Gogate, M., Farooq, S., Ieracitano, C., Dashtipour, K., Larijani, H., & Hussain, A. (2019). A survey on the role of wireless sensor networks and IoT in disaster management. *Geological disaster monitoring based on sensor networks*, 57-66.

Adeleke, A. Q., & Bamidele, A. (2020). Application of IoT in Flood Monitoring and Early Warning Systems: A Review. *International Journal of Advanced Computer Science and Applications*, *11*(3), 352–356.

Adnan, A., Ramli, M. Z., & Abd Razak, S. K. M. (2015, November). Disaster management and mitigation for earthquakes: are we ready. In *9th Asia Pacific structural engineering and construction conference (APSEC2015)* (pp. 34-44). IEEE.

Aggarwal, S., & Kumar, N. (2020). Path planning techniques for unmanned aerial vehicles: A review, solutions, and challenges. *Computer Communications, 149*, 270–299. doi:10.1016/j.comcom.2019.10.014

Agustiyara, P., Purnomo, E. P., & Ramdani, R. (2021). Using Artificial Intelligence Technique in Estimating Fire Hotspots of Forest Fires. *IOP Conference Series. Earth and Environmental Science, 717*(1), 012019. doi:10.1088/1755-1315/717/1/012019

Ahmad, K., Pogorelov, K., Riegler, M., Conci, N., & Halvorsen, P. (2017, September). CNN and GAN Based Satellite and Social Media Data Fusion for Disaster Detection. [Dublin, Ireland.]. *MediaEval, 17*, 13–15.

Aid, A., & Rassoul, I. (2017). Context-aware framework to support situation-awareness for disaster management. *International Journal of Ad Hoc and Ubiquitous Computing, 25*(3), 120–132. doi:10.1504/IJAHUC.2017.083597

Aisha, T. S., Wok, S., Manaf, A. M. A., & Ismail, R. (2015). Exploring the Use of Social Media during the 2014 Flood in Malaysia. *Procedia: Social and Behavioral Sciences, 211*, 931–937. doi:10.1016/j.sbspro.2015.11.123

Aitsi-Selmi, A., Egawa, S., Sasaki, H., Wannous, C., & Murray, V. (2015). The Sendai framework for disaster risk reduction: Renewing the global commitment to people's resilience, health, and well-being. *International Journal of Disaster Risk Science, 6*(2), 164–176. doi:10.1007/s13753-015-0050-9

Aitsi-Selmi, A., Murray, V., Wannous, C., Dickinson, C., Johnston, D., Kawasaki, A., Stevance, A.-S., & Yeung, T. (2016). Reflections on a science and technology agenda for 21st century disaster risk reduction: Based on the scientific content of the 2016 UNISDR science and technology conference on the implementation of the Sendai framework for disaster risk reduction 2015–2030. *International Journal of Disaster Risk Science, 7*(1), 1–29. doi:10.1007/s13753-016-0081-x

Akter, R., Golam, M., Doan, V.-S., Lee, J.-M., & Kim, D.-S. (2022). Iomt-net: Blockchain-integrated unauthorized uav localization using lightweight convolution neural network for internet of military things. *IEEE Internet of Things Journal, 10*(8), 6634–6651. doi:10.1109/JIOT.2022.3176310

Alavi, S. A., & Singh, S. K. (2021). A Review of Internet of Things (IoT)-Based Flood Monitoring and Alert System. *IEEE Access : Practical Innovations, Open Solutions, 9*, 57349–57367.

Ali, K., Nguyen, H. X., Shah, P., Vien, Q. T., & Ever, E. (2019). Internet of things (IoT) considerations, requirements, and architectures for disaster management system. *Performability in internet of things*, 111-125.

Alkhatib, R., Sahwan, W., Alkhatieb, A., & Schütt, B. (2023). A Brief Review of Machine Learning Algorithms in Forest Fires Science. In Applied Sciences (Switzerland), 13(14). doi:10.3390/app13148275

Allouch, A., Cheikhrouhou, O., Koubâa, A., Toumi, K., Khalgui, M., & Nguyen Gia, T. (2021). Utm-chain: Blockchain-based secure unmanned traffic management for internet of drones. *Sensors (Basel)*, *21*(9), 3049. doi:10.3390/s21093049 PMID:33925489

Alsumayt, A., El-Haggar, N., Amouri, L., Alfawaer, Z. M., & Aljameel, S. S. (2023). Smart flood detection with AI and blockchain integration in Saudi Arabia using drones. *Sensors (Basel)*, *23*(11), 5148. doi:10.3390/s23115148 PMID:37299876

Alzaylaee, M., Yerima, S., & Sezer, S. (2020). DL-Droid: Deep learning based android malware detection using real devices. *Computers & Security*, *89*, 101663. doi:10.1016/j.cose.2019.101663

Amit, S. N. K. B., & Aoki, Y. (2017). Disaster detection from aerial imagery with convolutional neural network. In *2017 International Electronics Symposium on Knowledge Creation and Intelligent Computing (IES-KCIC)*, (pp. 239–245). IEEE. 10.1109/KCIC.2017.8228593

Anisha, P. R., Kishor Kumar Reddy, C., & Marlia, M. (2023). *An intelligent deep feature based metabolism syndrome prediction system for sleep disorder diseases*. Springer Multimedia Tools and Applications., doi:10.1007/s11042-023-17296-4

Anyoha, R. (2017). The history of artificial intelligence: Can machine think? HMS. https://sitn.hms.harvard.edu/flash/2017/history-artificial-intelligence/

Arksey, H., & O'malley, L. (2005). Scoping studies: Towards a methodological framework. *International Journal of Social Research Methodology*, *8*(1), 19–32. doi:10.1080/1364557032000119616

Arslan, M., Roxin, A. M., Cruz, C., & Ginhac, D. (2017). *A review on applications of big data for disaster management*. IEEE. doi:10.1109/SITIS.2017.67

Ashraf, M. A., & Azad, M. A. K. (2015). Gender issues in disaster: Understanding the relationships of vulnerability, preparedness and capacity. *Environment and Ecology Research*, *3*(5), 136–142. doi:10.13189/eer.2015.030504

Assery, N., Xiaohong, Y., Almalki, S., Kaushik, R., & Xiuli, Q. (2019). *Comparing learning-based methods for identifying disaster-related tweets*. doi:10.1109/ICMLA.2019.00295

Ayamga, M., Akaba, S., & Nyaaba, A. A. (2021). Multifaceted applicability of drones: A review. *Technological Forecasting and Social Change*, *167*, 120677. doi:10.1016/j.techfore.2021.120677

Bagrow, J. P., Wang, D., & Barabási, A. L. (2011). Collective response of human populations to large-scale emergencies. *PLoS One*, *6*(3), e17680. doi:10.1371/journal.pone.0017680 PMID:21479206

Bail, R. D. F., Kovaleski, J. L., da Silva, V. L., Pagani, R. N., & Chiroli, D. M. D. G. (2021). Internet of things in disaster management: Technologies and uses. *Environmental Hazards*, *20*(5), 493–513. doi:10.1080/17477891.2020.1867493

Balaji, T. K., Annavarapu, C. S. R., & Bablani, A. (2021). Machine learning algorithms for social media analysis: A survey. *Computer Science Review*, *40*, 100395. doi:10.1016/j.cosrev.2021.100395

Barnabas, G., Citrawati, H., Santoso, T., & Surantha, N. (2023). Internet of Things in disaster management systems: A systematic review. *Paper presented at the AIP Conference Proceedings.* AIP. 10.1063/5.0109493

Barnes, J.M. (2021). *Theoretical foundations of community disaster resilience.* [Submitted in partial fulfillment of the requirements for the degree of Master of Arts in Security Studies (Homeland Security and Defense) from the Naval Postgraduate School].

Barzegar, A., & Roshan, G. (2019). An IoT-based Wireless Sensor Network Framework for Flood Monitoring. *International Journal of Advanced Computer Science and Applications, 10*(1), 63–68.

Basyah, N. A., Syukri, M., Fahmi, I., Ali, I., Rusli, Z., & Putri, E. S. (2023). Disaster prevention and management: A critical review of the literature. *Jurnal Penelitian Pendidikan IPA, 9*(11), 1045–1051. doi:10.29303/jppipa.v9i11.4486

Battisti, M., Belloc, F., & Del Gatto, M. (2023). *COVID-19, innovative firms and resilience.* (Economic Research Working Paper No. 73). https://www.wipo.int/edocs/pubdocs/en/wipo-pub-econstat-wp-73-en-covid-19-innovative-firms-and-resilience.pdf

Beekharry, D., & Baroudi, B. (2015). *Emergency risk management: Decision making factors and challenges when planning for disaster events.* Research Gate. https://www.researchgate.net/publication/280644351

Bengtsson, L., Lu, X., Thorson, A., Garfeld, R., & von Schreeb, J. (2011). Improved response to disasters and outbreaks by tracking population movements with mobile phone network data: A post-earthquake geospatial study in Haiti. *PLoS Medicine, 8*(8), e1001083. doi:10.1371/journal.pmed.1001083 PMID:21918643

Benzekri, W., El Moussati, A., Moussaoui, O., & Berrajaa, M. (2020). Early forest fire detection system using wireless sensor network and deep learning. *International Journal of Advanced Computer Science and Applications, 11*(5). doi:10.14569/IJACSA.2020.0110564

Bera, B., Das, A. K., & Sutrala, A. K. (2021). Private blockchain-based access control mechanism for unauthorized UAV detection and mitigation in Internet of Drones environment. *Computer Communications, 166*, 91–109. doi:10.1016/j.comcom.2020.12.005

Bera, B., Wazid, M., Das, A. K., & Rodrigues, J. J. (2021). Securing internet of drones networks using ai-envisioned smart-contract-based blockchain. *IEEE Internet of Things Magazine, 4*(4), 68–73. doi:10.1109/IOTM.001.2100044

Berawi, M. A. (2018). The role of technology in building a resilient city: Managing natural disasters. *International Journal of Technology, 5*(5), 862–865. doi:10.14716/ijtech.v9i5.2530

Bevington, J. S., Eguchi, R. T., Gill, S., Ghosh, S., & Huyck, C. K. (2015). A comprehensive analysis of building damage in the 2010 Haiti earthquake using high-resolution imagery and crowdsourcing. *Time-sensitive remote sensing*, 131-145.

Bhushan, B., Kumar, A., Agarwal, A. K., Kumar, A., Bhattacharya, P., & Kumar, A. (2023). Towards a Secure and Sustainable Internet of Medical Things (IoMT): Requirements, Design Challenges, Security Techniques, and Future Trends. *Sustainability (Basel)*, *15*(7), 6177. doi:10.3390/su15076177

Bine, L. M., Boukerche, A., Ruiz, L. B., & Loureiro, A. A. (2023). Connecting Internet of Drones and Urban Computing: Methods, protocols and applications. *Computer Networks*, 110136.

Blaikie, P. (2003). *At Risk – Natural hazards, people's vulnerability, and disasters*. Research Gate.

Borah, A. R., Ravi, N., Narayan, N., & D'Souza, P. J. (2024). *A Methodology for Forest Fire Detection and Notification Using AI and IoT Approaches*. 2024 2nd International Conference on Intelligent Data Communication Technologies and Internet of Things (IDCIoT), Bengaluru, India. 10.1109/IDCIoT59759.2024.10467597

Born, C. T., Briggs, S. M., Ciraulo, D. L., Frykberg, E. R., Hammond, J. S., Hirshberg, A., Lhowe, D. W., & O'Neill, P. A. (2007). Disasters and Mass Casualties: I. General Principles of Response and Management. *The Journal of the American Academy of Orthopaedic Surgeons*, *15*(7), 388–396. doi:10.5435/00124635-200707000-00004 PMID:17602028

Botzen, W. W., Deschenes, O., & Sanders, M. (2019). The economic impacts of natural disasters: A review of models and empirical studies. *Review of Environmental Economics and Policy*, *13*(2), 167–188. doi:10.1093/reep/rez004

Boulouard, Z., Ouaissa, M., Ouaissa, M., Krichen, M., Almutiq, M., & Algarni, M. (2023, December). Streamlining River Flood Prevention with an Integrated AIoT Framework. In *2023 20th ACS/IEEE International Conference on Computer Systems and Applications (AICCSA)* (pp. 1-6). IEEE. 10.1109/AICCSA59173.2023.10479264

Boulouard, Z., Ouaissa, M., Ouaissa, M., Krichen, M., Almutiq, M., & Gasmi, K. (2022). Detecting hateful and offensive speech in arabic social media using transfer learning. *Applied Sciences (Basel, Switzerland)*, *12*(24), 12823. doi:10.3390/app122412823

Boulouard, Z., Ouaissa, M., Ouaissa, M., Siddiqui, F., Almutiq, M., & Krichen, M. (2022). An integrated artificial intelligence of things environment for river flood prevention. *Sensors (Basel)*, *22*(23), 9485. doi:10.3390/s22239485 PMID:36502187

Bradshaw, S. (2004). *Socio-economic impacts of natural disasters: A gender analysis*.

Britannica. (2021). *Kōbe earthquake of 1995*. Encyclopedia Britannica. https://www.britannica.com/event/Kobe-earthquake-of-1995

Buchholz, M. (2019). Deep reinforcement learning. introduction. deep q network (dqn) algorithm. Research Gate.

Burger, A., Kennedy, W. G., & Crooks, A. (2021). Organizing theories for disasters into a complex adaptive system framework. *Urban Science (Basel, Switzerland)*, *5*(61), 61. Advance online publication. doi:10.3390/urbansci5030061

Bushnaq, O. M., Mishra, D., Natalizio, E., & Akyildiz, I. F. (2022). Unmanned aerial vehicles (UAVs) for disaster management. In *Nanotechnology-Based Smart Remote Sensing Networks for Disaster Prevention* (pp. 159–188). Elsevier. doi:10.1016/B978-0-323-91166-5.00013-6

Caragea, C., Squicciarini, A., Stehle, S., Neppalli, K., & Tapia, A. (2014). Mapping moods: geo-mapped sentiment analysis during Hurricane Sandy. In: *Proceedings of the 11th international conference on information systems for crisis response and management (ISCRAM 2014)*. ISCRAM.

Castelli, M., Vanneschi, L., & Popovič, A. (2015). Predicting burned areas of forest fires: An artificial intelligence approach. *Fire Ecology*, *11*(1), 106–118. doi:10.4996/fireecology.1101106

Chai, J., & Wu, H. Z. (2023). Prevention/mitigation of natural disasters in urban areas. *Smart Construction and Sustainable Cities*, *1*(1), 4. doi:10.1007/s44268-023-00002-6

Chamola, V., Hassija, V., Gupta, S., Goyal, A., Guizani, M., & Sikdar, B. (2020). Disaster and pandemic management using machine learning: A survey. *IEEE Internet of Things Journal*, *8*(21), 16047–16071. doi:10.1109/JIOT.2020.3044966 PMID:35782181

Cheema-Fox, A., LaPerla, B. R., Serafeim, G., & Wang, H. S. (2020). *Corporate resilience and response during COVID-19*. (Working Paper 20-108).

Chen, K., Wu, Z., Liu, H., & Guo, X. (2020). A Novel Distributed IoT Framework for Flood Early Warning Based on Edge Computing and Deep Learning. *IEEE Access : Practical Innovations, Open Solutions*, *8*, 20713–20724.

Chitra, U. T. (2023). *Forest Fire Detection System: 2023 Intelligent Computing and Control for Engineering and Business Systems*. ICCEBS., doi:10.1109/ICCEBS58601.2023.10448648

Ci, T., Liu, Z., & Wang, Y. (2019). Assessment of the degree of building damage caused by disaster using convolutional neural networks in combination with ordinal regression. *Remote Sensing (Basel)*, *11*(23), 2858. doi:10.3390/rs11232858

Clark, D. G., Ford, J. D., & Tabish, T. (2018). What role can unmanned aerial vehicles play in emergency response in the Arctic: A case study from Canada. *PLoS One*, *13*(12), e0205299. doi:10.1371/journal.pone.0205299 PMID:30562340

Coastal Resilience Centre. (2021). *Support strategies for socially marginalized neighborhoods likely impacted by natural hazards*. The University of North Carolina at Chapel Hill. https://coastalresiliencecenter.unc.edu/wp-content/uploads/sites/845/2021/07/Support-Strategies-for-Socially-Marginalized-Neighborhoods.pdf

Coman, C.-M., Toma, B., Constantin, M.-A., & Florescu, A. (2022). Ground Level Lidar as a Contributing Indicator in an Environmental Protection Application. SSRN Electronic Journal. doi:10.2139/ssrn.4096563

Cooner, A., Shao, Y., & Campbell, J. (2016). Detection of urban damage using remote sensing and machine learning algorithms: Revisiting the 2010 Haiti earthquake. *Remote Sensing (Basel)*, *8*(10), 868–885. doi:10.3390/rs8100868

Copeland, B. J. (2020). *What is artificial intelligence?* Alan Turning. https://www.alanturing.net/turing_archive/pages/Reference%20Articles/What%20is%20AI.html

Coppola, D. (2015). Introduction to International Disaster Management. Butterworth-Heinemann. Elsevier.

Costa-Saura, J. M., Balaguer-Beser, Á., Ruiz, L. A., Pardo-Pascual, J. E., & Soriano-Sancho, J. L. (2021). Empirical Models for Spatio-Temporal Live Fuel Moisture Content Estimation in Mixed Mediterranean Vegetation Areas Using Sentinel-2 Indices and Meteorological Data. *Remote Sensing (Basel)*, *13*(18), 3726. doi:10.3390/rs13183726

Costine, K. J. (2015). *The four fundamental theories of disasters*. Research Gate. https://www.researchgate.net/publication/273694762_the_four_fundamental_theories_of_disasters

Curtis, A., Duval-Diop, D., & Novak, J. (2010). Identifying spatial patterns of recovery and abandonment in the post-Katrina Holy Cross neighborhood of New Orleans. *Cartography and Geographic Information Science*, *37*(1), 45–56. doi:10.1559/152304010790588043

Cuthbertson, J., & Penney, G. (2023). Ethical decision making in disaster and emergency management: A systematic review of the literature. *Prehospital and Disaster Medicine*, *00*(00), 1–6. doi:10.1017/S1049023X23006325 PMID:37675490

Das, A., & Singh, S. K. (2021). IoT-Based Smart Water Management System for Flood Monitoring and Early Warning. *IEEE Sensors Journal*, *21*(10), 12595–12604.

Daud, S. M. S. M., Yusof, M. Y. P. M., Heo, C. C., Khoo, L. S., Singh, M. K. C., Mahmood, M. S., & Nawawi, H. (2022). Applications of drone in disaster management: A scoping review. *Science & Justice*, *62*(1), 30–42. doi:10.1016/j.scijus.2021.11.002 PMID:35033326

de la Llera, J. C., Rivera, F., Gil, M., & Schwarzhaupt, Ú. (2018). Mitigating risk through R&D+ innovation: Chile's national strategy for disaster resilience. *16ᵗʰ European Conference on Earthquake Engineering*. Research Gate.

Deo, R. C., & Sahin, M. (2015). An extreme learning machine model for the simulation of monthly effective drought index in eastern Australia. *Environmental Modelling & Software*, *67*, 144–159.

Derhab, A., Cheikhrouhou, O., Allouch, A., Koubaa, A., Qureshi, B., Ferrag, M. A., Maglaras, L., & Khan, F. A. (2023). Internet of drones security: Taxonomies, open issues, and future directions. *Vehicular Communications*, *39*, 100552. doi:10.1016/j.vehcom.2022.100552

Dhall, A., Dhasade, A., & Nalwade, A., VK, M. R., & Kulkarni, V. (2020). A survey on systematic approaches in managing forest fires. *Applied Geography (Sevenoaks, England)*, *121*, 102266. doi:10.1016/j.apgeog.2020.102266

Duarte, D., Nex, F., Kerle, N., & Vosselman, G. (2018). Satellite image classification of building damages using airborne and satellite image samples in a deep learning approach. *ISPRS Annals of the Photogrammetry, Remote Sensing and Spatial Information Sciences*, *4*(2), 89–96. doi:10.5194/isprs-annals-IV-2-89-2018

Ebi, K. L., Vanos, J., Baldwin, J. W., Bell, J. E., Hondula, D. M., Errett, N. A., Hayes, K., Reid, C. E., Saha, S., Spector, J., & Berry, P. (2021). Extreme weather and climate change: Population health and health system implications. *Annual Review of Public Health*, *42*(1), 293–315. doi:10.1146/annurev-publhealth-012420-105026 PMID:33406378

Edureka, (2020). *Top 15 hot artificial intelligence technologies*. Edureka. https://www.edureka.co/blog/top-15-hot-artificial-intelligence-technologies/

Ejaz, W., Ahmed, A., Mushtaq, A., & Ibnkahla, M. (2020). Energy-efficient task scheduling and physiological assessment in disaster management using UAV-assisted networks. *Computer Communications*, *155*, 150–157. doi:10.1016/j.comcom.2020.03.019

Ejaz, W., Azam, M. A., Saadat, S., Iqbal, F., & Hanan, A. (2019). Unmanned aerial vehicles enabled IoT platform for disaster management. *Energies*, *12*(14), 2706. doi:10.3390/en12142706

Emergency Database (EM-DAT). (2024). *Inventorying hazards and disasters worldwide since 1988*. EM-DAT. https://www.emdat.be

Erdelj, M., Natalizio, E., Chowdhury, K. R., & Akyildiz, I. F. (2017). Help from the sky: Leveraging UAVs for disaster management. *IEEE Pervasive Computing*, *16*(1), 24–32. doi:10.1109/MPRV.2017.11

Fan, C., Zhang, C., Yahja, A., & Mostafavi, A. (2021). Disaster City Digital Twin: A vision for integrating artificial and human intelligence for disaster management. *International Journal of Information Management*, *56*, 102049. doi:10.1016/j.ijinfomgt.2019.102049

Feizizadeh, B., Omarzadeh, D., Mohammadnejad, V., Khallaghi, H., Sharifi, A., & Karkarg, B. G. (2023). An integrated approach of artificial intelligence and geoinformation techniques applied to forest fire risk modeling in Gachsaran, Iran. *Journal of Environmental Planning and Management*, *66*(6), 1369–1391. doi:10.1080/09640568.2022.2027747

Feng, L., & Hu, F. (2019). Flood early warning model based on LSTM neural network and IoT. In *2019 9th International Conference on Electronics Information and Emergency Communication (ICEIEC)* (pp. 279-282). IEEE.

Finucane, M. L., Acosta, J., Wicker, A., & Whipkey, K. (2020). Short-Term Solutions to a Long-Term Challenge: Rethinking Disaster Recovery Planning to Reduce Vulnerabilities and Inequities. *International Journal of Environmental Research and Public Health*, *2020*(17), 482. doi:10.3390/ijerph17020482 PMID:31940859

Floreano, D., & Wood, R. (2015). Science, technology and the future of small autonomous drones. *Nature*, *521*(7553), 460–466. doi:10.1038/nature14542 PMID:26017445

Gagne, D. J. II, Haupt, S. E., Nychka, D. W., & Thompson, G. (2019). Interpretable deep learning for spatial analysis of severe hailstorms. *Monthly Weather Review*, *147*(8), 2827–2845. doi:10.1175/MWR-D-18-0316.1

Gagne, D. J. II, McGovern, A., Haupt, S. E., Sobash, R. A., Williams, J. K., & Xue, M. (2017). Storm-based probabilistic hail forecasting with machine learning applied to convection-allowing ensembles. *Weather and Forecasting*, *32*(5), 1819–1840. doi:10.1175/WAF-D-17-0010.1

Gao, X., Wang, J., Li, L., Shen, C., & Xu, J. (2020). A Framework for Internet of Things-Based Flood Monitoring and Emergency Response in Rural Areas. *IEEE Access : Practical Innovations, Open Solutions*, *8*, 194876–194886.

Gao, Y., Chen, Y. X., Ding, Y. S., & Tang, B. Y. (2006). Immune genetic algorithm based on network model for food disaster evaluation. *J Nat Disaster*, *15*, 110–114.

Gao, Z., Chen, H., Cao, Y., Li, J., & Li, C. (2021). An Integrated Flood Monitoring System Based on IoT, Satellite Remote Sensing, and Data Fusion Techniques. *IEEE Access : Practical Innovations, Open Solutions*, *9*, 29952–29966.

García-Magariño, I., Lacuesta, R., Rajarajan, M., & Lloret, J. (2019). Security in networks of unmanned aerial vehicles for surveillance with an agent-based approach inspired by the principles of blockchain. *Ad Hoc Networks*, *86*, 72–82. doi:10.1016/j.adhoc.2018.11.010

Gharibi, M., Boutaba, R., & Waslander, S. L. (2016). Internet of drones. *IEEE Access : Practical Innovations, Open Solutions*, *4*, 1148–1162. doi:10.1109/ACCESS.2016.2537208

Ghasemi, P., & Karimian, N. (2020, April). A qualitative study of various aspects of the application of IoT in disaster management. In *2020 6th International Conference on Web Research (ICWR)* (pp. 77-83). IEEE. 10.1109/ICWR49608.2020.9122323

Ghorbanzadeh, O., Liaghat, A. M., Nazari, B., Bahmanyar, M. A., & Bahiraei, M. (2020). A review on water level sensors in irrigation: Challenges and perspectives. *Computers and Electronics in Agriculture*, *175*, 105593.

Gilani, S. M., Anjum, A., Khan, A., Syed, M. H., Moqurrab, S. A., & Srivastava, G. (2024). A robust Internet of Drones security surveillance communication network based on IOTA. *Internet of Things : Engineering Cyber Physical Human Systems*, *25*, 101066. doi:10.1016/j.iot.2024.101066

Gilbert, S. W. (2010). *Disaster resilience: A guide to the literature*. National Institute of Standards and Technology Building and Fire Research Laboratory. https://www.govinfo.gov/content/pkg/GOVPUB-C13-724c6a83323886beabf2a2910a8fb81e/pdf/GOVPUB-C13-724c6a83323886beabf2a2910a8fb81e.pdf

Glantz, E. J., Ritter, F. E., Gilbreath, D., Stager, S. J., Anton, A., & Emani, R. (2020). UAV use in disaster management. In *Proceedings of the 17th ISCRAM Conference* (pp. 914-921). IEEE.

Goodchild, M. F., & Glennon, J. A. (2010). Crowdsourcing geographic information for disaster response: A research frontier. *International Journal of Digital Earth*, *3*(3), 231–241. doi:10.1080/17538941003759255

Goyal, H. R., Ghanshala, K. K., & Sharma, S. (2021). Post flood management system based on smart IoT devices using AI approach. *Materials Today: Proceedings*, *46*, 10411–10417. doi:10.1016/j.matpr.2020.12.947

Greifeneder, R., Bless, H., & Pham, M. (2011). When do people rely on afective and cognitive feelings in judgment? A review. *Personality and Social Psychology Review, 15*(2), 107–141. doi:10.1177/1088868310367640 PMID:20495111

Groumpos, P. P. (2023). A Critical Historic Overview of Artificial Intelligence: Issues, Challenges, Opportunities, and Threats. In *Artificial Intelligence and Applications, 1*(4). https://ojs.bonviewpress.com/index.php/AIA/article/view/689/580

Gubbi, J., Buyya, R., Marusic, S., & Palaniswami, M. (2013). Internet of Things (IoT): A vision, architectural elements, and future directions. *Future Generation Computer Systems, 29*(7), 1645–1660. doi:10.1016/j.future.2013.01.010

Guo, C., & Lin, X. (2019). A Deep Learning-Based Flood Early Warning System Using Internet of Things Sensors. In *2019 6th International Conference on Systems and Informatics (ICSAI)* (pp. 44-49). IEEE.

Gupta, S. G., Ghonge, D. M., & Jawandhiya, P. M. (2013). Review of unmanned aircraft system (UAS). [IJARCET]. *International Journal of Advanced Research in Computer Engineering and Technology, 2.*

Hafeez, S., Manzoor, H. U., Mohjazi, L., Zoha, A., Imran, M. A., & Sun, Y. (2024). Blockchain-Empowered Immutable and Reliable Delivery Service (BIRDS) Using UAV Networks. *arXiv preprint arXiv:2403.12060.*

Hafeez, S., Khan, A. R., Al-Quraan, M., Mohjazi, L., Zoha, A., Imran, M. A., & Sun, Y. (2023). Blockchain-assisted UAV communication systems: A comprehensive survey. *IEEE Open Journal of Vehicular Technology, 4*, 558–580. doi:10.1109/OJVT.2023.3295208

Harbi, Y., Medani, K., Gherbi, C., Senouci, O., Aliouat, Z., & Harous, S. (2023). A systematic literature review of blockchain technology for internet of drones security. *Arabian Journal for Science and Engineering, 48*(2), 1053–1074. doi:10.1007/s13369-022-07380-6 PMID:36337772

Hassanalian, M., & Abdelkefi, A. (2017). Classifications, applications, and design challenges of drones: A review. *Progress in Aerospace Sciences, 91*, 99–131. doi:10.1016/j.paerosci.2017.04.003

Hayajneh, A. M., Zaidi, S. A. R., McLernon, D. C., Di Renzo, M., & Ghogho, M. (2018). Performance analysis of UAV enabled disaster recovery networks: A stochastic geometric framework based on cluster processes. *IEEE Access : Practical Innovations, Open Solutions, 6*, 26215–26230. doi:10.1109/ACCESS.2018.2835638

Hecker, N., & Domres, B. D. (2018). The German emergency and disaster medicine and management system—history and present. *Chinese Journal of Traumatology, 21*(2), 64–72. doi:10.1016/j.cjtee.2017.09.003 PMID:29622286

Heidari, A., Jafari Navimipour, N., Unal, M., & Zhang, G. (2023). Machine learning applications in internet-of-drones: Systematic review, recent deployments, and open issues. *ACM Computing Surveys, 55*(12), 1–45. doi:10.1145/3571728

Hu, G., Rao, K., & Sun, Z. (2006). A preliminary framework to measure public health emergency response capacity. *Zeitschrift für Gesundheitswissenschaften, 14*(1), 43–47. doi:10.1007/s10389-005-0008-2

Hughes, A. A. (2023). Unmanned aerial vehicle use in humanitarian activities. *University of Lynchburg DMSc Doctoral Project Assignment Repository, 5*(1), 137.

Husman, M. A., Albattah, W., Abidin, Z. Z., Mustafah, Y. M., Kadir, K., Habib, S., Islam, M., & Khan, S. (2021). Unmanned aerial vehicles for crowd monitoring and analysis. *Electronics (Basel), 10*(23), 2974. doi:10.3390/electronics10232974

Hu, Z., Chen, L., & Sun, X. (2020). An Intelligent Flood Disaster Prediction Model Based on the Internet of Things and Deep Learning. *IEEE Access : Practical Innovations, Open Solutions, 8*, 12741–12752.

Inan, D. I., Beydoun, G., & Othman, S. H. (2023). Risk assessment and sustainable disaster management. *Sustainability (Basel), 15*(6), 1–5. doi:10.3390/su15065254

Independent Commission Multilateralism (ICM) & International Peace Institute (IPI) (2016). *Discussion paper: Humanitarian engagements-Independent Commission on Multilateralism.* https://reliefweb.int/report/world/discussion-paper-humanitarian-engagements-independent-commission-multilateralism

Islam, A., & Shin, S. Y. (2019). BHMUS: Blockchain based secure outdoor health monitoring scheme using UAV in smart city. *Paper presented at the 2019 7th international conference on information and communication technology (ICoICT).* IEEE. 10.1109/ICoICT.2019.8835373

Islam, A., & Shin, S. Y. (2019). Bus: A blockchain-enabled data acquisition scheme with the assistance of uav swarm in internet of things. *IEEE Access : Practical Innovations, Open Solutions, 7*, 103231–103249. doi:10.1109/ACCESS.2019.2930774

Israel, D. C., & Briones, R. M. (2012). Impacts of Natural Disasters on Agriculture, Food Security, and Natural Resources and Environment in the Philippines. In Y. Sawada & S. Oum (Eds.), *Economic and Welfare Impacts of Disasters in East Asia and Policy Responses. ERIA Research Project Report 2011-8* (pp. 553–599). ERIA.

Isser, R. (2022). *Disaster management: Theory, planning and preparedness.* CAPS. https://capsindia.org/wp-content/uploads/2022/08/Rajesh-Isser-1.pdf

Jameii, S. M., Zamirnaddafi, R. S., & Rezabakhsh, R. (2022). Internet of Flying Things security: A systematic review. *Concurrency and Computation, 34*(24), e7213. doi:10.1002/cpe.7213

Ji, H.-K., Kim, S.-W., & Kil, G.-S. (2020). Phase Analysis of Series Arc Signals for Low-Voltage Electrical Devices. *Energies, 13*(20), 5481. doi:10.3390/en13205481

Jiménez López, J., & Mulero-Pázmány, M. (2019). Drones for conservation in protected areas: Present and future. *Drones (Basel), 3*(1), 10. doi:10.3390/drones3010010

Ji, Y., Wang, D., Li, Q., Liu, T., & Bai, Y. (2024). Global Wildfire Danger Predictions Based on Deep Learning Taking into Account Static and Dynamic Variables. *Forests*, *15*(1), 216. doi:10.3390/f15010216

Jo, K., Heo, J., Jung, J., Kim, B., & Min, H. (2017). A rendezvous point estimation considering drone speed and data collection delay. *Paper presented at the 2017 4th International Conference on Computer Applications and Information Processing Technology (CAIPT)*. IEEE. 10.1109/CAIPT.2017.8320706

Justino, F. B., Bromwich, D. H., Schumacher, V., daSilva, A., & Wang, S. (2022). Arctic Oscillation and Pacific-North American pattern dominated-modulation of fire danger and wildfire occurrence. *NPJ Climate and Atmospheric Science*, *5*(1), 1–13. doi:10.1038/s41612-022-00274-2

Kalinaki, K., Malik, O. A., & Lai, D. T. C. (2023). FCD-AttResU-Net: An improved forest change detection in Sentinel-2 satellite images using attention residual U-Net. *International Journal of Applied Earth Observation and Geoinformation*, *122*, 103453. doi:10.1016/j.jag.2023.103453

Kalinaki, K., Malik, O. A., Lai, D. T. C., Sukri, R. S., & Wahab, R. B. H. A. (2023). Spatial-temporal mapping of forest vegetation cover changes along highways in Brunei using deep learning techniques and Sentinel-2 images. *Ecological Informatics*, *77*, 102193. doi:10.1016/j.ecoinf.2023.102193

Kamangir, H., Collins, W., Tissot, P., & King, S. A. (2020). A deep-learning model to predict thunderstorms within 400 km^2 South Texas domains. *Meteorological Applications*, *27*(2), e1905. doi:10.1002/met.1905

Kamilaris, A., Fonts, A., & Prenafeta-Boldú, F. X. (2019). The rise of blockchain technology in agriculture and food supply chains. *Trends in Food Science & Technology*, *91*, 640–652. doi:10.1016/j.tifs.2019.07.034

Kamruzzaman, M. D., Sarkar, N. I., Gutierrez, J., & Ray, S. K. (2017, January). A study of IoT-based post-disaster management. In *2017 international conference on information networking (ICOIN)* (pp. 406-410). IEEE.

Kapucu, N., & Liou, K. T. (2014). Disaster and development: Examining global issues and cases. In N. Kapucu & K. T. Liou (Eds.), *Disaster and development, environmental hazards*. Springer International Publishing Switzerland. doi:10.1007/978-3-319-04468-2

Karaca, Y., Cicek, M., Tatli, O., Sahin, A., Pasli, S., Beser, M. F., & Turedi, S. (2018). The potential use of unmanned aircraft systems (drones) in mountain search and rescue operations. *The American Journal of Emergency Medicine*, *36*(4), 583–588. doi:10.1016/j.ajem.2017.09.025 PMID:28928001

Karpatne, R. (2019). Machine learning for the geosciences: Challenges and opportunities. *IEEE Trans. Knowledge. Data Eng.*

Kaur, K., Singh, S., & Kaur, R. (2021). A review of IoT-based systems for monitoring and controlling soil irrigation. *Environmental Science and Pollution Research International, 28*(25), 31851–31866.

Keating, A., Campbell, K., Mechler, R., Magnuszewski, P., Mochizuki, J., Liu, W., Szoenyi, M., & McQuistan, C. (2017). Disaster resilience: What it is and how it can engender a meaningful change in development policy. *Development Policy Review, 35*(1), 65–91. doi:10.1111/dpr.12201

Kedys, J., Tchappi, I., & Najjar, A. (2024). UAVs for Disaster Management-An Exploratory Review. *Procedia Computer Science, 231*, 129–136. doi:10.1016/j.procs.2023.12.184

Kettridge, N., Turetsky, M., Sherwood, J., Thompson, D. K., Miller, C. A., Benscoter, B. W., Flannigan, M. D., Wotton, B. M., & Waddington, J. M. (2015). Moderate drop in water table increases peatland vulnerability to post-fire regime shift. *Scientific Reports, 5*(1), 8063. doi:10.1038/srep08063 PMID:25623290

Khaloo, A., Lattanzi, D., Cunningham, K., Dell'Andream, R., & Riley, M. (2017). Unmanned aerial vehicle inspection of the Placer River trail bridge through image-based 3D modelling. *Structure and Infrastructure Engineering, 14*(1), 124–136. doi:10.1080/15732479.2017.1330891

Khan, A., Hassan, B., Khan, S., Ahmed, R., & Abuassba, A. (2022). DeepFire: A Novel Dataset and Deep Transfer Learning Benchmark for Forest Fire Detection. *Mobile Information Systems, 2022*, 1–14. doi:10.1155/2022/5358359

Khan, H., Vasilescu, L. G., & Khan, A. (2008). Disaster management cycle – A theoretical approach. *Management and Marketing Journal, 6*(1), 43–50.

Khirani, S., Souahlia, A., & Rabehi, A. (2023). *Forest Fire Detection Using a Low Complex Convolutional Neural Network.* 2023 2nd International Conference on Electronics, Energy and Measurement (IC2EM), Medea, Algeria. 10.1109/IC2EM59347.2023.10453317

Kia, M. B., Pirasteh, S., Pradhan, B., Mahmud, A. R., Sulaiman, W. N. A., & Moradi, A. (2012). An artificial neural network model for flood simulation using GIS: Johor River Basin, Malaysia. *Environmental Earth Sciences, 67*(1), 251–264. doi:10.1007/s12665-011-1504-z

Kieslich, K., Lünich, M., & Marcinkowski, F. (2021). The threats of artificial intelligence scale (TAI) development, measurement and test over three application domains. *International Journal of Social Robotics, 13*(7), 1563–1577. doi:10.1007/s12369-020-00734-w

Kim, S. S., Shin, D. Y., Lim, E. T., Jung, Y. H., & Cho, S. B. (2022). Disaster Damage Investigation using Artificial Intelligence and Drone Mapping. *The International Archives of the Photogrammetry, Remote Sensing and Spatial Information Sciences, 43*, 1109–1114. doi:10.5194/isprs-archives-XLIII-B3-2022-1109-2022

Kim, Y.-k., & Sohn, H.-G. (2018). Disaster theory. In Y.-k. Kim & H.-G. Sohn (Eds.), *Disaster risk management in the Republic of Korea.* Disaster Risk Reduction., doi:10.1007/978-981-10-4789-3_2

Kinaneva, D., Hristov, G., Raychev, J., & Zahariev, P. (2019). Early forest fire detection using drones and artificial intelligence. *2019 42nd International Convention on Information and Communication Technology, Electronics and Microelectronics, MIPRO 2019 - Proceedings*. IEEE. 10.23919/MIPRO.2019.8756696

Kishor Kumar Reddy, C., Anisha, P. R., & Marlia Mohd Hanafah, Y. V. S. S. (2023). *An intelligent optimized cyclone intensity prediction framework using satellite images*. Springer Earth Science Informatics. doi:10.1007/s12145-023-00983-z

Konings, A. G., Saatchi, S. S., Frankenberg, C., Keller, M., Leshyk, V., Anderegg, W. R. L., Humphrey, V., Matheny, A. M., Trugman, A., Sack, L., Agee, E., Barnes, M. L., Binks, O., Cawse-Nicholson, K., Christoffersen, B. O., Entekhabi, D., Gentine, P., Holtzman, N. M., Katul, G. G., & Zuidema, P. A. (2021). Detecting forest response to droughts with global observations of vegetation water content. *Global Change Biology, 27*(23), 6005–6024. doi:10.1111/gcb.15872 PMID:34478589

Kousky, C. (2016). Impacts of natural disasters on children. *The Future of Children, 26*(1), 73–92. doi:10.1353/foc.2016.0004

Kucuk, O., & Sevinc, V. (2023). Fire behavior prediction with artificial intelligence in thinned black pine (Pinus nigra Arnold) stand. *Forest Ecology and Management, 529*, 120707. doi:10.1016/j.foreco.2022.120707

Kuglitsch, M. M., Pelivan, I., Ceola, S., Menon, M., & Xoplaki, E. (2022). Facilitating adoption of AI in natural disaster management through collaboration. *Nature Communications, 13*(1), 1579. doi:10.1038/s41467-022-29285-6 PMID:35332147

Kumar, M. S., Vimal, S., Jhanjhi, N., Dhanabalan, S. S., & Alhumyani, H. A. (2021). Blockchain based peer to peer communication in autonomous drone operation. *Energy Reports, 7*, 7925–7939. doi:10.1016/j.egyr.2021.08.073

Kumar, R. L., Pham, Q.-V., Khan, F., Piran, M. J., & Dev, K. (2021). Blockchain for securing aerial communications: Potentials, solutions, and research directions. *Physical Communication, 47*, 101390. doi:10.1016/j.phycom.2021.101390

Kumar, S., Jadon, P., Sharma, L., Bhushan, B., & Obaid, A. J. (2023). Decentralized Blockchain Technology for the Development of IoT-Based Smart City Applications. In D. K. Sharma, R. Sharma, G. Jeon, & Z. Polkowski (Eds.), *Low Power Architectures for IoT Applications. Springer Tracts in Electrical and Electronics Engineering*. Springer. doi:10.1007/978-981-99-0639-0_13

Labib, N. S., Brust, M. R., Danoy, G., & Bouvry, P. (2021). The rise of drones in internet of things: A survey on the evolution, prospects and challenges of unmanned aerial vehicles. *IEEE Access : Practical Innovations, Open Solutions, 9*, 115466–115487. doi:10.1109/ACCESS.2021.3104963

Le, P., Wang, C., & Chai, K. (2018). Using wireless sensor networks for flash flood early warning system: challenges and opportunities. In *2018 10th International Conference on Information Technology in Medicine and Education (ITME)* (pp. 55-59). IEEE.

Lee, S., & Kim, S. (2020). A Flood Early Warning System Based on IoT and Deep Learning Technologies. *Sustainability*, *12*(6), 2432.

Levine, N. M., & Spencer, B. F. Jr. (2022). Post-Earthquake Building Evaluation Using UAVs: A BIM-Based Digital Twin Framework. *Sensors (Basel)*, *22*(3), 873. doi:10.3390/s22030873 PMID:35161619

Li, Y., Liu, Y., & Lv, Y. (2018). Flood monitoring and early warning system based on internet of things (IoT) and machine learning. In *2018 IEEE 3rd Advanced Information Technology, Electronic and Automation Control Conference (IAEAC)* (pp. 1494-1498). IEEE.

Li, D., & Zhang, C. (2021). A Novel Framework for Flood Forecasting Based on LSTM Neural Network and IoT Sensors. *IEEE Transactions on Industrial Informatics*, *17*(1), 532–540.

Li, J., & Rao, H. (2010). Twitter as a rapid response news service: An exploration in the context of the 2008 China earthquake. *The Electronic Journal on Information Systems in Developing Countries*, *42*(1), 1–22. doi:10.1002/j.1681-4835.2010.tb00300.x

Li, K., Lau, W. F., Au, M. H., Ho, I. W.-H., & Wang, Y. (2020). Efficient message authentication with revocation transparency using blockchain for vehicular networks. *Computers & Electrical Engineering*, *86*, 106721. doi:10.1016/j.compeleceng.2020.106721

Lim, G. J., Kim, S., Cho, J., Gong, Y., & Khodaei, A. (2016). Multi-UAV pre-positioning and routing for power network damage assessment. *IEEE Transactions on Smart Grid*, *9*(4), 3643–3651. doi:10.1109/TSG.2016.2637408

Lin, T. Z. (2022). Suppressing Forest Fires in Global Climate Change Through Artificial Intelligence: A Case Study on British Columbia. *Proceedings - 2022 International Conference on Big Data, Information and Computer Network, BDICN 2022*. IEEE. 10.1109/BDICN55575.2022.00086

Linardos, V., Drakaki, M., Tzionas, P., & Karnavas, Y. L. (2022). Machine learning in disaster management: recent developments in methods and applications. *Machine Learning and Knowledge Extraction*, *4*(2).

Lin, C., He, D., Kumar, N., Choo, K.-K. R., Vinel, A., & Huang, X. (2018). Security and privacy for the internet of drones: Challenges and solutions. *IEEE Communications Magazine*, *56*(1), 64–69. doi:10.1109/MCOM.2017.1700390

Lin, X., Li, Z., Chen, W., Sun, X., & Gao, D. (2023). Forest Fire Prediction Based on Long- and Short-Term Time-Series Network. *Forests*, *14*(4), 778. doi:10.3390/f14040778

Lin, Y. R. (2015). Event-related crowd activities on social media. In B. Gonçalves & N. Perra (Eds.), *Social phenomena* (pp. 235–250). Springer. doi:10.1007/978-3-319-14011-7_12

Liu, X., & Ansari, N. (2018). Resource allocation in UAV-assisted M2M communications for disaster rescue. *IEEE Wireless Communications Letters*, *8*(2), 580–583. doi:10.1109/LWC.2018.2880467

Liu, Y., Liu, S., Sun, X., Wang, K., Wang, X., & Li, J. (2020). A Deep Learning-Based IoT Framework for Flood Monitoring and Prediction. *IEEE Transactions on Industrial Informatics*, *16*(3), 1932–1941.

Li, X., Sun, X., Gu, W., Lin, Z., Liu, Y., & Yang, W. (2021). An IoT-enabled Flood Monitoring System for Disaster Early Warning. *IEEE Access : Practical Innovations, Open Solutions*, *9*, 7865–7875.

Li, Y., Li, K., Wang, K., Li, G., & Wang, Z. (2023). An Improved Stacking Model for Forest Fire Susceptibility Prediction in Chongqing City, China. *2023 IEEE Smart World Congress (SWC)*, Portsmouth, United Kingdom. 10.1109/SWC57546.2023.10449165

Li, Z., Chen, Q., Li, J., Huang, J., Mo, W., Wong, D. S., & Jiang, H. (2024). A secure and efficient UAV network defense strategy: Convergence of blockchain and deep learning. *Computer Standards & Interfaces*, *90*, 103844. doi:10.1016/j.csi.2024.103844

Lord, F. M. (1951). *A theory of test scores and their relation to the trait measured (Research Bulletin No. RB-51-13)*. Princeton: Educational Testing Service. doi:10.1002/j.2333-8504.1951.tb00922.x

Lord, F. M. (1952). *A theory of test scores (Psychometric Monograph No. 7)*. Psychometric Corporation.

Lord, F. M. (1953). The relation of test score to the trait underlying the test. *Educational and Psychological Measurement*, *13*(4), 517–549. doi:10.1177/001316445301300401

Lu, J., Zhan, Q., Zou, Z., & Wu, D. (2020). Development of a Low-Cost IoT-Based System for Real-Time Flood Monitoring in an Agricultural Field. *IEEE Access : Practical Innovations, Open Solutions*, *8*, 97636–97645.

Macias, M., Barrado, C., Pastor, E., & Royo, P. (2019). The future of drones and their public acceptance. In 2019 IEEE/AIAA 38th Digital Avionics Systems Conference (DASC) (pp. 1-8). IEEE. 10.1109/DASC43569.2019.9081623

Madichetty, S., & Sridevi, M. (2019). Detecting informative tweets during disaster using deep neural networks. In *2019 11th international conference on communication systems & networks (COMSNETS)* (pp. 709-713). IEEE. 10.1109/COMSNETS.2019.8711095

Mahaveerakannan, R., Anitha, C., & Aby, K. (2023). An IoT based forest fire detection system using integration of cat swarm with LSTM model. *Computer Communications*, *211*, 37–45. doi:10.1016/j.comcom.2023.08.020

Mandirola, M., Casarotti, C., Peloso, S., Lanese, I., Brunesi, E., & Senaldi, I. (2022). Use of UAS for damage inspection and assessment of bridge infrastructures. *International Journal of Disaster Risk Reduction*, *72*, 102824. doi:10.1016/j.ijdrr.2022.102824

Manoharan, G., Razak, A., Rao, B. S., Singh, R., Ashtikar, S. P., & Nivedha, M. (2024). Navigating the Crescendo of Challenges in Harnessing Artificial Intelligence for Disaster Management. In D. Satishkumar & M. Sivaraja (Eds.), *Predicting Natural Disasters With AI and Machine Learning* (pp. 64–94). IGI Global. doi:10.4018/979-8-3693-2280-2.ch003

Marini, L., Ferraris, L., Mancini, M., & Meucci, L. (2017). Rainfall estimates from low-cost rain gauges: Evaluation and application in the context of a citizen observatory. *Atmospheric Research, 197,* 179–189.

Maurya, S., Rauthan, M., Verma, R., & Ahmad, R. B. (2022). Blockchain based Secure System for the Internet of Drones (IoD). *Paper presented at the 2022 6th International Conference on Electronics, Communication and Aerospace Technology.*

Mavroulis, S., Spyrou, N. I., & Emmanuel, A. (2019). UAV and GIS based rapid earthquake-induced building damage assessment and methodology for EMS-98 isoseismal map drawing: The June 12, 2017 Mw 6.3 Lesvos (Northeastern Aegean, Greece) earthquake. *International Journal of Disaster Risk Reduction, 37,* 101169. doi:10.1016/j.ijdrr.2019.101169

Ma, W., Feng, Z., Cheng, Z., Chen, S., & Wang, F. (2020). Identifying Forest Fire Driving Factors and Related Impacts in China Using Random Forest Algorithm. *Forests, 11*(5), 507. doi:10.3390/f11050507

McKinsey & Company. (2020). *COVID-19 implications for life sciences R&D: Recovery and the next normal.* Pharmaceuticals & Medical Products Practice. https://www.mckinsey.de/~/media/McKinsey/Industries/Pharmaceuticals%20and%20Medical%20Products/Our%20Insights/COVID%2019%20implications%20for%20life%20sciences%20R%20and%20D%20Recovery%20and%20the%20next%20normal/COVID19implicationsforlifesciencesRDRecoveryandthenextnormalvF.pdf

McWilliams, J. C. (2019). A perspective on the legacy of Edward Lorenz. *Earth and Space Science (Hoboken, N.J.), 6*(3), 336–350. doi:10.1029/2018EA000434

Mehta, P., Gupta, R., & Tanwar, S. (2020). Blockchain envisioned UAV networks: Challenges, solutions, and comparisons. *Computer Communications, 151,* 518–538. doi:10.1016/j.comcom.2020.01.023

Merkert, R., & Bushell, J. (2020). Managing the drone revolution: A systematic literature review into the current use of airborne drones and future strategic directions for their effective control. *Journal of Air Transport Management, 89,* 101929. doi:10.1016/j.jairtraman.2020.101929 PMID:32952321

Mhaisen, N., Fetais, N., Erbad, A., Mohamed, A., & Guizani, M. (2020). To chain or not to chain: A reinforcement learning approach for blockchain-enabled IoT monitoring applications. *Future Generation Computer Systems, 111,* 39–51. doi:10.1016/j.future.2020.04.035

Michailidis, E. T., Maliatsos, K., Skoutas, D. N., Vouyioukas, D., & Skianis, C. (2022). Secure UAV-aided mobile edge computing for IoT: A review. *IEEE Access : Practical Innovations, Open Solutions, 10,* 86353–86383. doi:10.1109/ACCESS.2022.3199408

Milanović, S., Kaczmarowski, J., Ciesielski, M., Trailović, Z., Mielcarek, M., Szczygieł, R., Kwiatkowski, M., Bałazy, R., Zasada, M., & Milanović, S. D. (2023). Modeling and Mapping of Forest Fire Occurrence in the Lower Silesian Voivodeship of Poland Based on Machine Learning Methods. *Forests*, *14*(1), 46. doi:10.3390/f14010046

Miller, L., Zhu, L., Yebra, M., Rüdiger, C., & Webb, G. I. (2022). Multi-modal temporal CNNs for live fuel moisture content estimation. *Environmental Modelling & Software*, *156*(105467), 105467. doi:10.1016/j.envsoft.2022.105467

Miyano, K., Shinkuma, R., Mandayam, N. B., Sato, T., & Oki, E. (2019). Utility based scheduling for multi-UAV search systems in disaster-hit areas. *IEEE Access : Practical Innovations, Open Solutions*, *7*, 26810–26820. doi:10.1109/ACCESS.2019.2900865

Mohapatra, S., & Shrimali, B. (2023). *Fault Detection in Forest Fire Monitoring Using Negative Selection Approach*. 2023 IEEE 11th Region 10 Humanitarian Technology Conference (R10-HTC), Rajkot, India. 10.1109/R10-HTC57504.2023.10461878

Mohsin, B., Steinhäusler, F., Madl, P., & Kiefel, M. (2016). An innovative system to enhance situational awareness in disaster response. *Journal of Homeland Security and Emergency Management*, *13*(3), 301–327. doi:10.1515/jhsem-2015-0079

Mojtahedi, S. M. H., & Lan-Oo, B. (2013). *Theoretical framework for stakeholders' disaster response index in the built environment*. IRBnet. https://www.irbnet.de/daten/iconda/CIB_DC27260.pdf

Moloi, T., & Marwala, T. (2021). A High-Level Overview of Artificial Intelligence: Historical Overview and Emerging Developments. In *Artificial Intelligence and the Changing Nature of Corporations. Future of Business and Finance*. Springer., doi:10.1007/978-3-030-76313-8_2

Munawar, H. S., Hammad, A., Ullah, F., & Ali, T. H. (2019). *After the flood: A novel application of image processing and machine learning for post-flood disaster management*.

Munawar, H. S., Qayyum, S., Ullah, F., & Sepasgozar, S. (2020). Big data and its applications in smart real estate and the disaster management life cycle: A systematic analysis. *Big Data and Cognitive Computing*, *4*(2), 4. doi:10.3390/bdcc4020004

Musa, A. I. (2013). Understanding the intersections of paradigm, meta-theory, and theory in library and information science research: A social constructionist perspective. *Samaru Journal of Information Studies*, *13*(1 & 2). https://www.researchgate.net/publication/309479829

Nair, B. B., & Dileep, M. R. (2020). A study on the role of tourism in destination's disaster and resilience management. *Journal of Environmental Management and Tourism*, *11*(6), 1496–1507. doi:10.14505/jemt.11.6(46).20

Natekar, S., Patil, S., Nair, A., & Roychowdhury, S. (2021). *Forest Fire Prediction using LSTM*. 2021 2nd International Conference for Emerging Technology (INCET), Belagavi, India. 10.1109/INCET51464.2021.9456113

National Institute of Disaster Management. (n.d.). *Understanding Disasters.* https://nidm.gov.in/PDF/Disaster_about.pdf

Naveen, L., & Mohan, H. S. (2019). Atmospheric weather prediction using various machine learning techniques: A survey. In *Proceedings of the IEEE 3rd International Conference on Computing Methodologies and Communication (ICCMC)*, Erode, India.

Nebot, À., & Mugica, F. (2021). Forest fire forecasting using fuzzy logic models. *Forests, 12*(8), 1005. doi:10.3390/f12081005

Nguyen, T., Katila, R., & Gia, T. N. (2023). An advanced Internet-of-Drones System with Blockchain for improving quality of service of Search and Rescue: A feasibility study. *Future Generation Computer Systems, 140*, 36–52. doi:10.1016/j.future.2022.10.002

Nguyen, T., Tran, D., Nguyen, D., & Dinh, H. (2020). Flood Early Warning System Using IoT-Based Environmental Monitoring. In *2020 International Conference on Advanced Computing and Applications (ACOMP)* (pp. 246-249). IEEE.

Nia, K., Prasetya, D., & Arifin, D. (2023). Detection of the post-earthquake damage in Mamuju Regency in January 2021 using Sentinel-1 satellite imagery. *Buletin Poltanesa, 24*(1).

Nihad, B., & Nadjat, A. (2023). *Drones Use in Disaster Management* [Doctoral dissertation, university center of abdalhafid boussouf-MILA].

Ni, X., Liu, C., Cecil, D. J., & Zhang, Q. (2017). On the detection of hail using satellite passive microwave radiometers and precipitation radar. *Journal of Applied Meteorology and Climatology, 56*(10), 2693–2709. doi:10.1175/JAMC-D-17-0065.1

Nouri, M., & Shafie-Khah, M. (2020). An IoT-Based Framework for Flood Early Warning Systems: An Overview. *IEEE Internet of Things Journal, 8*(6), 4547–4556.

Nunavath, V. & Goodwin, M. (2019). The Use of Artificial Intelligence in Disaster Management Systematic Literature Review. (pp. 1–8). IEEE.

Nunavath, V. (2018). *The role of artificial intelligence in social media big data analytics for disaster management-initial results of a systematic literature review.* (pp. 1–4). IEEE.

Nweke, H. F., Teh, Y. W., Mujtaba, G., & Al-Garadi, M. A. (2019). Data fusion and multiple classifier systems for human activity detection and health monitoring: Review and open research directions. *Information Fusion, 46*, 147–170. doi:10.1016/j.inffus.2018.06.002

O'Reilly, A. M., O'Sullivan, J. J., & Deering, C. T. (2020). Data-driven ensemble machine learning to improve the predictive accuracy of early warning flood forecasting systems. *Journal of Hydrology (Amsterdam), 583*, 124608.

Oláh, N., Molnár, B., & Huszti, A. (2023). Secure Registration Protocol for the Internet of Drones Using Blockchain and Physical Unclonable Function Technology. *Symmetry, 15*(10), 1886. doi:10.3390/sym15101886

Ouaissa, M., Rhattoy, A., & Chana, I. (2018, October). New security level of authentication and key agreement protocol for the IoT on LTE mobile networks. In *2018 6th International Conference on Wireless Networks and Mobile Communications (WINCOM)* (pp. 1-6). IEEE. 10.1109/WINCOM.2018.8629767

Pancholi, D., Jadav, N. K., Tanwar, S., Garg, D., & Zanjani, S. M. (2023). *Blockchain-based Secure UAV-assisted Battlefield Operation underlying 5G.* Paper presented at the 2023 14th International Conference on Information and Knowledge Technology (IKT).

Pang, G. (2022). Artificial intelligence for natural disaster management. *IEEE Intelligent Systems*, *37*(6), 3–6. doi:10.1109/MIS.2022.3220061

Pang, Y., Li, Y., Feng, Z., Feng, Z., Zhao, Z., Chen, S., & Zhang, H. (2022). Forest Fire Occurrence Prediction in China Based on Machine Learning Methods. *Remote Sensing (Basel)*, *14*(21), 5546. doi:10.3390/rs14215546

Park, S., Won Han, K., & Lee, K. (2020). *A study on fire detection technology through spectrum analysis of smoke particles.* 2020 International Conference on Information and Communication Technology Convergence (ICTC), Jeju, Korea (South). 10.1109/ICTC49870.2020.9289272

Park, S., Park, S. H., Park, L. W., Park, S., Lee, S., Lee, T., Lee, S., Jang, H., Kim, S., Chang, H., & Park, S. (2018). Design and implementation of a smart IoT based building and town disaster management system in smart city infrastructure. *Applied Sciences (Basel, Switzerland)*, *8*(11), 2239. doi:10.3390/app8112239

Pathak, P., Damle, M., Pal, P. R., & Yadav, V. (2019). Humanitarian impact of drones in healthcare and disaster management. *Int. J. Recent Technol. Eng*, *7*(5), 201–205.

Peng, W., Zhu, Y., & Song, J. (2020). An intelligent flood forecasting model based on LSTM network and GA-SVM. *Journal of Hydrology (Amsterdam)*, *587*, 125045.

Pepinsky, T. B. (2022). *Theory, comparison, and design: A review essay. Book review essay.* Cambridge University Press.

Perry, R. W. (2018). Defining disaster: An evolving concept. In H. Rodriguez, W. Donner, & J. Trainor (Eds.), *Handbook of disaster research.* Springer International Publishing. doi:10.1007/978-3-319-63254-4_1

Pettorru, G., Fadda, M., Girau, R., Sole, M., Anedda, M., & Giusto, D. (2023). Using Artificial Intelligence and IoT Solution for Forest Fire Prevention. *2023 International Conference on Computing, Networking and Communications, ICNC 2023.* IEEE. 10.1109/ICNC57223.2023.10074289

Pham, B. T., Luu, C., Van Phong, T., Nguyen, H. D., Van Le, H., Tran, T. Q., & Prakash, I. (2021). Flood risk assessment using hybrid artificial intelligence models integrated with multi-criteria decision analysis in Quang Nam Province, Vietnam. *Journal of Hydrology (Amsterdam)*, *592*, 125815. doi:10.1016/j.jhydrol.2020.125815

Pouyanfar, S., Sadiq, S., Yan, Y., Tian, H., Tao, Y., Reyes, M. P., Shyu, M.-L., Chen, S.-C., & Iyengar, S. S. (2018). A survey on deep learning: Algorithms, techniques, and applications. *ACM Computing Surveys*, *51*(5), 1–36. doi:10.1145/3234150

Praveen Kumar, B., Kalpana, A. V., & Nalini, S. (2023). Gated Attention Based Deep Learning Model for Analyzing the Influence of Social Media on Education. *Journal of Experimental & Theoretical Artificial Intelligence*, 1–15. doi:10.1080/0952813X.2023.2188262

Purbahapsari, A. F., & Batoarung, I. B. (2022). *Geospatial Artificial Intelligence for Early Detection of Forest and Land Fires*. KnE Social Sciences. doi:10.18502/kss.v7i9.10947

Puthal, D., Malik, N., Mohanty, S. P., Kougianos, E., & Yang, C. (2018). The blockchain as a decentralized security framework [future directions]. *IEEE Consumer Electronics Magazine*, *7*(2), 18–21. doi:10.1109/MCE.2017.2776459

Qiang, Y. (2019). Disparities of population exposed to flood hazards in the United States. *Journal of Environmental Management*, *232*, 295–304. doi:10.1016/j.jenvman.2018.11.039 PMID:30481643

Radhi, A. A., & Ibrahim, A. A. (2023). *Forest Fire Detection Techniques Based on IoT Technology: Review*. 2023 1st IEEE International Conference on Smart Technology (ICE-SMARTec), Bandung, Indonesia.10.1109/ICE-SMARTECH59237.2023.10461943

Raimi, L., Kah, J. M., & Tariq, M. U. (2022). The Discourse of Blue Economy Definitions, Measurements, and Theories: Implications for Strengthening Academic Research and Industry Practice. In L. Raimi & J. Kah (Eds.), *Implications for Entrepreneurship and Enterprise Development in the Blue Economy* (pp. 1–17). IGI Global. doi:10.4018/978-1-6684-3393-5.ch001

Raimi, L., Tariq, M. U., & Kah, J. M. (2022). Diversity, Equity, and Inclusion as the Future Workplace Ethics: Theoretical Review. In L. Raimi & J. Kah (Eds.), *Mainstreaming Diversity, Equity, and Inclusion as Future Workplace Ethics* (pp. 1–27). IGI Global. doi:10.4018/978-1-6684-3657-8.ch001

Ramadan, M., Hilles, S. M., & Alkhedher, M. (2023). Design and Study of an AI-Powered Autonomous Stair Climbing Robot. *El-Cezeri*, *10*(3), 571–585. doi:10.31202/ecjse.1272769

Ramchurn, S. D., Huynh, T. D., Ikuno, Y., Flann, J., Wu, F., Moreau, L., Jennings, N. R., Fischer, J., Jiang, W., Rodden, T., Simpson, E., Reece, S., & Roberts, S. J. (2015). HAC-ER: a disaster response system based on humanagent collectives. In: *Proceedings of the 2015 international conference on autonomous agents and multiagent systems (AAMAS 2015)*, (pp. 533–541). IEEE.

Rana, T., Shankar, A., Sultan, M. K., Patan, R., & Balusamy, B. (2019). *An intelligent approach for UAV and drone privacy security using blockchain methodology*. Paper presented at the 2019 9th International Conference on Cloud Computing, Data Science & Engineering (Confluence). 10.1109/CONFLUENCE.2019.8776613

Rao, T. V. N., Jakkam, P., & Medipally, S. (2024). Future Trends and Innovations in Natural Disaster Detection Using AI and ML. In *Predicting Natural Disasters With AI and Machine Learning* (pp. 110–134). IGI Global.

Ray, P. P., Mukherjee, M., & Shu, L. (2017). Internet of things for disaster management: State-of-the-art and prospects. *IEEE Access : Practical Innovations, Open Solutions, 5*, 18818–18835. doi:10.1109/ACCESS.2017.2752174

Raza, M., & Rehman, S. (2021). A Comprehensive Survey on IoT-Based Smart Flood Detection and Monitoring Systems. *IEEE Access : Practical Innovations, Open Solutions, 9*, 7954–7967.

Razavi-Termeh, S. V., Sadeghi-Niaraki, A., & Choi, S. M. (2020). Ubiquitous GIS-based forest fire susceptibility mapping using artificial intelligence methods. *Remote Sensing (Basel), 12*(10), 1689. doi:10.3390/rs12101689

Ren, D., Zhang, Y., Wang, L., Sun, H., Ren, S., & Gu, J. (2024). FCLGYOLO: Feature Constraint and Local Guided Global Feature for Fire Detection in Unmanned Aerial Vehicle Imagery. *IEEE Journal of Selected Topics in Applied Earth Observations and Remote Sensing, 17*, 5864–5875. doi:10.1109/JSTARS.2024.3358544

Renugadevi, R., & Medida, L. H. (2024). Artificial Intelligence and IoT-Based Disaster Management System. In Predicting Natural Disasters With AI and Machine Learning (pp. 135-146). IGI Global. doi:10.4018/979-8-3693-2280-2.ch006

Renugadevi, R., Vyshnavi, T., Reddy, T. P., & Lahari, P. S. (2023). Air Quality Prediction using Random Forest algorithm. *International Conference on Research Methodologies in Knowledge Management, Artificial Intelligence and Telecommunication Engineering (RMKMATE)*, Chennai, India. 10.1109/RMKMATE59243.2023.10369180

Restas, A. (2017). Disaster management supported by unmanned aerial systems (UAS) focusing especially on natural disasters. *Zeszyty Naukowe SGSP/Szkoła Główna Służby Pożarniczej, 61*.

Ribeiro, F. N., Castro, P. M., Azevedo, J. R., & Silva, A. M. (2018). A Review on the Use of IoT to Monitor Water Quality Through Aquatic Ecosystems. *IEEE Internet of Things Journal, 6*(4), 6373–6385.

Rodríguez, R. M., Alarcón, F., Rubio, D. S., & Ollero, A. (2013). *Autonomous management of an UAV Airfield*. In proceedings of the 3rd international conference on application and theory of automation in command and control systems, Naples, Italy.

Roper, S., & Turner, J. (2020). R&D and innovation after COVID-19: What can we expect? A review of prior research and data trends after the great financial crisis. *International Small Business Journal, 38*(6), 504–514. doi:10.1177/0266242620947946

Rubí, J. N., Carvalho, P., & Gondim, P. R. (2023). Application of machine learning models in the behavioral study of forest fires in the Brazilian Federal District region. *Engineering Applications of Artificial Intelligence, 118*, 105649. doi:10.1016/j.engappai.2022.105649

Russell, S. J., & Norvig, P. (2022). *Artificial intelligence: A modern approach*. Pearson.

Rust, J., Stillwell, D., Loe, A., & Sun, L. (2017). Introduction to item response theory. https://www.psychometrics.cam.ac.uk/system/files/documents/SSRMCIRT2017.pdf

Sahebjamnia, N., Torabi, S. A., & Mansouri, S. A. (2015). Integrated business continuity and disaster recovery planning: Towards organizational resilience. *European Journal of Operational Research*, *242*(1), 261–273. doi:10.1016/j.ejor.2014.09.055

Sakr, G. E., Elhajj, I. H., Mitri, G., & Wejinya, U. C. (2010). Artificial intelligence for forest fire prediction. *IEEE/ASME International Conference on Advanced Intelligent Mechatronics, AIM*. IEEE. 10.1109/AIM.2010.5695809

Salamati, S. P. N., & Kulatunga, U. (2017). *The importance of disaster management & impact of natural disasters on hospitals*. The 6th World Construction Symposium 2017: What's New and What's Next in the Built Environment Sustainability Agenda? Colombo, Sri Lanka.

Salamati, S. P. N., & Kulatunga, U. (2017). The challenges of hospital disaster managers in natural disaster events. *5th International Conference on Disaster Management and Human Health: Reducing Risk, Improving Outcomes*, Seville, Spain.

Samarakkody, A., Amaratunga, D., & Haigh, R. (2023). Technological innovations for enhancing disaster resilience in Smart Cities: A comprehensive urban scholar's analysis. *Sustainability (Basel)*, *15*(15), 12036. doi:10.3390/su151512036

SAMHSA. (2017). *Greater impact: How disasters affect people of low socioeconomic status*. Disaster Technical Assistance Centre, Supplemental Research Bulletin. https://www.samhsa. gov/sites/default/files/dtac/srb-low-ses_2.pdf

Sandhu, D., & Kaur, S. (2013). Psychological impacts of natural disasters. *Indian Journal of Health and Wellbeing*, *4*(6), 1317.

Seema, S., Kumar, M., & Singh, R. (2020). Seismic vulnerability assessment of NCT of Delhi using GIS-based multiple criteria decision analysis. In *Lecture Notes in Civil Engineering* (Vol. 118, pp. 11–20). Springer.

Sevinç, V. (2023). Mapping the forest fire risk zones using artificial intelligence with risk factors data. *Environmental Science and Pollution Research International*, *30*(2), 4721–4732. doi:10.1007/s11356-022-22515-w PMID:35974271

Shafik, W. (2023a). A Comprehensive Cybersecurity Framework for Present and Future Global Information Technology Organizations. In *Effective Cybersecurity Operations for Enterprise-Wide Systems* (pp. 56–79). IGI Global. doi:10.4018/978-1-6684-9018-1.ch002

Shafik, W. (2023b). *IoT-Based Energy Harvesting and Future Research Trends in Wireless Sensor Networks. Handbook of Research on Network-Enabled IoT Applications for Smart City Services*. IGI Global. doi:10.4018/979-8-3693-0744-1.ch016

Shafik, W. (2024a). *Blockchain-Based Internet of Things (B-IoT): Challenges, Solutions, Opportunities, Open Research Questions, and Future Trends. Blockchain-based Internet of Things: Opportunities, Challenges and Solutions*. Chapman and Hall/CRC. doi:10.1201/9781003407096-3

Shafik, W. (2024b). *Navigating Emerging Challenges in Robotics and Artificial Intelligence in Africa. Examining the Rapid Advance of Digital Technology in Africa.* IGI Global. doi:10.4018/978-1-6684-9962-7.ch007

Shafik, W. (2024c). *Toward a More Ethical Future of Artificial Intelligence and Data Science. The Ethical Frontier of AI and Data Analysis.* IGI Global. doi:10.4018/979-8-3693-2964-1.ch022

Shah, S. A., Seker, D. Z., Hameed, S., & Draheim, D. (2019). The rising role of big data analytics and IoT in disaster management: Recent advances, taxonomy and prospects. *IEEE Access : Practical Innovations, Open Solutions, 7,* 54595–54614. doi:10.1109/ACCESS.2019.2913340

ShamimM. A. B.NasreenM. (2016). Theoretical approach to disaster management. doi:10.13140/RG.2.2.25708.97927

Shankar, K., Zhang, Y., Liu, Y., Wu, L., & Che, C. (2020). Hyperparameter tuning deep learning for diabetic retinopathy fundus image classification. *IEEE Access : Practical Innovations, Open Solutions, 8,* 118164–118173. doi:10.1109/ACCESS.2020.3005152

Shao, Y., Feng, Z., Cao, M., Wang, W., Sun, L., Yang, X., Ma, T., Guo, Z., Fahad, S., Liu, X., & Wang, Z. (2023). An Ensemble Model for Forest Fire Occurrence Mapping in China. *Forests, 14*(4), 704. doi:10.3390/f14040704

Sharma, K., & Anand, D. (2023). AI and IoT in Supply Chain Management and Disaster Management. In Artificial Intelligence in Cyber-Physical Systems (pp. 275-289). CRC Press. doi:10.1201/9781003248750-16

Sharma, M., & Kaur, J. (2019). Disaster management using internet of things. In *Handbook of research on big data and the IoT* (pp. 211–222). IGI Global. doi:10.4018/978-1-5225-7432-3.ch012

Shavarani, S. M. (2019). Multi-level facility location-allocation problem for post-disaster humanitarian relief distribution: A case study. *Journal of Humanitarian Logistics and Supply Chain Management, 9*(1), 70–81. doi:10.1108/JHLSCM-05-2018-0036

Shi, W., Cao, J., Zhang, Q., Li, Y., & Xu, L. (2016). Edge computing: Vision and challenges. *IEEE Internet of Things Journal, 3*(5), 637–646. doi:10.1109/JIOT.2016.2579198

Shi, W., Zhang, M., Zhang, R., Chen, S., & Zhan, Z. (2020). Change detection based on artificial intelligence: State-of-the-art and challenges. *Remote Sensing (Basel), 12*(10), 1688. doi:10.3390/rs12101688

Singh, M., Aujla, G. S., Bali, R. S., Vashisht, S., Singh, A., & Jindal, A. (2020). Blockchain-enabled secure communication for drone delivery: A case study in COVID-like scenarios. *Proceedings of the 2nd ACM MobiCom Workshop on Drone Assisted Wireless Communications for 5G and beyond.* ACM.

Singh, R. (2023). *Turkey Earthquake 2023: Reducing Risks in the Indian Context.* MP-IDSA. https://idsa.in/idsacomments/turkey-earthquake-2023-rsingh-120423

Singh, C. R., & Manoharan, G. (2024). Strengthening Resilience: AI and Machine Learning in Emergency Decision-Making for Natural Disasters. In D. Satishkumar & M. Sivaraja (Eds.), *Internet of Things and AI for Natural Disaster Management and Prediction* (pp. 249–278). IGI Global., doi:10.4018/979-8-3693-4284-8.ch012

Singh, N. (2007). Expert system prototype of food aid distribution. *Asia Pacific Journal of Clinical Nutrition*, *16*, 116–121. PMID:17392088

Singh, S., & Chockalingam, J. (2023). Using Ensemble Machine Learning Algorithm to Predict Forest Fire Occurrence Probability in Madhya Pradesh and Chhattisgarh, India. *Advances in Space Research*, *73*(6), 2969–2987. doi:10.1016/j.asr.2023.12.054

Singh, Z. (2020). Disasters: Implications, mitigation, and preparedness. *Indian Journal of Public Health*, *64*(1), 1–3. doi:10.4103/ijph.IJPH_40_20 PMID:32189674

Sinha, A., Kumar, P., Rana, N. P., Islam, R., & Dwivedi, Y. K. (2019). Impact of internet of things (IoT) in disaster management: A task-technology fit perspective. *Annals of Operations Research*, *283*(1-2), 759–794. doi:10.1007/s10479-017-2658-1

Sitorus, F. P., Sunyoto, A., & Setiaji, B. (2023). *Forest Fire Disaster Classification Using Artificial Neural Network Method*. 2023 6th International Conference on Information and Communications Technology (ICOIACT), Yogyakarta, Indonesia. 10.1109/ICOIACT59844.2023.10455926

Soulakellis, N., Vasilakos, C., Chatzistamatis, S., Kavroudakis, D., Tataris, G., Papadopoulou, E.-E., Papakonstantinou, A., Roussou, O., & Kontos, T. (2020). Post-Earthquake Recovery Phase Monitoring and Mapping Based on UAS Data. *ISPRS International Journal of Geo-Information*, *9*(7), 447. doi:10.3390/ijgi9070447

Southard, N. (2017). *The socio-political and economic causes of natural disasters*. [Thesis, CMC]. https://scholarship.claremont.edu/cmc_theses/1720

Sowah, R. A., Ofoli, A. R., Krakani, S., & Fiawoo, S. (2017). Hardware Design and Web-Based Communication Modules of a Real-Time Multisensor Fire Detection and Notification System Using Fuzzy Logic. *IEEE Transactions on Industry Applications*, *53*(1), 559–566. doi:10.1109/TIA.2016.2613075

Srivastava, M., Abdelzaher, T., & Szymanski, B. (2012). Human-centric sensing. Philos Trans R Soc A Math Phys. *Engineering and Science*, *370*, 176–197.

Strzelecka, A., Kurdyś-Kujawska, A., & Zawadzka, D. (2020). Application of logistic regression models to assess household financial decisions regarding debt. *Procedia Computer Science*, *176*, 3418–3427. doi:10.1016/j.procs.2020.09.055

Subramani, J., Maria, A., Rajasekaran, A. S., & Lloret, J. (2024). Physically secure and privacy-preserving blockchain enabled authentication scheme for internet of drones. *Security and Privacy*, *7*(3), 364. doi:10.1002/spy2.364

Suleyman, M., & Bhaskar, M. (2023). *The coming wave: Technology, power, and the 21st century greatest dilemma*. New York: Crown Publishing Group. https://pdfget.com/please-wait-for-few-moments-6/

Sulistijono, I. A., & Risnumawan, A. (2016). From concrete to abstract: Multilayer neural networks for disaster victims detection. In *2016 International electronics symposium (IES)* (pp. 93-98). IEEE.

Sulthana, S. F., Wise, C. T. A., Ravikumar, C. V., Anbazhagan, R., Idayachandran, G., & Pau, G. (2023). Review Study on Recent Developments in Fire Sensing Methods. *IEEE Access : Practical Innovations, Open Solutions*, *11*, 90269–90282. doi:10.1109/ACCESS.2023.3306812

Sun, R., Gao, G., Gong, Z., & Wu, J. (2020). A review of risk analysis methods for natural disasters. *Natural Hazards*, *100*(2), 571–593. doi:10.1007/s11069-019-03826-7

Sun, W., Bocchini, P., & Davison, B. D. (2020). Applications of artificial intelligence for disaster management. *Natural Hazards*, *103*(3), 2631–2689. doi:10.1007/s11069-020-04124-3

Supriya, Y., & Gadekallu, T. R. (2023). Particle Swarm-Based Federated Learning Approach for Early Detection of Forest Fires. *Sustainability (Basel)*, *15*(2), 964. doi:10.3390/su15020964

Suteris, M. S., Rahman, F. A., & Ismail, A. (2018). Route schedule optimization method of unmanned aerial vehicle implementation for maritime surveillance in monitoring trawler activities in Kuala Kedah, Malaysia. *Int. J. Supply Chain Manag*, *7*(5), 245–249.

Syifa, M., Kadavi, P. R., & Lee, C. W. (2019). An Artificial Intelligence Application for Post-Earthquake Damage Mapping in Palu, Central Sulawesi, Indonesia. *Sensors (Basel)*, *19*(3), 542–560. doi:10.3390/s19030542 PMID:30696050

Tabish, S. A., & Syed, N. (2015). Disaster Preparedness: Current Trends and Future Directions. [IJSR]. *International Journal of Scientific Research*, 438.

Tan, L., Guo, J., Mohanarajah, S., & Zhou. (2021). Can we detect trends in natural disaster management with artificial intelligence? *A review of modeling practices, 107*(3), 2389-2417

Tanaka, G., Yamane, T., Héroux, J. B., Nakane, R., Kanazawa, N., Takeda, S., Numata, H., Nakano, D., & Hirose, A. (2019). Recent advances in physical reservoir computing: A review. *Neural Networks*, *115*, 100–123. doi:10.1016/j.neunet.2019.03.005 PMID:30981085

Tan, D., Qu, W., & Tu, J. (2010). The damage detection based on the fuzzy clustering and support vector machine. In *2010 International Conference on Intelligent System Design and Engineering Application* (Vol. 2, pp. 598-601). IEEE. 10.1109/ISDEA.2010.404

Tan, L., Guo, J., Mohanarajah, S., & Zhou, K. (2021). Can we detect trends in natural disaster management with artificial intelligence? A review of modeling practices. *Natural Hazards*, *107*(3), 2389–2417. doi:10.1007/s11069-020-04429-3

Tanzi, T. J., Chandra, M., Isnard, J., Camara, D., Sebastien, O., & Harivelo, F. (2016). Towards" drone-borne" disaster management: future application scenarios. In *XXIII ISPRS Congress, Commission VIII* (Volume III-8) (Vol. 3, pp. 181-189). Copernicus GmbH.

Tariq, M. U. (2024). Multidisciplinary Service Learning in Higher Education: Concepts, Implementation, and Impact. In S. Watson (Ed.), Applications of Service Learning in Higher Education (pp. 1-19). IGI Global. doi:10.4018/979-8-3693-2133-1.ch001

Tariq, M. U. (2024). Advanced Wearable Medical Devices and Their Role in Transformative Remote Health Monitoring. In M. Garcia & R. de Almeida (Eds.), *Transformative Approaches to Patient Literacy and Healthcare Innovation* (pp. 308–326). IGI Global., doi:10.4018/979-8-3693-3661-8.ch015

Tariq, M. U. (2024). Emerging Trends and Innovations in Blockchain-Digital Twin Integration for Green Investments: A Case Study Perspective. In S. Jafar, R. Rodriguez, H. Kannan, S. Akhtar, & P. Plugmann (Eds.), *Harnessing Blockchain-Digital Twin Fusion for Sustainable Investments* (pp. 148–175). IGI Global., doi:10.4018/979-8-3693-1878-2.ch007

Tariq, M. U. (2024). Emotional Intelligence in Understanding and Influencing Consumer Behavior. In T. Musiolik, R. Rodriguez, & H. Kannan (Eds.), *AI Impacts in Digital Consumer Behavior* (pp. 56–81). IGI Global., doi:10.4018/979-8-3693-1918-5.ch003

Tariq, M. U. (2024). Empowering Educators in the Learning Ecosystem. In F. Al Husseiny & A. Munna (Eds.), *Preparing Students for the Future Educational Paradigm* (pp. 232–255). IGI Global., doi:10.4018/979-8-3693-1536-1.ch010

Tariq, M. U. (2024). Empowering Student Entrepreneurs: From Idea to Execution. In G. Cantafio & A. Munna (Eds.), *Empowering Students and Elevating Universities With Innovation Centers* (pp. 83–111). IGI Global. doi:10.4018/979-8-3693-1467-8.ch005

Tariq, M. U. (2024). Enhancing Cybersecurity Protocols in Modern Healthcare Systems: Strategies and Best Practices. In M. Garcia & R. de Almeida (Eds.), *Transformative Approaches to Patient Literacy and Healthcare Innovation* (pp. 223–241). IGI Global., doi:10.4018/979-8-3693-3661-8.ch011

Tariq, M. U. (2024). Equity and Inclusion in Learning Ecosystems. In F. Al Husseiny & A. Munna (Eds.), *Preparing Students for the Future Educational Paradigm* (pp. 155–176). IGI Global., doi:10.4018/979-8-3693-1536-1.ch007

Tariq, M. U. (2024). Fintech Startups and Cryptocurrency in Business: Revolutionizing Entrepreneurship. In K. Kankaew, P. Nakpathom, A. Chnitphattana, K. Pitchayadejanant, & S. Kunnapapdeelert (Eds.), *Applying Business Intelligence and Innovation to Entrepreneurship* (pp. 106–124). IGI Global., doi:10.4018/979-8-3693-1846-1.ch006

Tariq, M. U. (2024). Leveraging Artificial Intelligence for a Sustainable and Climate-Neutral Economy in Asia. In P. Ordóñez de Pablos, M. Almunawar, & M. Anshari (Eds.), *Strengthening Sustainable Digitalization of Asian Economy and Society* (pp. 1–21). IGI Global., doi:10.4018/979-8-3693-1942-0.ch001

Tariq, M. U. (2024). Metaverse in Business and Commerce. In J. Kumar, M. Arora, & G. Erkol Bayram (Eds.), *Exploring the Use of Metaverse in Business and Education* (pp. 47–72). IGI Global., doi:10.4018/979-8-3693-5868-9.ch004

Tariq, M. U. (2024). Revolutionizing Health Data Management With Blockchain Technology: Enhancing Security and Efficiency in a Digital Era. In M. Garcia & R. de Almeida (Eds.), *Emerging Technologies for Health Literacy and Medical Practice* (pp. 153–175). IGI Global., doi:10.4018/979-8-3693-1214-8.ch008

Tariq, M. U. (2024). The Role of Emerging Technologies in Shaping the Global Digital Government Landscape. In Y. Guo (Ed.), *Emerging Developments and Technologies in Digital Government* (pp. 160–180). IGI Global. doi:10.4018/979-8-3693-2363-2.ch009

Tariq, M. U. (2024). The Transformation of Healthcare Through AI-Driven Diagnostics. In A. Sharma, N. Chanderwal, S. Tyagi, P. Upadhyay, & A. Tyagi (Eds.), *Enhancing Medical Imaging with Emerging Technologies* (pp. 250–264). IGI Global. doi:10.4018/979-8-3693-5261-8.ch015

Tariq, M. U., & Ismail, M. U. S. B. (2024). AI-powered COVID-19 forecasting: A comprehensive comparison of advanced deep learning methods. *Osong Public Health and Research Perspectives*, 2210–9099. doi:10.24171/j.phrp.2023.0287 PMID:38621765

Tatsubori, M., Watanabe, H., Shibayama, A., Sato, S., & Imamura, F. (2012). Social web in disaster archives. In: *The proceedings of the 21st international conference companion on world wide web—WWW'12 Companion*. ACM. 10.1145/2187980.2188190

Tchappi, I., Mualla, Y., Galland, S., Bottaro, A., Kamla, V. C., & Kamgang, J. C. (2022). Multilevel and holonic model for dynamic holarchy management: Application to large-scale road traffic. *Engineering Applications of Artificial Intelligence*, *109*, 104622. doi:10.1016/j.engappai.2021.104622

Teixeira, A. H. C., Wendland, E., Reichert, J. M., Montenegro, S. M. G. L., & Leitão, M. M. V. B. R. (2019). Soil moisture monitoring using low-cost sensors in precision agriculture. *Computers and Electronics in Agriculture*, *138*, 200–211.

Thomas, M. (2023). Dangerous risks of artificial intelligence. *Built In.* https://builtin.com/artificial-intelligence/risks-of-artificial-intelligence

Thurstone, L. L. (1927). A law of comparative judgment. *Psychological Review*, *34*(4), 273–286. doi:10.1037/h0070288

Tien Bui, D., Hoang, N. D., & Samui, P. (2019). Spatial pattern analysis and prediction of forest fire using new machine learning approach of Multivariate Adaptive Regression Splines and Differential Flower Pollination optimization: A case study at Lao Cai province (Viet Nam). *Journal of Environmental Management*, *237*, 476–487. doi:10.1016/j.jenvman.2019.01.108 PMID:30825780

Tissera, P. C., Printista, A. M., & Luque, E. (2012). A hybrid simulation model to test behaviour designs in an emergency evacuation. *Procedia Computer Science*, *9*, 266–275. doi:10.1016/j.procs.2012.04.028

Tonti, I., Lingua, A. M., Piccinini, F., Pierdicca, R., & Malinverni, E. S. (2023). Digitalization and Spatial Documentation of Post-Earthquake Temporary Housing in Central Italy: An Integrated Geomatic Approach Involving UAV and a GIS-Based System. *Drones (Basel)*, *7*(7), 438. doi:10.3390/drones7070438

Tubis, A. A., Poturaj, H., Dereń, K., & Żurek, A. (2024). Risks of Drone Use in Light of Literature Studies. *Sensors (Basel)*, *24*(4), 1205. doi:10.3390/s24041205 PMID:38400363

UNDP. (2023). *Innovation in disaster management: Leveraging technology to save more lives.* One United Nations Plaza, New York. https://www.undp.org/sites/g/files/zskgke326/files/2024-03/innovation_in_disaster_management_web_final_compressed.pdf

UNDRR. (n.d.). *Sendai Framework Terminology on Disaster Risk Reduction.* UNDRR. https://www.undrr.org/terminology/disaster-risk-management

United Nations Office for Disaster Risk Reduction (UNISDR). (2005). Hyogo Framework of Action for 2005-2015: Building the Resilience of Nations and Communities to Disasters. *World Conference on Disaster Reduction; Hyogo, Japan: International Strategy for Disaster Reduction (ISDR).* UN. https://www.unisdr.org/2005/wcdr/intergover/officialdoc/L-docs/Hyogo-framework-for-action-english.pdf

United Nations. (1994). *Yokohoma Strategy and Plan of Action for a Safer World guidelines for Natural Disaster Prevention, Preparedness and Mitigation.* Prevention Web. https://www.preventionweb.net/files/8241_doc6841contenido1.pdf

United Nations. (1999). *Activities of the International Decade for Natural Disaster Reduction; Report of the General Secretary.* UN. https://www.un.org/esa/documents/ecosoc/docs/1999/e1999-80.htm#:~:text=Faced%20with%20an%20increasing%20frequency,United%20Nations%2C%20would%20pay%20special

van Niekerk, D. (2011). *USAID disaster risk reduction training course for Southern Africa.* Prevention Web. https://www.preventionweb.net/files/26081_kp1concepdisasterrisk1.pdf

Varalakshmi, M. S. (2019). Evolution and Significance of Drones in Modern Technology. *8 IJRAR, 6*(1). https://ijrar.org/papers/IJRAR19J1225.pdf

Vashishth, T. K., Kumar, B., Sharma, V., Chaudhary, S., Kumar, S., & Sharma, K. K. (2023). The Evolution of AI and Its Transformative Effects on Computing: A Comparative Analysis. In *Intelligent Engineering Applications and Applied Sciences for Sustainability* (pp. 425–442). IGI Global. doi:10.4018/979-8-3693-0044-2.ch022

Velev, D., & Zlateva, P. (2023). Challenges of artificial intelligence application for disaster risk Management. *The International Archives of the Photogrammetry, Remote Sensing and Spatial Information Sciences*, *48*, 387–394. doi:10.5194/isprs-archives-XLVIII-M-1-2023-387-2023

Vermiglio, C., Noto, G., Rodríguez Bolívar, M. P., & Zarone, V. (2022). Disaster management and emerging technologies: A performance-based perspective. *Meditari Accountancy Research*, *30*(4), 1093–1117. doi:10.1108/MEDAR-02-2021-1206

Vetrivel, A., Gerke, M., Kerle, N., Nex, F., & Vosselman, G. (2018). Disaster damage detection through synergistic use of deep learning and 3D point cloud features derived from very high-resolution oblique aerial images, and multiple-kernel-learning. *ISPRS Journal of Photogrammetry and Remote Sensing, 140*, 45–59. doi:10.1016/j.isprsjprs.2017.03.001

Vieweg, S. (2012). *Situational awareness in mass emergency: a behavioral and linguistic analysis of microblogged communications.* [PhD dissertation, University of Colorado at Boulder].

Wamsler, C., & Johannessen, A. (2020). Meeting at the crossroads? Developing national strategies for disaster risk reduction and resilience: Relevance, scope for, and challenges to, integration. *International Journal of Disaster Risk Reduction, 45*, 101452. doi:10.1016/j.ijdrr.2019.101452

Wandelt, S., Wang, S., Zheng, C., & Sun, X. (2023). AERIAL: A Meta Review and Discussion of Challenges Toward Unmanned Aerial Vehicle Operations in Logistics, Mobility, and Monitoring. *IEEE Transactions on Intelligent Transportation Systems*, 1–14. doi:10.1109/TITS.2023.3343713

Wang, G., Zhang, Y., Qu, Y., Chen, Y., & Maqsood, H. (2019). Early Forest Fire Region Segmentation Based on Deep Learning. *Proceedings of the 31st Chinese Control and Decision Conference, CCDC 2019.* IEEE. 10.1109/CCDC.2019.8833125

Wang, J., Zhang, J., Gong, L., Li, Q., & Zhou, D. (2018). Indirect seismic economic loss assessment and recovery evaluation using nighttime light images—Application for Wenchuan earthquake. *Natural Hazards and Earth System Sciences, 18*(12), 3253–3266. doi:10.5194/nhess-18-3253-2018

Wang, J., Zhou, K., Xing, W., Li, H., & Yang, Z. (2023). Applications, evolutions, and challenges of drones in maritime transport. *Journal of Marine Science and Engineering, 11*(11), 2056. doi:10.3390/jmse11112056

Wang, L., Zhao, Q., Wen, Z., & Qu, J. (2018). RAFFIA: Short-term forest fire danger rating prediction via multiclass logistic regression. *Sustainability (Basel), 10*(12), 4620. doi:10.3390/su10124620

Wang, P., Lv, W., Wang, C., & Hou, J. (2018). *Hail storms recognition based on convolutional neural network.* In *Proceedings of the 13th World Congress on Intelligent Control and Automation (WCICA)*, Changsha, China. 10.1109/WCICA.2018.8630701

Wang, S., Wan, J., Li, D., & Zhang, C. (2020). Cloud-based IoT for real-time hydrological monitoring and flood forecasting. *Computer Networks, 168*, 107047.

Wang, S., Wan, J., Zhang, D., Li, D., & Zhang, C. (2019). Towards smart factory for industry 4.0: A self-organized multi-agent system with big data based feedback and coordination. *Computer Networks, 151*, 264–276.

Weston, R., & Gore, P. A. Jr. (2006). Brief guide to structural equation modeling. *The Counseling Psychologist, 34*(5), 719–751. doi:10.1177/0011000006286345

Wisner, B., Blaikie, P., Cannon, T., & Davis, I. (2004). *At risk: Natural hazards, people's vulnerability, and disasters* (2nd ed.). Routledge.

World Bank. (2021). *Overlooked: Examining the impact of disasters and climate shocks on poverty in the Europe and Central Asia region*. World Bank. https://documents1.worldbank. org/curated/en/493181607687673440/pdf/Overlooked-Examining-the-Impact-of-Disasters-and-Climate-Shocks-on-Poverty-in-the-Europe-and-Central-Asia-Region.pdf

World Health Organization. (2019). *Health emergency and disaster risk management framework*. WHO.

Wu, Y., Song, P., & Wang, F. (2020). Hybrid consensus algorithm optimization: A mathematical method based on POS and PBFT and its application in blockchain. *Mathematical Problems in Engineering, 2020*, 2020. doi:10.1155/2020/7270624

Wu, Z., Li, M., Wang, B., Quan, Y., & Liu, J. (2021). Using artificial intelligence to estimate the probability of forest fires in heilongjiang, Northeast China. *Remote Sensing (Basel), 13*(9), 1813. doi:10.3390/rs13091813

Xiao, W., Li, M., Alzahrani, B., Alotaibi, R., Barnawi, A., & Ai, Q. (2021). A blockchain-based secure crowd monitoring system using UAV swarm. *IEEE Network, 35*(1), 108–115. doi:10.1109/MNET.011.2000210

Xie, J., Qi, T., Hu, W., Huang, H., Chen, B., & Zhang, J. (2022). Retrieval of Live Fuel Moisture Content Based on Multi-Source Remote Sensing Data and Ensemble Deep Learning Model. *Remote Sensing (Basel), 14*(17), 4378. doi:10.3390/rs14174378

Xie, L., Chen, J., & Lv, J. (2021). An IoT-enabled system for flood monitoring and early warning based on artificial intelligence algorithms. *Environmental Monitoring and Assessment, 193*(7), 452. PMID:34181101

Xue, Z., Zheng, Z., Yi, Z., Han, Y., Liu, W., & Peng, J. (2023). A Fire Detection and Assessment Method based on YOLOv8. 2023 China Automation Congress. CAC. doi:10.1109/CAC59555.2023.10451727

Xu, R., Wei, S., Chen, Y., Chen, G., & Pham, K. (2022). Lightman: A lightweight microchained fabric for assurance-and resilience-oriented urban air mobility networks. *Drones (Basel), 6*(12), 421. doi:10.3390/drones6120421

Yang, T., Xie, J., Li, G., Mou, N., Li, Z., Tian, C., & Zhao, J. (2019). Social media big data mining and spatio-temporal analysis on public emotions for disaster mitigation. *ISPRS International Journal of Geo-Information, 8*(1), 29. doi:10.3390/ijgi8010029

Yang, W., Wang, S., Yin, X., Wang, X., & Hu, J. (2022). A review on security issues and solutions of the internet of drones. *IEEE Open Journal of the Computer Society, 3*, 96–110. doi:10.1109/OJCS.2022.3183003

Yang, Z., Yu, X., Dedman, S., Rosso, M., Zhu, J., Yang, J., Xia, Y., Tian, Y., Zhang, G., & Wang, J. (2022). UAV remote sensing applications in marine monitoring: Knowledge visualization and review. *The Science of the Total Environment, 838*, 155939. doi:10.1016/j.scitotenv.2022.155939 PMID:35577092

Yan, L., & Suryadi, H. (2021). An Overview of IoT Applications in Water Management: Challenges and Opportunities. *IEEE Internet of Things Journal, 8*(16), 12807–12825.

Ye, L., & Abe, M. (2012). *The impacts of natural disasters on global supply chains (No. 115).* ARTNeT working paper series.

Yim, J., Park, H., Kwon, E., Kim, S., & Lee, Y.-T. (2018). Low-power image stitching management for reducing power consumption of UAVs for disaster management system. *Paper presented at the 2018 IEEE International Conference on Consumer Electronics (ICCE).* IEEE. 10.1109/ICCE.2018.8326248

Yong, H., & Zhu, Z. (2023). Rapid assessment of seismic risk for railway bridges based on machine learning. *International Journal of Structural Stability and Dynamics.*

Yuan, H., Zhan, S., & Tan, W. (2018). Application of wireless sensor networks and IoT technology in flood warning. In *2018 International Conference on Smart Grid and Electrical Automation (ICSGEA)* (pp. 113-116). IEEE.

Zafar, U., Shah, M. A., Wahid, A., Akhunzada, A., & Arif, S. (2019). Exploring IoT applications for disaster management: identifying key factors and proposing future directions. *Recent trends and advances in wireless and IoT-enabled networks*, 291-309.

Zakari, R. Y., Shafik, W., Kalinaki, K., & Iheaturu, C. J. (2024). Internet of Forestry Things (IoFT) Technologies and Applications in Forest Management. In *Advanced IoT Technologies and Applications in the Industry 4.0 Digital Economy* (pp. 275-295). CRC Press.

Zaré M, & Afrouz SG. (2011). Crisis management of Tohoku; Japan earthquake and tsunami. *Iran J Public Health, 41*(6), 12-20

Zhang, C., Yu, S., Zhou, J., Cheng, H., & Liu, J. (2018). Blockchain-based cloud data storage: A survey. *IEEE Access : Practical Innovations, Open Solutions, 6*, 9365–9375.

Zhang, D., Lin, T., Wang, K., Wang, L., & Li, H. (2021). An improved LSTM-based multi-model prediction method for air quality forecasting. *Journal of Cleaner Production, 320*, 128956.

Zhang, Y., & Peacock, W. G. (2009). Planning for housing recovery? Lessons learned from Hurricane Andrew. *Journal of the American Planning Association, 76*(1), 5–24. doi:10.1080/01944360903294556

Zhu, X., Zhang, G. & Sun, B. (2019). *A comprehensive literature review of the demand forecasting methods of emergency resources from the perspective of artificial intelligence.* Research Gate.

Zhu, J., Zhang, D., & Xia, J. (2020). CNN-based spatio-temporal feature learning for flood inundation mapping with SAR and optical remote sensing images. *Remote Sensing, 12*(15), 2365.

Zulch, H. (2019). *Psychological preparedness for natural hazards – improving disaster preparedness policy and practice.* Contributing Paper to GAR 2019. UNDRR. https://www.undrr.org/publication/psychological-preparedness-natural-hazards-improving-disaster-preparedness-policy-and

Related References

To continue our tradition of advancing information science and technology research, we have compiled a list of recommended IGI Global readings. These references will provide additional information and guidance to further enrich your knowledge and assist you with your own research and future publications.

Aasi, P., Rusu, L., & Vieru, D. (2017). The Role of Culture in IT Governance Five Focus Areas: A Literature Review. *International Journal of IT/Business Alignment and Governance, 8*(2), 42-61. https://doi.org/ doi:10.4018/IJITBAG.2017070103

Abdrabo, A. A. (2018). Egypt's Knowledge-Based Development: Opportunities, Challenges, and Future Possibilities. In A. Alraouf (Ed.), *Knowledge-Based Urban Development in the Middle East* (pp. 80–101). Hershey, PA: IGI Global. doi:10.4018/978-1-5225-3734-2.ch005

Abu Doush, I., & Alhami, I. (2018). Evaluating the Accessibility of Computer Laboratories, Libraries, and Websites in Jordanian Universities and Colleges. *International Journal of Information Systems and Social Change, 9*(2), 44–60. doi:10.4018/IJISSC.2018040104

Adegbore, A. M., Quadri, M. O., & Oyewo, O. R. (2018). A Theoretical Approach to the Adoption of Electronic Resource Management Systems (ERMS) in Nigerian University Libraries. In A. Tella & T. Kwanya (Eds.), *Handbook of Research on Managing Intellectual Property in Digital Libraries* (pp. 292–311). Hershey, PA: IGI Global. doi:10.4018/978-1-5225-3093-0.ch015

Afolabi, O. A. (2018). Myths and Challenges of Building an Effective Digital Library in Developing Nations: An African Perspective. In A. Tella & T. Kwanya (Eds.), *Handbook of Research on Managing Intellectual Property in Digital Libraries* (pp. 51–79). Hershey, PA: IGI Global. doi:10.4018/978-1-5225-3093-0.ch004

Agarwal, P., Kurian, R., & Gupta, R. K. (2022). Additive Manufacturing Feature Taxonomy and Placement of Parts in AM Enclosure. In S. Salunkhe, H. Hussein, & J. Davim (Eds.), *Applications of Artificial Intelligence in Additive Manufacturing* (pp. 138–176). IGI Global. https://doi.org/10.4018/978-1-7998-8516-0.ch007

Al-Alawi, A. I., Al-Hammam, A. H., Al-Alawi, S. S., & AlAlawi, E. I. (2021). The Adoption of E-Wallets: Current Trends and Future Outlook. In Y. Albastaki, A. Razzaque, & A. Sarea (Eds.), *Innovative Strategies for Implementing FinTech in Banking* (pp. 242–262). IGI Global. https://doi.org/10.4018/978-1-7998-3257-7.ch015

Alsharo, M. (2017). Attitudes Towards Cloud Computing Adoption in Emerging Economies. *International Journal of Cloud Applications and Computing*, 7(3), 44–58. doi:10.4018/IJCAC.2017070102

Amer, T. S., & Johnson, T. L. (2017). Information Technology Progress Indicators: Research Employing Psychological Frameworks. In A. Mesquita (Ed.), *Research Paradigms and Contemporary Perspectives on Human-Technology Interaction* (pp. 168–186). Hershey, PA: IGI Global. doi:10.4018/978-1-5225-1868-6.ch008

Andreeva, A., & Yolova, G. (2021). Liability in Labor Legislation: New Challenges Related to the Use of Artificial Intelligence. In B. Vassileva & M. Zwilling (Eds.), *Responsible AI and Ethical Issues for Businesses and Governments* (pp. 214–232). IGI Global. https://doi.org/10.4018/978-1-7998-4285-9.ch012

Anohah, E. (2017). Paradigm and Architecture of Computing Augmented Learning Management System for Computer Science Education. *International Journal of Online Pedagogy and Course Design*, 7(2), 60–70. doi:10.4018/IJOPCD.2017040105

Anohah, E., & Suhonen, J. (2017). Trends of Mobile Learning in Computing Education from 2006 to 2014: A Systematic Review of Research Publications. *International Journal of Mobile and Blended Learning*, 9(1), 16–33. doi:10.4018/IJMBL.2017010102

Arbaiza, C. S., Huerta, H. V., & Rodriguez, C. R. (2021). Contributions to the Technological Adoption Model for the Peruvian Agro-Export Sector. *International Journal of E-Adoption*, 13(1), 1–17. https://doi.org/10.4018/IJEA.2021010101

Bailey, E. K. (2017). Applying Learning Theories to Computer Technology Supported Instruction. In M. Grassetti & S. Brookby (Eds.), *Advancing Next-Generation Teacher Education through Digital Tools and Applications* (pp. 61–81). Hershey, PA: IGI Global. doi:10.4018/978-1-5225-0965-3.ch004

Baker, J. D. (2021). Introduction to Machine Learning as a New Methodological Framework for Performance Assessment. In M. Bocarnea, B. Winston, & D. Dean (Eds.), *Handbook of Research on Advancements in Organizational Data Collection and Measurements: Strategies for Addressing Attitudes, Beliefs, and Behaviors* (pp. 326–342). IGI Global. https://doi.org/10.4018/978-1-7998-7665-6.ch021

Banerjee, S., Sing, T. Y., Chowdhury, A. R., & Anwar, H. (2018). Let's Go Green: Towards a Taxonomy of Green Computing Enablers for Business Sustainability. In M. Khosrow-Pour (Ed.), *Green Computing Strategies for Competitive Advantage and Business Sustainability* (pp. 89–109). Hershey, PA: IGI Global. doi:10.4018/978-1-5225-5017-4.ch005

Basham, R. (2018). Information Science and Technology in Crisis Response and Management. In M. Khosrow-Pour, D.B.A. (Ed.), Encyclopedia of Information Science and Technology, Fourth Edition (pp. 1407-1418). Hershey, PA: IGI Global. doi:10.4018/978-1-5225-2255-3.ch121

Batyashe, T., & Iyamu, T. (2018). Architectural Framework for the Implementation of Information Technology Governance in Organisations. In M. Khosrow-Pour, D.B.A. (Ed.), Encyclopedia of Information Science and Technology, Fourth Edition (pp. 810-819). Hershey, PA: IGI Global. doi:10.4018/978-1-5225-2255-3.ch070

Bekleyen, N., & Çelik, S. (2017). Attitudes of Adult EFL Learners towards Preparing for a Language Test via CALL. In D. Tafazoli & M. Romero (Eds.), *Multiculturalism and Technology-Enhanced Language Learning* (pp. 214–229). Hershey, PA: IGI Global. doi:10.4018/978-1-5225-1882-2.ch013

Bergeron, F., Croteau, A., Uwizeyemungu, S., & Raymond, L. (2017). A Framework for Research on Information Technology Governance in SMEs. In S. De Haes & W. Van Grembergen (Eds.), *Strategic IT Governance and Alignment in Business Settings* (pp. 53–81). Hershey, PA: IGI Global. doi:10.4018/978-1-5225-0861-8.ch003

Bhardwaj, M., Shukla, N., & Sharma, A. (2021). Improvement and Reduction of Clustering Overhead in Mobile Ad Hoc Network With Optimum Stable Bunching Algorithm. In S. Kumar, M. Trivedi, P. Ranjan, & A. Punhani (Eds.), *Evolution of Software-Defined Networking Foundations for IoT and 5G Mobile Networks* (pp. 139–158). IGI Global. https://doi.org/10.4018/978-1-7998-4685-7.ch008

Bhatt, G. D., Wang, Z., & Rodger, J. A. (2017). Information Systems Capabilities and Their Effects on Competitive Advantages: A Study of Chinese Companies. *Information Resources Management Journal, 30*(3), 41–57. doi:10.4018/IRMJ.2017070103

Bhattacharya, A. (2021). Blockchain, Cybersecurity, and Industry 4.0. In A. Tyagi, G. Rekha, & N. Sreenath (Eds.), *Opportunities and Challenges for Blockchain Technology in Autonomous Vehicles* (pp. 210–244). IGI Global. https://doi.org/10.4018/978-1-7998-3295-9.ch013

Bhyan, P., Shrivastava, B., & Kumar, N. (2022). Requisite Sustainable Development Contemplating Buildings: Economic and Environmental Sustainability. In A. Hussain, K. Tiwari, & A. Gupta (Eds.), *Addressing Environmental Challenges Through Spatial Planning* (pp. 269–288). IGI Global. https://doi.org/10.4018/978-1-7998-8331-9.ch014

Boido, C., Davico, P., & Spallone, R. (2021). Digital Tools Aimed to Represent Urban Survey. In M. Khosrow-Pour D.B.A. (Ed.), *Encyclopedia of Information Science and Technology, Fifth Edition* (pp. 1181-1195). IGI Global. https://doi.org/10.4018/978-1-7998-3479-3.ch082

Borkar, P. S., Chanana, P. U., Atwal, S. K., Londe, T. G., & Dalal, Y. D. (2021). The Replacement of HMI (Human-Machine Interface) in Industry Using Single Interface Through IoT. In R. Raut & A. Mihovska (Eds.), *Examining the Impact of Deep Learning and IoT on Multi-Industry Applications* (pp. 195–208). IGI Global. https://doi.org/10.4018/978-1-7998-7511-6.ch011

Brahmane, A. V., & Krishna, C. B. (2021). Rider Chaotic Biography Optimization-driven Deep Stacked Auto-encoder for Big Data Classification Using Spark Architecture: Rider Chaotic Biography Optimization. *International Journal of Web Services Research*, *18*(3), 42–62. https://doi.org/10.4018/ijwsr.2021070103

Burcoff, A., & Shamir, L. (2017). Computer Analysis of Pablo Picasso's Artistic Style. *International Journal of Art, Culture and Design Technologies*, *6*(1), 1–18. doi:10.4018/IJACDT.2017010101

Byker, E. J. (2017). I Play I Learn: Introducing Technological Play Theory. In C. Martin & D. Polly (Eds.), *Handbook of Research on Teacher Education and Professional Development* (pp. 297–306). Hershey, PA: IGI Global. doi:10.4018/978-1-5225-1067-3.ch016

Calongne, C. M., Stricker, A. G., Truman, B., & Arenas, F. J. (2017). Cognitive Apprenticeship and Computer Science Education in Cyberspace: Reimagining the Past. In A. Stricker, C. Calongne, B. Truman, & F. Arenas (Eds.), *Integrating an Awareness of Selfhood and Society into Virtual Learning* (pp. 180–197). Hershey, PA: IGI Global. doi:10.4018/978-1-5225-2182-2.ch013

Carneiro, A. D. (2017). Defending Information Networks in Cyberspace: Some Notes on Security Needs. In M. Dawson, D. Kisku, P. Gupta, J. Sing, & W. Li (Eds.), Developing Next-Generation Countermeasures for Homeland Security Threat Prevention (pp. 354-375). Hershey, PA: IGI Global. https://doi.org/ doi:10.4018/978-1-5225-0703-1.ch016

Carvalho, W. F., & Zarate, L. (2021). Causal Feature Selection. In A. Azevedo & M. Santos (Eds.), *Integration Challenges for Analytics, Business Intelligence, and Data Mining* (pp. 145-160). IGI Global. https://doi.org/10.4018/978-1-7998-5781-5.ch007

Chase, J. P., & Yan, Z. (2017). Affect in Statistics Cognition. In *Assessing and Measuring Statistics Cognition in Higher Education Online Environments: Emerging Research and Opportunities* (pp. 144–187). Hershey, PA: IGI Global. doi:10.4018/978-1-5225-2420-5.ch005

Chatterjee, A., Roy, S., & Shrivastava, R. (2021). A Machine Learning Approach to Prevent Cancer. In G. Rani & P. Tiwari (Eds.), *Handbook of Research on Disease Prediction Through Data Analytics and Machine Learning* (pp. 112–141). IGI Global. https://doi.org/10.4018/978-1-7998-2742-9.ch007

Cifci, M. A. (2021). Optimizing WSNs for CPS Using Machine Learning Techniques. In A. Luhach & A. Elçi (Eds.), *Artificial Intelligence Paradigms for Smart Cyber-Physical Systems* (pp. 204–228). IGI Global. https://doi.org/10.4018/978-1-7998-5101-1.ch010

Cimermanova, I. (2017). Computer-Assisted Learning in Slovakia. In D. Tafazoli & M. Romero (Eds.), *Multiculturalism and Technology-Enhanced Language Learning* (pp. 252–270). Hershey, PA: IGI Global. doi:10.4018/978-1-5225-1882-2.ch015

Cipolla-Ficarra, F. V., & Cipolla-Ficarra, M. (2018). Computer Animation for Ingenious Revival. In F. Cipolla-Ficarra, M. Ficarra, M. Cipolla-Ficarra, A. Quiroga, J. Alma, & J. Carré (Eds.), *Technology-Enhanced Human Interaction in Modern Society* (pp. 159–181). Hershey, PA: IGI Global. doi:10.4018/978-1-5225-3437-2.ch008

Cockrell, S., Damron, T. S., Melton, A. M., & Smith, A. D. (2018). Offshoring IT. In M. Khosrow-Pour, D.B.A. (Ed.), Encyclopedia of Information Science and Technology, Fourth Edition (pp. 5476-5489). Hershey, PA: IGI Global. https://doi.org/ doi:10.4018/978-1-5225-2255-3.ch476

Coffey, J. W. (2018). Logic and Proof in Computer Science: Categories and Limits of Proof Techniques. In J. Horne (Ed.), *Philosophical Perceptions on Logic and Order* (pp. 218–240). Hershey, PA: IGI Global. doi:10.4018/978-1-5225-2443-4.ch007

Dale, M. (2017). Re-Thinking the Challenges of Enterprise Architecture Implementation. In M. Tavana (Ed.), *Enterprise Information Systems and the Digitalization of Business Functions* (pp. 205–221). Hershey, PA: IGI Global. doi:10.4018/978-1-5225-2382-6.ch009

Das, A., & Mohanty, M. N. (2021). An Useful Review on Optical Character Recognition for Smart Era Generation. In A. Tyagi (Ed.), *Multimedia and Sensory Input for Augmented, Mixed, and Virtual Reality* (pp. 1–41). IGI Global. https://doi.org/10.4018/978-1-7998-4703-8.ch001

Dash, A. K., & Mohapatra, P. (2021). A Survey on Prematurity Detection of Diabetic Retinopathy Based on Fundus Images Using Deep Learning Techniques. In S. Saxena & S. Paul (Eds.), *Deep Learning Applications in Medical Imaging* (pp. 140–155). IGI Global. https://doi.org/10.4018/978-1-7998-5071-7.ch006

De Maere, K., De Haes, S., & von Kutzschenbach, M. (2017). CIO Perspectives on Organizational Learning within the Context of IT Governance. *International Journal of IT/Business Alignment and Governance, 8*(1), 32-47. https://doi.org/doi:10.4018/IJITBAG.2017010103

Demir, K., Çaka, C., Yaman, N. D., İslamoğlu, H., & Kuzu, A. (2018). Examining the Current Definitions of Computational Thinking. In H. Ozcinar, G. Wong, & H. Ozturk (Eds.), *Teaching Computational Thinking in Primary Education* (pp. 36–64). Hershey, PA: IGI Global. doi:10.4018/978-1-5225-3200-2.ch003

Deng, X., Hung, Y., & Lin, C. D. (2017). Design and Analysis of Computer Experiments. In S. Saha, A. Mandal, A. Narasimhamurthy, S. V, & S. Sangam (Eds.), Handbook of Research on Applied Cybernetics and Systems Science (pp. 264-279). Hershey, PA: IGI Global. doi:10.4018/978-1-5225-2498-4.ch013

Denner, J., Martinez, J., & Thiry, H. (2017). Strategies for Engaging Hispanic/Latino Youth in the US in Computer Science. In Y. Rankin & J. Thomas (Eds.), *Moving Students of Color from Consumers to Producers of Technology* (pp. 24–48). Hershey, PA: IGI Global. doi:10.4018/978-1-5225-2005-4.ch002

Devi, A. (2017). Cyber Crime and Cyber Security: A Quick Glance. In R. Kumar, P. Pattnaik, & P. Pandey (Eds.), *Detecting and Mitigating Robotic Cyber Security Risks* (pp. 160–171). Hershey, PA: IGI Global. doi:10.4018/978-1-5225-2154-9.ch011

Dhaya, R., & Kanthavel, R. (2022). Futuristic Research Perspectives of IoT Platforms. In D. Jeya Mala (Ed.), *Integrating AI in IoT Analytics on the Cloud for Healthcare Applications* (pp. 258–275). IGI Global. doi:10.4018/978-1-7998-9132-1.ch015

Doyle, D. J., & Fahy, P. J. (2018). Interactivity in Distance Education and Computer-Aided Learning, With Medical Education Examples. In M. Khosrow-Pour, D.B.A. (Ed.), Encyclopedia of Information Science and Technology, Fourth Edition (pp. 5829-5840). Hershey, PA: IGI Global. https://doi.org/ doi:10.4018/978-1-5225-2255-3.ch507

Eklund, P. (2021). Reinforcement Learning in Social Media Marketing. In B. Christiansen & T. Škrinjarić (Eds.), *Handbook of Research on Applied AI for International Business and Marketing Applications* (pp. 30–48). IGI Global. https://doi.org/10.4018/978-1-7998-5077-9.ch003

El Ghandour, N., Benaissa, M., & Lebbah, Y. (2021). An Integer Linear Programming-Based Method for the Extraction of Ontology Alignment. *International Journal of Information Technology and Web Engineering, 16*(2), 25–44. https://doi.org/10.4018/IJITWE.2021040102

Elias, N. I., & Walker, T. W. (2017). Factors that Contribute to Continued Use of E-Training among Healthcare Professionals. In F. Topor (Ed.), *Handbook of Research on Individualism and Identity in the Globalized Digital Age* (pp. 403–429). Hershey, PA: IGI Global. doi:10.4018/978-1-5225-0522-8.ch018

Fisher, R. L. (2018). Computer-Assisted Indian Matrimonial Services. In M. Khosrow-Pour, D.B.A. (Ed.), Encyclopedia of Information Science and Technology, Fourth Edition (pp. 4136-4145). Hershey, PA: IGI Global. doi:10.4018/978-1-5225-2255-3.ch358

Galiautdinov, R. (2021). Nonlinear Filtering in Artificial Neural Network Applications in Business and Engineering. In Q. Do (Ed.), *Artificial Neural Network Applications in Business and Engineering* (pp. 1–23). IGI Global. https://doi.org/10.4018/978-1-7998-3238-6.ch001

Gardner-McCune, C., & Jimenez, Y. (2017). Historical App Developers: Integrating CS into K-12 through Cross-Disciplinary Projects. In Y. Rankin & J. Thomas (Eds.), *Moving Students of Color from Consumers to Producers of Technology* (pp. 85–112). Hershey, PA: IGI Global. doi:10.4018/978-1-5225-2005-4.ch005

Garg, P. K. (2021). The Internet of Things-Based Technologies. In S. Kumar, M. Trivedi, P. Ranjan, & A. Punhani (Eds.), *Evolution of Software-Defined Networking Foundations for IoT and 5G Mobile Networks* (pp. 37–65). IGI Global. https://doi.org/10.4018/978-1-7998-4685-7.ch003

Related References

Garg, T., & Bharti, M. (2021). Congestion Control Protocols for UWSNs. In N. Goyal, L. Sapra, & J. Sandhu (Eds.), *Energy-Efficient Underwater Wireless Communications and Networking* (pp. 85–100). IGI Global. https://doi.org/10.4018/978-1-7998-3640-7.ch006

Gauttier, S. (2021). A Primer on Q-Method and the Study of Technology. In M. Khosrow-Pour D.B.A. (Eds.), *Encyclopedia of Information Science and Technology, Fifth Edition* (pp. 1746-1756). IGI Global. https://doi.org/10.4018/978-1-7998-3479-3.ch120

Ghafele, R., & Gibert, B. (2018). Open Growth: The Economic Impact of Open Source Software in the USA. In M. Khosrow-Pour (Ed.), *Optimizing Contemporary Application and Processes in Open Source Software* (pp. 164–197). Hershey, PA: IGI Global. doi:10.4018/978-1-5225-5314-4.ch007

Ghobakhloo, M., & Azar, A. (2018). Information Technology Resources, the Organizational Capability of Lean-Agile Manufacturing, and Business Performance. *Information Resources Management Journal, 31*(2), 47–74. doi:10.4018/IRMJ.2018040103

Gikandi, J. W. (2017). Computer-Supported Collaborative Learning and Assessment: A Strategy for Developing Online Learning Communities in Continuing Education. In J. Keengwe & G. Onchwari (Eds.), *Handbook of Research on Learner-Centered Pedagogy in Teacher Education and Professional Development* (pp. 309–333). Hershey, PA: IGI Global. doi:10.4018/978-1-5225-0892-2.ch017

Gokhale, A. A., & Machina, K. F. (2017). Development of a Scale to Measure Attitudes toward Information Technology. In L. Tomei (Ed.), *Exploring the New Era of Technology-Infused Education* (pp. 49–64). Hershey, PA: IGI Global. doi:10.4018/978-1-5225-1709-2.ch004

Goswami, J. K., Jalal, S., Negi, C. S., & Jalal, A. S. (2022). A Texture Features-Based Robust Facial Expression Recognition. *International Journal of Computer Vision and Image Processing, 12*(1), 1–15. https://doi.org/10.4018/IJCVIP.2022010103

Hafeez-Baig, A., Gururajan, R., & Wickramasinghe, N. (2017). Readiness as a Novel Construct of Readiness Acceptance Model (RAM) for the Wireless Handheld Technology. In N. Wickramasinghe (Ed.), *Handbook of Research on Healthcare Administration and Management* (pp. 578–595). Hershey, PA: IGI Global. doi:10.4018/978-1-5225-0920-2.ch035

Hanafizadeh, P., Ghandchi, S., & Asgarimehr, M. (2017). Impact of Information Technology on Lifestyle: A Literature Review and Classification. *International Journal of Virtual Communities and Social Networking*, 9(2), 1–23. doi:10.4018/IJVCSN.2017040101

Haseski, H. İ., Ilic, U., & Tuğtekin, U. (2018). Computational Thinking in Educational Digital Games: An Assessment Tool Proposal. In H. Ozcinar, G. Wong, & H. Ozturk (Eds.), *Teaching Computational Thinking in Primary Education* (pp. 256–287). Hershey, PA: IGI Global. doi:10.4018/978-1-5225-3200-2.ch013

Hee, W. J., Jalleh, G., Lai, H., & Lin, C. (2017). E-Commerce and IT Projects: Evaluation and Management Issues in Australian and Taiwanese Hospitals. *International Journal of Public Health Management and Ethics*, 2(1), 69–90. doi:10.4018/IJPHME.2017010104

Hernandez, A. A. (2017). Green Information Technology Usage: Awareness and Practices of Philippine IT Professionals. *International Journal of Enterprise Information Systems*, 13(4), 90–103. doi:10.4018/IJEIS.2017100106

Hernandez, M. A., Marin, E. C., Garcia-Rodriguez, J., Azorin-Lopez, J., & Cazorla, M. (2017). Automatic Learning Improves Human-Robot Interaction in Productive Environments: A Review. *International Journal of Computer Vision and Image Processing*, 7(3), 65–75. doi:10.4018/IJCVIP.2017070106

Hirota, A. (2021). Design of Narrative Creation in Innovation: "Signature Story" and Two Types of Pivots. In T. Ogata & J. Ono (Eds.), *Bridging the Gap Between AI, Cognitive Science, and Narratology With Narrative Generation* (pp. 363–376). IGI Global. https://doi.org/10.4018/978-1-7998-4864-6.ch012

Hond, D., Asgari, H., Jeffery, D., & Newman, M. (2021). An Integrated Process for Verifying Deep Learning Classifiers Using Dataset Dissimilarity Measures. *International Journal of Artificial Intelligence and Machine Learning*, 11(2), 1–21. https://doi.org/10.4018/IJAIML.289536

Horne-Popp, L. M., Tessone, E. B., & Welker, J. (2018). If You Build It, They Will Come: Creating a Library Statistics Dashboard for Decision-Making. In L. Costello & M. Powers (Eds.), *Developing In-House Digital Tools in Library Spaces* (pp. 177–203). Hershey, PA: IGI Global. doi:10.4018/978-1-5225-2676-6.ch009

Hu, H., Hu, P. J., & Al-Gahtani, S. S. (2017). User Acceptance of Computer Technology at Work in Arabian Culture: A Model Comparison Approach. In M. Khosrow-Pour (Ed.), *Handbook of Research on Technology Adoption, Social Policy, and Global Integration* (pp. 205–228). Hershey, PA: IGI Global. doi:10.4018/978-1-5225-2668-1.ch011

Huang, C., Sun, Y., & Fuh, C. (2022). Vehicle License Plate Recognition With Deep Learning. In C. Chen, W. Yang, & L. Chen (Eds.), *Technologies to Advance Automation in Forensic Science and Criminal Investigation* (pp. 161-219). IGI Global. https://doi.org/10.4018/978-1-7998-8386-9.ch009

Ifinedo, P. (2017). Using an Extended Theory of Planned Behavior to Study Nurses' Adoption of Healthcare Information Systems in Nova Scotia. *International Journal of Technology Diffusion*, 8(1), 1–17. doi:10.4018/IJTD.2017010101

Ilie, V., & Sneha, S. (2018). A Three Country Study for Understanding Physicians' Engagement With Electronic Information Resources Pre and Post System Implementation. *Journal of Global Information Management*, 26(2), 48–73. doi:10.4018/JGIM.2018040103

Ilo, P. I., Nkiko, C., Ugwu, C. I., Ekere, J. N., Izuagbe, R., & Fagbohun, M. O. (2021). Prospects and Challenges of Web 3.0 Technologies Application in the Provision of Library Services. In M. Khosrow-Pour D.B.A. (Ed.), *Encyclopedia of Information Science and Technology, Fifth Edition* (pp. 1767-1781). IGI Global. https://doi.org/10.4018/978-1-7998-3479-3.ch122

Inoue-Smith, Y. (2017). Perceived Ease in Using Technology Predicts Teacher Candidates' Preferences for Online Resources. *International Journal of Online Pedagogy and Course Design*, 7(3), 17–28. doi:10.4018/IJOPCD.2017070102

Islam, A. Y. (2017). Technology Satisfaction in an Academic Context: Moderating Effect of Gender. In A. Mesquita (Ed.), *Research Paradigms and Contemporary Perspectives on Human-Technology Interaction* (pp. 187–211). Hershey, PA: IGI Global. doi:10.4018/978-1-5225-1868-6.ch009

Jagdale, S. C., Hable, A. A., & Chabukswar, A. R. (2021). Protocol Development in Clinical Trials for Healthcare Management. In M. Khosrow-Pour D.B.A. (Ed.), *Encyclopedia of Information Science and Technology, Fifth Edition* (pp. 1797-1814). IGI Global. https://doi.org/10.4018/978-1-7998-3479-3.ch124

Jamil, G. L., & Jamil, C. C. (2017). Information and Knowledge Management Perspective Contributions for Fashion Studies: Observing Logistics and Supply Chain Management Processes. In G. Jamil, A. Soares, & C. Pessoa (Eds.), *Handbook of Research on Information Management for Effective Logistics and Supply Chains* (pp. 199–221). Hershey, PA: IGI Global. doi:10.4018/978-1-5225-0973-8.ch011

Jamil, M. I., & Almunawar, M. N. (2021). Importance of Digital Literacy and Hindrance Brought About by Digital Divide. In M. Khosrow-Pour D.B.A. (Ed.), *Encyclopedia of Information Science and Technology, Fifth Edition* (pp. 1683-1698). IGI Global. https://doi.org/10.4018/978-1-7998-3479-3.ch116

Janakova, M. (2018). Big Data and Simulations for the Solution of Controversies in Small Businesses. In M. Khosrow-Pour, D.B.A. (Ed.), Encyclopedia of Information Science and Technology, Fourth Edition (pp. 6907-6915). Hershey, PA: IGI Global. doi:10.4018/978-1-5225-2255-3.ch598

Jhawar, A., & Garg, S. K. (2018). Logistics Improvement by Investment in Information Technology Using System Dynamics. In A. Azar & S. Vaidyanathan (Eds.), *Advances in System Dynamics and Control* (pp. 528–567). Hershey, PA: IGI Global. doi:10.4018/978-1-5225-4077-9.ch017

Kalelioğlu, F., Gülbahar, Y., & Doğan, D. (2018). Teaching How to Think Like a Programmer: Emerging Insights. In H. Ozcinar, G. Wong, & H. Ozturk (Eds.), *Teaching Computational Thinking in Primary Education* (pp. 18–35). Hershey, PA: IGI Global. doi:10.4018/978-1-5225-3200-2.ch002

Kamberi, S. (2017). A Girls-Only Online Virtual World Environment and its Implications for Game-Based Learning. In A. Stricker, C. Calongne, B. Truman, & F. Arenas (Eds.), *Integrating an Awareness of Selfhood and Society into Virtual Learning* (pp. 74–95). Hershey, PA: IGI Global. doi:10.4018/978-1-5225-2182-2.ch006

Kamel, S., & Rizk, N. (2017). ICT Strategy Development: From Design to Implementation – Case of Egypt. In C. Howard & K. Hargiss (Eds.), *Strategic Information Systems and Technologies in Modern Organizations* (pp. 239–257). Hershey, PA: IGI Global. doi:10.4018/978-1-5225-1680-4.ch010

Kamel, S. H. (2018). The Potential Role of the Software Industry in Supporting Economic Development. In M. Khosrow-Pour, D.B.A. (Ed.), Encyclopedia of Information Science and Technology, Fourth Edition (pp. 7259-7269). Hershey, PA: IGI Global. doi:10.4018/978-1-5225-2255-3.ch631

Kang, H., Kang, Y., & Kim, J. (2022). Improved Fall Detection Model on GRU Using PoseNet. *International Journal of Software Innovation*, 10(2), 1–11. https://doi.org/10.4018/IJSI.289600

Kankam, P. K. (2021). Employing Case Study and Survey Designs in Information Research. *Journal of Information Technology Research*, 14(1), 167–177. https://doi.org/10.4018/JITR.2021010110

Karas, V., & Schuller, B. W. (2021). Deep Learning for Sentiment Analysis: An Overview and Perspectives. In F. Pinarbasi & M. Taskiran (Eds.), *Natural Language Processing for Global and Local Business* (pp. 97–132). IGI Global. https://doi.org/10.4018/978-1-7998-4240-8.ch005

Kaufman, L. M. (2022). Reimagining the Magic of the Workshop Model. In T. Driscoll III, (Ed.), *Designing Effective Distance and Blended Learning Environments in K-12* (pp. 89–109). IGI Global. https://doi.org/10.4018/978-1-7998-6829-3.ch007

Kawata, S. (2018). Computer-Assisted Parallel Program Generation. In M. Khosrow-Pour, D.B.A. (Ed.), Encyclopedia of Information Science and Technology, Fourth Edition (pp. 4583-4593). Hershey, PA: IGI Global. doi:10.4018/978-1-5225-2255-3.ch398

Kharb, L., & Singh, P. (2021). Role of Machine Learning in Modern Education and Teaching. In S. Verma & P. Tomar (Ed.), *Impact of AI Technologies on Teaching, Learning, and Research in Higher Education* (pp. 99-123). IGI Global. https://doi.org/10.4018/978-1-7998-4763-2.ch006

Khari, M., Shrivastava, G., Gupta, S., & Gupta, R. (2017). Role of Cyber Security in Today's Scenario. In R. Kumar, P. Pattnaik, & P. Pandey (Eds.), *Detecting and Mitigating Robotic Cyber Security Risks* (pp. 177–191). Hershey, PA: IGI Global. doi:10.4018/978-1-5225-2154-9.ch013

Khekare, G., & Sheikh, S. (2021). Autonomous Navigation Using Deep Reinforcement Learning in ROS. *International Journal of Artificial Intelligence and Machine Learning, 11*(2), 63–70. https://doi.org/10.4018/IJAIML.20210701.oa4

Khouja, M., Rodriguez, I. B., Ben Halima, Y., & Moalla, S. (2018). IT Governance in Higher Education Institutions: A Systematic Literature Review. *International Journal of Human Capital and Information Technology Professionals, 9*(2), 52–67. doi:10.4018/IJHCITP.2018040104

Kiourt, C., Pavlidis, G., Koutsoudis, A., & Kalles, D. (2017). Realistic Simulation of Cultural Heritage. *International Journal of Computational Methods in Heritage Science, 1*(1), 10–40. doi:10.4018/IJCMHS.2017010102

Köse, U. (2017). An Augmented-Reality-Based Intelligent Mobile Application for Open Computer Education. In G. Kurubacak & H. Altinpulluk (Eds.), *Mobile Technologies and Augmented Reality in Open Education* (pp. 154–174). Hershey, PA: IGI Global. doi:10.4018/978-1-5225-2110-5.ch008

Lahmiri, S. (2018). Information Technology Outsourcing Risk Factors and Provider Selection. In M. Gupta, R. Sharman, J. Walp, & P. Mulgund (Eds.), *Information Technology Risk Management and Compliance in Modern Organizations* (pp. 214–228). Hershey, PA: IGI Global. doi:10.4018/978-1-5225-2604-9.ch008

Lakkad, A. K., Bhadaniya, R. D., Shah, V. N., & Lavanya, K. (2021). Complex Events Processing on Live News Events Using Apache Kafka and Clustering Techniques. *International Journal of Intelligent Information Technologies*, *17*(1), 39–52. https://doi.org/10.4018/IJIIT.2021010103

Landriscina, F. (2017). Computer-Supported Imagination: The Interplay Between Computer and Mental Simulation in Understanding Scientific Concepts. In I. Levin & D. Tsybulsky (Eds.), *Digital Tools and Solutions for Inquiry-Based STEM Learning* (pp. 33–60). Hershey, PA: IGI Global. doi:10.4018/978-1-5225-2525-7.ch002

Lara López, G. (2021). Virtual Reality in Object Location. In A. Negrón & M. Muñoz (Eds.), *Latin American Women and Research Contributions to the IT Field* (pp. 307–324). IGI Global. https://doi.org/10.4018/978-1-7998-7552-9.ch014

Lee, W. W. (2018). Ethical Computing Continues From Problem to Solution. In M. Khosrow-Pour, D.B.A. (Ed.), Encyclopedia of Information Science and Technology, Fourth Edition (pp. 4884-4897). Hershey, PA: IGI Global. doi:10.4018/978-1-5225-2255-3.ch423

Lin, S., Chen, S., & Chuang, S. (2017). Perceived Innovation and Quick Response Codes in an Online-to-Offline E-Commerce Service Model. *International Journal of E-Adoption*, *9*(2), 1–16. doi:10.4018/IJEA.2017070101

Liu, M., Wang, Y., Xu, W., & Liu, L. (2017). Automated Scoring of Chinese Engineering Students' English Essays. *International Journal of Distance Education Technologies*, *15*(1), 52–68. doi:10.4018/IJDET.2017010104

Ma, X., Li, X., Zhong, B., Huang, Y., Gu, Y., Wu, M., Liu, Y., & Zhang, M. (2021). A Detector and Evaluation Framework of Abnormal Bidding Behavior Based on Supplier Portrait. *International Journal of Information Technology and Web Engineering*, *16*(2), 58–74. https://doi.org/10.4018/IJITWE.2021040104

Mabe, L. K., & Oladele, O. I. (2017). Application of Information Communication Technologies for Agricultural Development through Extension Services: A Review. In T. Tossy (Ed.), *Information Technology Integration for Socio-Economic Development* (pp. 52–101). Hershey, PA: IGI Global. doi:10.4018/978-1-5225-0539-6.ch003

Mahboub, S. A., Sayed Ali Ahmed, E., & Saeed, R. A. (2021). Smart IDS and IPS for Cyber-Physical Systems. In A. Luhach & A. Elçi (Eds.), *Artificial Intelligence Paradigms for Smart Cyber-Physical Systems* (pp. 109–136). IGI Global. https://doi.org/10.4018/978-1-7998-5101-1.ch006

Manogaran, G., Thota, C., & Lopez, D. (2018). Human-Computer Interaction With Big Data Analytics. In D. Lopez & M. Durai (Eds.), *HCI Challenges and Privacy Preservation in Big Data Security* (pp. 1–22). Hershey, PA: IGI Global. doi:10.4018/978-1-5225-2863-0.ch001

Margolis, J., Goode, J., & Flapan, J. (2017). A Critical Crossroads for Computer Science for All: "Identifying Talent" or "Building Talent," and What Difference Does It Make? In Y. Rankin & J. Thomas (Eds.), *Moving Students of Color from Consumers to Producers of Technology* (pp. 1–23). Hershey, PA: IGI Global. doi:10.4018/978-1-5225-2005-4.ch001

Mazzù, M. F., Benetton, A., Baccelloni, A., & Lavini, L. (2022). A Milk Blockchain-Enabled Supply Chain: Evidence From Leading Italian Farms. In P. De Giovanni (Ed.), *Blockchain Technology Applications in Businesses and Organizations* (pp. 73–98). IGI Global. https://doi.org/10.4018/978-1-7998-8014-1.ch004

Mbale, J. (2018). Computer Centres Resource Cloud Elasticity-Scalability (CRECES): Copperbelt University Case Study. In S. Aljawarneh & M. Malhotra (Eds.), *Critical Research on Scalability and Security Issues in Virtual Cloud Environments* (pp. 48–70). Hershey, PA: IGI Global. doi:10.4018/978-1-5225-3029-9.ch003

McKee, J. (2018). The Right Information: The Key to Effective Business Planning. In *Business Architectures for Risk Assessment and Strategic Planning: Emerging Research and Opportunities* (pp. 38–52). Hershey, PA: IGI Global. doi:10.4018/978-1-5225-3392-4.ch003

Meddah, I. H., Remil, N. E., & Meddah, H. N. (2021). Novel Approach for Mining Patterns. *International Journal of Applied Evolutionary Computation*, *12*(1), 27–42. https://doi.org/10.4018/IJAEC.2021010103

Mensah, I. K., & Mi, J. (2018). Determinants of Intention to Use Local E-Government Services in Ghana: The Perspective of Local Government Workers. *International Journal of Technology Diffusion*, *9*(2), 41–60. doi:10.4018/IJTD.2018040103

Mohamed, J. H. (2018). Scientograph-Based Visualization of Computer Forensics Research Literature. In J. Jeyasekar & P. Saravanan (Eds.), *Innovations in Measuring and Evaluating Scientific Information* (pp. 148–162). Hershey, PA: IGI Global. doi:10.4018/978-1-5225-3457-0.ch010

Montañés-Del Río, M. Á., Cornejo, V. R., Rodríguez, M. R., & Ortiz, J. S. (2021). Gamification of University Subjects: A Case Study for Operations Management. *Journal of Information Technology Research*, *14*(2), 1–29. https://doi.org/10.4018/JITR.2021040101

Moore, R. L., & Johnson, N. (2017). Earning a Seat at the Table: How IT Departments Can Partner in Organizational Change and Innovation. *International Journal of Knowledge-Based Organizations*, 7(2), 1–12. doi:10.4018/IJKBO.2017040101

Mukul, M. K., & Bhattaharyya, S. (2017). Brain-Machine Interface: Human-Computer Interaction. In E. Noughabi, B. Raahemi, A. Albadvi, & B. Far (Eds.), *Handbook of Research on Data Science for Effective Healthcare Practice and Administration* (pp. 417–443). Hershey, PA: IGI Global. doi:10.4018/978-1-5225-2515-8.ch018

Na, L. (2017). Library and Information Science Education and Graduate Programs in Academic Libraries. In L. Ruan, Q. Zhu, & Y. Ye (Eds.), *Academic Library Development and Administration in China* (pp. 218–229). Hershey, PA: IGI Global. doi:10.4018/978-1-5225-0550-1.ch013

Nagpal, G., Bishnoi, G. K., Dhami, H. S., & Vijayvargia, A. (2021). Use of Data Analytics to Increase the Efficiency of Last Mile Logistics for Ecommerce Deliveries. In B. Patil & M. Vohra (Eds.), *Handbook of Research on Engineering, Business, and Healthcare Applications of Data Science and Analytics* (pp. 167–180). IGI Global. https://doi.org/10.4018/978-1-7998-3053-5.ch009

Nair, S. M., Ramesh, V., & Tyagi, A. K. (2021). Issues and Challenges (Privacy, Security, and Trust) in Blockchain-Based Applications. In A. Tyagi, G. Rekha, & N. Sreenath (Eds.), *Opportunities and Challenges for Blockchain Technology in Autonomous Vehicles* (pp. 196–209). IGI Global. https://doi.org/10.4018/978-1-7998-3295-9.ch012

Naomi, J. F. M., K., & V., S. (2021). Machine and Deep Learning Techniques in IoT and Cloud. In S. Velayutham (Ed.), *Challenges and Opportunities for the Convergence of IoT, Big Data, and Cloud Computing* (pp. 225-247). IGI Global. https://doi.org/10.4018/978-1-7998-3111-2.ch013

Nath, R., & Murthy, V. N. (2018). What Accounts for the Differences in Internet Diffusion Rates Around the World? In M. Khosrow-Pour, D.B.A. (Ed.), Encyclopedia of Information Science and Technology, Fourth Edition (pp. 8095-8104). Hershey, PA: IGI Global. https://doi.org/ doi:10.4018/978-1-5225-2255-3.ch705

Nedelko, Z., & Potocan, V. (2018). The Role of Emerging Information Technologies for Supporting Supply Chain Management. In M. Khosrow-Pour, D.B.A. (Ed.), Encyclopedia of Information Science and Technology, Fourth Edition (pp. 5559-5569). Hershey, PA: IGI Global. doi:10.4018/978-1-5225-2255-3.ch483

Negrini, L., Giang, C., & Bonnet, E. (2022). Designing Tools and Activities for Educational Robotics in Online Learning. In N. Eteokleous & E. Nisiforou (Eds.), *Designing, Constructing, and Programming Robots for Learning* (pp. 202–222). IGI Global. https://doi.org/10.4018/978-1-7998-7443-0.ch010

Ngafeeson, M. N. (2018). User Resistance to Health Information Technology. In M. Khosrow-Pour, D.B.A. (Ed.), Encyclopedia of Information Science and Technology, Fourth Edition (pp. 3816-3825). Hershey, PA: IGI Global. doi:10.4018/978-1-5225-2255-3.ch331

Nguyen, T. T., Giang, N. L., Tran, D. T., Nguyen, T. T., Nguyen, H. Q., Pham, A. V., & Vu, T. D. (2021). A Novel Filter-Wrapper Algorithm on Intuitionistic Fuzzy Set for Attribute Reduction From Decision Tables. *International Journal of Data Warehousing and Mining*, *17*(4), 67–100. https://doi.org/10.4018/IJDWM.2021100104

Nigam, A., & Dewani, P. P. (2022). Consumer Engagement Through Conditional Promotions: An Exploratory Study. *Journal of Global Information Management*, *30*(5), 1–19. https://doi.org/10.4018/JGIM.290364

Odagiri, K. (2017). Introduction of Individual Technology to Constitute the Current Internet. In *Strategic Policy-Based Network Management in Contemporary Organizations* (pp. 20–96). Hershey, PA: IGI Global. doi:10.4018/978-1-68318-003-6.ch003

Odia, J. O., & Akpata, O. T. (2021). Role of Data Science and Data Analytics in Forensic Accounting and Fraud Detection. In B. Patil & M. Vohra (Eds.), *Handbook of Research on Engineering, Business, and Healthcare Applications of Data Science and Analytics* (pp. 203–227). IGI Global. https://doi.org/10.4018/978-1-7998-3053-5.ch011

Okike, E. U. (2018). Computer Science and Prison Education. In I. Biao (Ed.), *Strategic Learning Ideologies in Prison Education Programs* (pp. 246–264). Hershey, PA: IGI Global. doi:10.4018/978-1-5225-2909-5.ch012

Olelewe, C. J., & Nwafor, I. P. (2017). Level of Computer Appreciation Skills Acquired for Sustainable Development by Secondary School Students in Nsukka LGA of Enugu State, Nigeria. In C. Ayo & V. Mbarika (Eds.), *Sustainable ICT Adoption and Integration for Socio-Economic Development* (pp. 214–233). Hershey, PA: IGI Global. doi:10.4018/978-1-5225-2565-3.ch010

Oliveira, M., Maçada, A. C., Curado, C., & Nodari, F. (2017). Infrastructure Profiles and Knowledge Sharing. *International Journal of Technology and Human Interaction*, *13*(3), 1–12. doi:10.4018/IJTHI.2017070101

Otarkhani, A., Shokouhyar, S., & Pour, S. S. (2017). Analyzing the Impact of Governance of Enterprise IT on Hospital Performance: Tehran's (Iran) Hospitals – A Case Study. *International Journal of Healthcare Information Systems and Informatics*, *12*(3), 1–20. doi:10.4018/IJHISI.2017070101

Otunla, A. O., & Amuda, C. O. (2018). Nigerian Undergraduate Students' Computer Competencies and Use of Information Technology Tools and Resources for Study Skills and Habits' Enhancement. In M. Khosrow-Pour, D.B.A. (Ed.), Encyclopedia of Information Science and Technology, Fourth Edition (pp. 2303-2313). Hershey, PA: IGI Global. https://doi.org/ doi:10.4018/978-1-5225-2255-3.ch200

Özçınar, H. (2018). A Brief Discussion on Incentives and Barriers to Computational Thinking Education. In H. Ozcinar, G. Wong, & H. Ozturk (Eds.), *Teaching Computational Thinking in Primary Education* (pp. 1–17). Hershey, PA: IGI Global. doi:10.4018/978-1-5225-3200-2.ch001

Pandey, J. M., Garg, S., Mishra, P., & Mishra, B. P. (2017). Computer Based Psychological Interventions: Subject to the Efficacy of Psychological Services. *International Journal of Computers in Clinical Practice*, *2*(1), 25–33. doi:10.4018/IJCCP.2017010102

Pandkar, S. D., & Paatil, S. D. (2021). Big Data and Knowledge Resource Centre. In S. Dhamdhere (Ed.), *Big Data Applications for Improving Library Services* (pp. 90–106). IGI Global. https://doi.org/10.4018/978-1-7998-3049-8.ch007

Patro, C. (2017). Impulsion of Information Technology on Human Resource Practices. In P. Ordóñez de Pablos (Ed.), *Managerial Strategies and Solutions for Business Success in Asia* (pp. 231–254). Hershey, PA: IGI Global. doi:10.4018/978-1-5225-1886-0.ch013

Patro, C. S., & Raghunath, K. M. (2017). Information Technology Paraphernalia for Supply Chain Management Decisions. In M. Tavana (Ed.), *Enterprise Information Systems and the Digitalization of Business Functions* (pp. 294–320). Hershey, PA: IGI Global. doi:10.4018/978-1-5225-2382-6.ch014

Paul, P. K. (2018). The Context of IST for Solid Information Retrieval and Infrastructure Building: Study of Developing Country. *International Journal of Information Retrieval Research*, *8*(1), 86–100. doi:10.4018/IJIRR.2018010106

Related References

Paul, P. K., & Chatterjee, D. (2018). iSchools Promoting "Information Science and Technology" (IST) Domain Towards Community, Business, and Society With Contemporary Worldwide Trend and Emerging Potentialities in India. In M. Khosrow-Pour, D.B.A. (Ed.), Encyclopedia of Information Science and Technology, Fourth Edition (pp. 4723-4735). Hershey, PA: IGI Global. https://doi.org/ doi:10.4018/978-1-5225-2255-3.ch410

Pessoa, C. R., & Marques, M. E. (2017). Information Technology and Communication Management in Supply Chain Management. In G. Jamil, A. Soares, & C. Pessoa (Eds.), *Handbook of Research on Information Management for Effective Logistics and Supply Chains* (pp. 23–33). Hershey, PA: IGI Global. doi:10.4018/978-1-5225-0973-8.ch002

Pineda, R. G. (2018). Remediating Interaction: Towards a Philosophy of Human-Computer Relationship. In M. Khosrow-Pour (Ed.), *Enhancing Art, Culture, and Design With Technological Integration* (pp. 75–98). Hershey, PA: IGI Global. doi:10.4018/978-1-5225-5023-5.ch004

Prabha, V. D., & R., R. (2021). Clinical Decision Support Systems: Decision-Making System for Clinical Data. In G. Rani & P. Tiwari (Eds.), *Handbook of Research on Disease Prediction Through Data Analytics and Machine Learning* (pp. 268-280). IGI Global. https://doi.org/10.4018/978-1-7998-2742-9.ch014

Pushpa, R., & Siddappa, M. (2021). An Optimal Way of VM Placement Strategy in Cloud Computing Platform Using ABCS Algorithm. *International Journal of Ambient Computing and Intelligence*, *12*(3), 16–38. https://doi.org/10.4018/IJACI.2021070102

Qian, Y. (2017). Computer Simulation in Higher Education: Affordances, Opportunities, and Outcomes. In P. Vu, S. Fredrickson, & C. Moore (Eds.), *Handbook of Research on Innovative Pedagogies and Technologies for Online Learning in Higher Education* (pp. 236–262). Hershey, PA: IGI Global. doi:10.4018/978-1-5225-1851-8.ch011

Rahman, N. (2017). Lessons from a Successful Data Warehousing Project Management. *International Journal of Information Technology Project Management*, *8*(4), 30–45. doi:10.4018/IJITPM.2017100103

Rahman, N. (2018). Environmental Sustainability in the Computer Industry for Competitive Advantage. In M. Khosrow-Pour (Ed.), *Green Computing Strategies for Competitive Advantage and Business Sustainability* (pp. 110–130). Hershey, PA: IGI Global. doi:10.4018/978-1-5225-5017-4.ch006

Rajh, A., & Pavetic, T. (2017). Computer Generated Description as the Required Digital Competence in Archival Profession. *International Journal of Digital Literacy and Digital Competence, 8*(1), 36–49. doi:10.4018/IJDLDC.2017010103

Raman, A., & Goyal, D. P. (2017). Extending IMPLEMENT Framework for Enterprise Information Systems Implementation to Information System Innovation. In M. Tavana (Ed.), *Enterprise Information Systems and the Digitalization of Business Functions* (pp. 137–177). Hershey, PA: IGI Global. doi:10.4018/978-1-5225-2382-6.ch007

Rao, A. P., & Reddy, K. S. (2021). Automated Soil Residue Levels Detecting Device With IoT Interface. In V. Sathiyamoorthi & A. Elci (Eds.), *Challenges and Applications of Data Analytics in Social Perspectives* (Vol. S, pp. 123–135). IGI Global. https://doi.org/10.4018/978-1-7998-2566-1.ch007

Rao, Y. S., Rauta, A. K., Saini, H., & Panda, T. C. (2017). Mathematical Model for Cyber Attack in Computer Network. *International Journal of Business Data Communications and Networking, 13*(1), 58–65. doi:10.4018/IJBDCN.2017010105

Rapaport, W. J. (2018). Syntactic Semantics and the Proper Treatment of Computationalism. In M. Danesi (Ed.), *Empirical Research on Semiotics and Visual Rhetoric* (pp. 128–176). Hershey, PA: IGI Global. doi:10.4018/978-1-5225-5622-0.ch007

Raut, R., Priyadarshinee, P., & Jha, M. (2017). Understanding the Mediation Effect of Cloud Computing Adoption in Indian Organization: Integrating TAM-TOE- Risk Model. *International Journal of Service Science, Management, Engineering, and Technology, 8*(3), 40–59. doi:10.4018/IJSSMET.2017070103

Rezaie, S., Mirabedini, S. J., & Abtahi, A. (2018). Designing a Model for Implementation of Business Intelligence in the Banking Industry. *International Journal of Enterprise Information Systems, 14*(1), 77–103. doi:10.4018/IJEIS.2018010105

Rezende, D. A. (2018). Strategic Digital City Projects: Innovative Information and Public Services Offered by Chicago (USA) and Curitiba (Brazil). In M. Lytras, L. Daniela, & A. Visvizi (Eds.), *Enhancing Knowledge Discovery and Innovation in the Digital Era* (pp. 204–223). Hershey, PA: IGI Global. doi:10.4018/978-1-5225-4191-2.ch012

Rodriguez, A., Rico-Diaz, A. J., Rabuñal, J. R., & Gestal, M. (2017). Fish Tracking with Computer Vision Techniques: An Application to Vertical Slot Fishways. In M. S., & V. V. (Eds.), Multi-Core Computer Vision and Image Processing for Intelligent Applications (pp. 74-104). Hershey, PA: IGI Global. https://doi.org/doi:10.4018/978-1-5225-0889-2.ch003

Romero, J. A. (2018). Sustainable Advantages of Business Value of Information Technology. In M. Khosrow-Pour, D.B.A. (Ed.), Encyclopedia of Information Science and Technology, Fourth Edition (pp. 923-929). Hershey, PA: IGI Global. doi:10.4018/978-1-5225-2255-3.ch079

Romero, J. A. (2018). The Always-On Business Model and Competitive Advantage. In N. Bajgoric (Ed.), *Always-On Enterprise Information Systems for Modern Organizations* (pp. 23–40). Hershey, PA: IGI Global. doi:10.4018/978-1-5225-3704-5.ch002

Rosen, Y. (2018). Computer Agent Technologies in Collaborative Learning and Assessment. In M. Khosrow-Pour, D.B.A. (Ed.), Encyclopedia of Information Science and Technology, Fourth Edition (pp. 2402-2410). Hershey, PA: IGI Global. doi:10.4018/978-1-5225-2255-3.ch209

Roy, D. (2018). Success Factors of Adoption of Mobile Applications in Rural India: Effect of Service Characteristics on Conceptual Model. In M. Khosrow-Pour (Ed.), *Green Computing Strategies for Competitive Advantage and Business Sustainability* (pp. 211–238). Hershey, PA: IGI Global. doi:10.4018/978-1-5225-5017-4.ch010

Ruffin, T. R., & Hawkins, D. P. (2018). Trends in Health Care Information Technology and Informatics. In M. Khosrow-Pour, D.B.A. (Ed.), Encyclopedia of Information Science and Technology, Fourth Edition (pp. 3805-3815). Hershey, PA: IGI Global. doi:10.4018/978-1-5225-2255-3.ch330

Sadasivam, U. M., & Ganesan, N. (2021). Detecting Fake News Using Deep Learning and NLP. In S. Misra, C. Arumugam, S. Jaganathan, & S. S. (Eds.), *Confluence of AI, Machine, and Deep Learning in Cyber Forensics* (pp. 117-133). IGI Global. https://doi.org/10.4018/978-1-7998-4900-1.ch007

Safari, M. R., & Jiang, Q. (2018). The Theory and Practice of IT Governance Maturity and Strategies Alignment: Evidence From Banking Industry. *Journal of Global Information Management, 26*(2), 127–146. doi:10.4018/JGIM.2018040106

Sahin, H. B., & Anagun, S. S. (2018). Educational Computer Games in Math Teaching: A Learning Culture. In E. Toprak & E. Kumtepe (Eds.), *Supporting Multiculturalism in Open and Distance Learning Spaces* (pp. 249–280). Hershey, PA: IGI Global. doi:10.4018/978-1-5225-3076-3.ch013

Sakalle, A., Tomar, P., Bhardwaj, H., & Sharma, U. (2021). Impact and Latest Trends of Intelligent Learning With Artificial Intelligence. In S. Verma & P. Tomar (Eds.), *Impact of AI Technologies on Teaching, Learning, and Research in Higher Education* (pp. 172-189). IGI Global. https://doi.org/10.4018/978-1-7998-4763-2.ch011

Sala, N. (2021). Virtual Reality, Augmented Reality, and Mixed Reality in Education: A Brief Overview. In D. Choi, A. Dailey-Hebert, & J. Estes (Eds.), *Current and Prospective Applications of Virtual Reality in Higher Education* (pp. 48–73). IGI Global. https://doi.org/10.4018/978-1-7998-4960-5.ch003

Salunkhe, S., Kanagachidambaresan, G., Rajkumar, C., & Jayanthi, K. (2022). Online Detection and Prediction of Fused Deposition Modelled Parts Using Artificial Intelligence. In S. Salunkhe, H. Hussein, & J. Davim (Eds.), *Applications of Artificial Intelligence in Additive Manufacturing* (pp. 194–209). IGI Global. https://doi.org/10.4018/978-1-7998-8516-0.ch009

Samy, V. S., Pramanick, K., Thenkanidiyoor, V., & Victor, J. (2021). Data Analysis and Visualization in Python for Polar Meteorological Data. *International Journal of Data Analytics*, 2(1), 32–60. https://doi.org/10.4018/IJDA.2021010102

Sanna, A., & Valpreda, F. (2017). An Assessment of the Impact of a Collaborative Didactic Approach and Students' Background in Teaching Computer Animation. *International Journal of Information and Communication Technology Education*, 13(4), 1–16. doi:10.4018/IJICTE.2017100101

Sarivougioukas, J., & Vagelatos, A. (2022). Fused Contextual Data With Threading Technology to Accelerate Processing in Home UbiHealth. *International Journal of Software Science and Computational Intelligence*, 14(1), 1–14. https://doi.org/10.4018/IJSSCI.285590

Scott, A., Martin, A., & McAlear, F. (2017). Enhancing Participation in Computer Science among Girls of Color: An Examination of a Preparatory AP Computer Science Intervention. In Y. Rankin & J. Thomas (Eds.), *Moving Students of Color from Consumers to Producers of Technology* (pp. 62–84). Hershey, PA: IGI Global. doi:10.4018/978-1-5225-2005-4.ch004

Shanmugam, M., Ibrahim, N., Gorment, N. Z., Sugu, R., Dandarawi, T. N., & Ahmad, N. A. (2022). Towards an Integrated Omni-Channel Strategy Framework for Improved Customer Interaction. In P. Lai (Ed.), *Handbook of Research on Social Impacts of E-Payment and Blockchain Technology* (pp. 409–427). IGI Global. https://doi.org/10.4018/978-1-7998-9035-5.ch022

Sharma, A., & Kumar, S. (2021). Network Slicing and the Role of 5G in IoT Applications. In S. Kumar, M. Trivedi, P. Ranjan, & A. Punhani (Eds.), *Evolution of Software-Defined Networking Foundations for IoT and 5G Mobile Networks* (pp. 172–190). IGI Global. https://doi.org/10.4018/978-1-7998-4685-7.ch010

Related References

Siddoo, V., & Wongsai, N. (2017). Factors Influencing the Adoption of ISO/IEC 29110 in Thai Government Projects: A Case Study. *International Journal of Information Technologies and Systems Approach, 10*(1), 22–44. doi:10.4018/IJITSA.2017010102

Silveira, C., Hir, M. E., & Chaves, H. K. (2022). An Approach to Information Management as a Subsidy of Global Health Actions: A Case Study of Big Data in Health for Dengue, Zika, and Chikungunya. In J. Lima de Magalhães, Z. Hartz, G. Jamil, H. Silveira, & L. Jamil (Eds.), *Handbook of Research on Essential Information Approaches to Aiding Global Health in the One Health Context* (pp. 219–234). IGI Global. https://doi.org/10.4018/978-1-7998-8011-0.ch012

Simões, A. (2017). Using Game Frameworks to Teach Computer Programming. In R. Alexandre Peixoto de Queirós & M. Pinto (Eds.), *Gamification-Based E-Learning Strategies for Computer Programming Education* (pp. 221–236). Hershey, PA: IGI Global. doi:10.4018/978-1-5225-1034-5.ch010

Simões de Almeida, R., & da Silva, T. (2022). AI Chatbots in Mental Health: Are We There Yet? In A. Marques & R. Queirós (Eds.), *Digital Therapies in Psychosocial Rehabilitation and Mental Health* (pp. 226–243). IGI Global. https://doi.org/10.4018/978-1-7998-8634-1.ch011

Singh, L. K., Khanna, M., Thawkar, S., & Gopal, J. (2021). Robustness for Authentication of the Human Using Face, Ear, and Gait Multimodal Biometric System. *International Journal of Information System Modeling and Design, 12*(1), 39–72. https://doi.org/10.4018/IJISMD.2021010103

Sllame, A. M. (2017). Integrating LAB Work With Classes in Computer Network Courses. In H. Alphin Jr, R. Chan, & J. Lavine (Eds.), *The Future of Accessibility in International Higher Education* (pp. 253–275). Hershey, PA: IGI Global. doi:10.4018/978-1-5225-2560-8.ch015

Smirnov, A., Ponomarev, A., Shilov, N., Kashevnik, A., & Teslya, N. (2018). Ontology-Based Human-Computer Cloud for Decision Support: Architecture and Applications in Tourism. *International Journal of Embedded and Real-Time Communication Systems, 9*(1), 1–19. doi:10.4018/IJERTCS.2018010101

Smith-Ditizio, A. A., & Smith, A. D. (2018). Computer Fraud Challenges and Its Legal Implications. In M. Khosrow-Pour, D.B.A. (Ed.), Encyclopedia of Information Science and Technology, Fourth Edition (pp. 4837-4848). Hershey, PA: IGI Global. doi:10.4018/978-1-5225-2255-3.ch419

Sosnin, P. (2018). Figuratively Semantic Support of Human-Computer Interactions. In *Experience-Based Human-Computer Interactions: Emerging Research and Opportunities* (pp. 244–272). Hershey, PA: IGI Global. doi:10.4018/978-1-5225-2987-3.ch008

Srilakshmi, R., & Jaya Bhaskar, M. (2021). An Adaptable Secure Scheme in Mobile Ad hoc Network to Protect the Communication Channel From Malicious Behaviours. *International Journal of Information Technology and Web Engineering*, *16*(3), 54–73. https://doi.org/10.4018/IJITWE.2021070104

Sukhwani, N., Kagita, V. R., Kumar, V., & Panda, S. K. (2021). Efficient Computation of Top-K Skyline Objects in Data Set With Uncertain Preferences. *International Journal of Data Warehousing and Mining*, *17*(3), 68–80. https://doi.org/10.4018/IJDWM.2021070104

Susanto, H., Yie, L. F., Setiana, D., Asih, Y., Yoganingrum, A., Riyanto, S., & Saputra, F. A. (2021). Digital Ecosystem Security Issues for Organizations and Governments: Digital Ethics and Privacy. In Z. Mahmood (Ed.), *Web 2.0 and Cloud Technologies for Implementing Connected Government* (pp. 204–228). IGI Global. https://doi.org/10.4018/978-1-7998-4570-6.ch010

Syväjärvi, A., Leinonen, J., Kivivirta, V., & Kesti, M. (2017). The Latitude of Information Management in Local Government: Views of Local Government Managers. *International Journal of Electronic Government Research*, *13*(1), 69–85. doi:10.4018/IJEGR.2017010105

Tanque, M., & Foxwell, H. J. (2018). Big Data and Cloud Computing: A Review of Supply Chain Capabilities and Challenges. In A. Prasad (Ed.), *Exploring the Convergence of Big Data and the Internet of Things* (pp. 1–28). Hershey, PA: IGI Global. doi:10.4018/978-1-5225-2947-7.ch001

Teixeira, A., Gomes, A., & Orvalho, J. G. (2017). Auditory Feedback in a Computer Game for Blind People. In T. Issa, P. Kommers, T. Issa, P. Isaías, & T. Issa (Eds.), *Smart Technology Applications in Business Environments* (pp. 134–158). Hershey, PA: IGI Global. doi:10.4018/978-1-5225-2492-2.ch007

Tewari, P., Tiwari, P., & Goel, R. (2022). Information Technology in Supply Chain Management. In V. Garg & R. Goel (Eds.), *Handbook of Research on Innovative Management Using AI in Industry 5.0* (pp. 165–178). IGI Global. https://doi.org/10.4018/978-1-7998-8497-2.ch011

Thompson, N., McGill, T., & Murray, D. (2018). Affect-Sensitive Computer Systems. In M. Khosrow-Pour, D.B.A. (Ed.), Encyclopedia of Information Science and Technology, Fourth Edition (pp. 4124-4135). Hershey, PA: IGI Global. doi:10.4018/978-1-5225-2255-3.ch357

Triberti, S., Brivio, E., & Galimberti, C. (2018). On Social Presence: Theories, Methodologies, and Guidelines for the Innovative Contexts of Computer-Mediated Learning. In M. Marmon (Ed.), *Enhancing Social Presence in Online Learning Environments* (pp. 20–41). Hershey, PA: IGI Global. doi:10.4018/978-1-5225-3229-3.ch002

Tripathy, B. K. T. R., S., & Mohanty, R. K. (2018). Memetic Algorithms and Their Applications in Computer Science. In S. Dash, B. Tripathy, & A. Rahman (Eds.), Handbook of Research on Modeling, Analysis, and Application of Nature-Inspired Metaheuristic Algorithms (pp. 73-93). Hershey, PA: IGI Global. https://doi.org/doi:10.4018/978-1-5225-2857-9.ch004

Turulja, L., & Bajgoric, N. (2017). Human Resource Management IT and Global Economy Perspective: Global Human Resource Information Systems. In M. Khosrow-Pour (Ed.), *Handbook of Research on Technology Adoption, Social Policy, and Global Integration* (pp. 377–394). Hershey, PA: IGI Global. doi:10.4018/978-1-5225-2668-1.ch018

Unwin, D. W., Sanzogni, L., & Sandhu, K. (2017). Developing and Measuring the Business Case for Health Information Technology. In K. Moahi, K. Bwalya, & P. Sebina (Eds.), *Health Information Systems and the Advancement of Medical Practice in Developing Countries* (pp. 262–290). Hershey, PA: IGI Global. doi:10.4018/978-1-5225-2262-1.ch015

Usharani, B. (2022). House Plant Leaf Disease Detection and Classification Using Machine Learning. In M. Mundada, S. Seema, S. K.G., & M. Shilpa (Eds.), *Deep Learning Applications for Cyber-Physical Systems* (pp. 17-26). IGI Global. https://doi.org/10.4018/978-1-7998-8161-2.ch002

Vadhanam, B. R. S., M., Sugumaran, V., V., V., & Ramalingam, V. V. (2017). Computer Vision Based Classification on Commercial Videos. In M. S., & V. V. (Eds.), Multi-Core Computer Vision and Image Processing for Intelligent Applications (pp. 105-135). Hershey, PA: IGI Global. https://doi.org/doi:10.4018/978-1-5225-0889-2.ch004

Vairinho, S. (2022). Innovation Dynamics Through the Encouragement of Knowledge Spin-Off From Touristic Destinations. In C. Ramos, S. Quinteiro, & A. Gonçalves (Eds.), *ICT as Innovator Between Tourism and Culture* (pp. 170–190). IGI Global. https://doi.org/10.4018/978-1-7998-8165-0.ch011

Valverde, R., Torres, B., & Motaghi, H. (2018). A Quantum NeuroIS Data Analytics Architecture for the Usability Evaluation of Learning Management Systems. In S. Bhattacharyya (Ed.), *Quantum-Inspired Intelligent Systems for Multimedia Data Analysis* (pp. 277–299). Hershey, PA: IGI Global. doi:10.4018/978-1-5225-5219-2.ch009

Vassilis, E. (2018). Learning and Teaching Methodology: "1:1 Educational Computing. In K. Koutsopoulos, K. Doukas, & Y. Kotsanis (Eds.), *Handbook of Research on Educational Design and Cloud Computing in Modern Classroom Settings* (pp. 122–155). Hershey, PA: IGI Global. doi:10.4018/978-1-5225-3053-4.ch007

Verma, S., & Jain, A. K. (2022). A Survey on Sentiment Analysis Techniques for Twitter. In B. Gupta, D. Peraković, A. Abd El-Latif, & D. Gupta (Eds.), *Data Mining Approaches for Big Data and Sentiment Analysis in Social Media* (pp. 57–90). IGI Global. https://doi.org/10.4018/978-1-7998-8413-2.ch003

Wang, H., Huang, P., & Chen, X. (2021). Research and Application of a Multidimensional Association Rules Mining Method Based on OLAP. *International Journal of Information Technology and Web Engineering*, *16*(1), 75–94. https://doi.org/10.4018/IJITWE.2021010104

Wexler, B. E. (2017). Computer-Presented and Physical Brain-Training Exercises for School Children: Improving Executive Functions and Learning. In B. Dubbels (Ed.), *Transforming Gaming and Computer Simulation Technologies across Industries* (pp. 206–224). Hershey, PA: IGI Global. doi:10.4018/978-1-5225-1817-4.ch012

Wimble, M., Singh, H., & Phillips, B. (2018). Understanding Cross-Level Interactions of Firm-Level Information Technology and Industry Environment: A Multilevel Model of Business Value. *Information Resources Management Journal*, *31*(1), 1–20. doi:10.4018/IRMJ.2018010101

Wimmer, H., Powell, L., Kilgus, L., & Force, C. (2017). Improving Course Assessment via Web-based Homework. *International Journal of Online Pedagogy and Course Design*, *7*(2), 1–19. doi:10.4018/IJOPCD.2017040101

Wong, S. (2021). Gendering Information and Communication Technologies in Climate Change. In M. Khosrow-Pour D.B.A. (Eds.), *Encyclopedia of Information Science and Technology, Fifth Edition* (pp. 1408-1422). IGI Global. https://doi.org/10.4018/978-1-7998-3479-3.ch096

Related References

Wong, Y. L., & Siu, K. W. (2018). Assessing Computer-Aided Design Skills. In M. Khosrow-Pour, D.B.A. (Ed.), Encyclopedia of Information Science and Technology, Fourth Edition (pp. 7382-7391). Hershey, PA: IGI Global. doi:10.4018/978-1-5225-2255-3.ch642

Wongsurawat, W., & Shrestha, V. (2018). Information Technology, Globalization, and Local Conditions: Implications for Entrepreneurs in Southeast Asia. In P. Ordóñez de Pablos (Ed.), *Management Strategies and Technology Fluidity in the Asian Business Sector* (pp. 163–176). Hershey, PA: IGI Global. doi:10.4018/978-1-5225-4056-4.ch010

Yamada, H. (2021). Homogenization of Japanese Industrial Technology From the Perspective of R&D Expenses. *International Journal of Systems and Service-Oriented Engineering, 11*(2), 24–51. doi:10.4018/IJSSOE.2021070102

Yang, Y., Zhu, X., Jin, C., & Li, J. J. (2018). Reforming Classroom Education Through a QQ Group: A Pilot Experiment at a Primary School in Shanghai. In H. Spires (Ed.), *Digital Transformation and Innovation in Chinese Education* (pp. 211–231). Hershey, PA: IGI Global. doi:10.4018/978-1-5225-2924-8.ch012

Yilmaz, R., Sezgin, A., Kurnaz, S., & Arslan, Y. Z. (2018). Object-Oriented Programming in Computer Science. In M. Khosrow-Pour, D.B.A. (Ed.), Encyclopedia of Information Science and Technology, Fourth Edition (pp. 7470-7480). Hershey, PA: IGI Global. doi:10.4018/978-1-5225-2255-3.ch650

Yu, L. (2018). From Teaching Software Engineering Locally and Globally to Devising an Internationalized Computer Science Curriculum. In S. Dikli, B. Etheridge, & R. Rawls (Eds.), *Curriculum Internationalization and the Future of Education* (pp. 293–320). Hershey, PA: IGI Global. doi:10.4018/978-1-5225-2791-6.ch016

Yuhua, F. (2018). Computer Information Library Clusters. In M. Khosrow-Pour, D.B.A. (Ed.), Encyclopedia of Information Science and Technology, Fourth Edition (pp. 4399-4403). Hershey, PA: IGI Global. doi:10.4018/978-1-5225-2255-3.ch382

Zakaria, R. B., Zainuddin, M. N., & Mohamad, A. H. (2022). Distilling Blockchain: Complexity, Barriers, and Opportunities. In P. Lai (Ed.), *Handbook of Research on Social Impacts of E-Payment and Blockchain Technology* (pp. 89–114). IGI Global. https://doi.org/10.4018/978-1-7998-9035-5.ch007

Zhang, Z., Ma, J., & Cui, X. (2021). Genetic Algorithm With Three-Dimensional Population Dominance Strategy for University Course Timetabling Problem. *International Journal of Grid and High Performance Computing, 13*(2), 56–69. https://doi.org/10.4018/IJGHPC.2021040104

About the Contributors

Mariyam Ouaissa is currently an Assistant Professor in Networks and Systems at ENSA, Chouaib Doukkali University, El Jadida, Morocco. She received her Ph.D. degree in 2019 from National Graduate School of Arts and Crafts, Meknes, Morocco and her Engineering Degree in 2013 from the National School of Applied Sciences, Khouribga, Morocco. She is a communication and networking researcher and practitioner with industry and academic experience. Dr Ouaissa's research is multidisciplinary that focuses on Internet of Things, M2M, WSN, vehicular communications and cellular networks, security networks, congestion overload problem and the resource allocation management and access control. She is serving as a reviewer for international journals and conferences including as IEEE Access, Wireless Communications and Mobile Computing. Since 2020, she is a member of "International Association of Engineers IAENG" and "International Association of Online Engineering", and since 2021, she is an "ACM Professional Member". She has published more than 40 research papers (this includes book chapters, peer-reviewed journal articles, and peer-reviewed conference manuscripts), 12 edited books, and 6 special issue as guest editor. She has served on Program Committees and Organizing Committees of several conferences and events and has organized many Symposiums / Workshops / Conferences as a General Chair.

Mariya Ouaissa is currently a Professor in Cybersecurity and Networks at Faculty of Sciences Semlalia, Cadi Ayyad University, Marrakech, Morocco. She is a Ph.D. graduated in 2019 in Computer Science and Networks, at the Laboratory of Modelisation of Mathematics and Computer Science from ENSAM-Moulay Ismail University, Meknes, Morocco. She is a Networks and Telecoms Engineer, graduated in 2013 from National School of Applied Sciences Khouribga, Morocco. She is a Co-Founder and IT Consultant at IT Support and Consulting Center. She was working for School of Technology of Meknes Morocco as a Visiting Professor from 2013 to 2021. She is member of the International Association of Engineers and International Association of Online Engineering, and since 2021, she is an "ACM Professional Member". She is Expert Reviewer with Academic Exchange Information Centre (AEIC) and Brand Ambassador with Bentham Science. She has served and continues to serve on technical program and organizer committees of several conferences and events and has organized many Symposiums/Workshops/Conferences as a General Chair also as a reviewer of numerous international journals. Dr. Ouaissa has made contributions in the fields of information security and privacy, Internet of Things security, and wireless and constrained networks security. Her main research topics are IoT, M2M, D2D, WSN, Cellular Networks, and Vehicular Networks. She has published over 50 papers (book chapters, international journals, and conferences/workshops), 12 edited books, and 8 special issue as guest editor .

Zakaria Boulouard is currently a Professor at Department of Computer Sciences at the "Faculty of Sciences and Techniques Mohammedia, Hassan II University, Casablanca, Morocco". In 2018, he joined the "Advanced Smart Systems" Research Team at the "Computer Sciences Laboratory of Mohammedia". He received his PhD degree in 2018 from "Ibn Zohr University, Morocco" and his Engineering Degree in 2013 from the "National School of Applied Sciences, Khouribga, Morocco". His research interests include Artificial Intelligence, Big Data Visualization and Analytics, Optimization and Competitive Intelligence. Since 2017, he is a member of "Draa-Tafilalet Foundation of Experts and Researchers", and since 2020, he is an "ACM Professional Member". He has served on Program Committees and Organizing Committees of several conferences and events and has organized many Symposiums/Workshops/Conferences as a General Chair. He has served and continues to serve as a reviewer of numerous international conferences and journals. He has published several research papers. This includes book chapters, peer-reviewed journal articles, peer-reviewed conference manuscripts, edited books and special issue journals.

Celestine Iwendi is an associate Professor (Senior Lecturer) with the School of Creative Technologies, University of Bolton, United Kingdom. He is an IEEE Brand Ambassador. He has a PhD in Electronics Engineering, ACM Distinguished Speaker, a Senior Member of IEEE, a Seasoned Lecturer and a Chartered Engineer. A highly motivated researcher and teacher with emphasis on communication, hands-on experience, willing-to-learn and a 21 years technical expertise. Celestine has developed operational, maintenance, and testing procedures for electronic products, components, equipment, and systems; provided technical support and instruction to staff and customers regarding equipment standards, assisting with specific, difficult in-service engineering; Inspected electronic and communication equipment, instruments, products, and systems to ensure conformance to specifications, safety standards, and regulations. He is a wireless sensor network Chief Evangelist, AI, ML and IoT expert and designer. Celestine is an Associate Professor (Senior Lecturer) at the School of Creative Technologies at the University of Bolton, United

Kingdom. He is also a Board Member of IEEE Sweden Section, a Fellow of The Higher Education Academy, United Kingdom and a fellow of Institute of Management Consultants to add to his teaching, managerial and professional experiences. Visiting Professor to three Universities and an IEEE Philanthropist.

Moez Krichen obtained his HDR (Ability to Conduct Researches) in Computer Science from the University of Sfax (Sfax, Tunisia) in 2018. He obtained his PhD in Computer Science in 2007. He is currently an Associate Professor at the University of Al-Baha (KSA) and a member of the Research Laboratory on Development and Control of Distributed Applications - REDCAD (Sfax, Tunisia). His main Research Interest is Model-Based Conformance/Load/Security Testing Methodologies for Real-Time, Distributed, and Dynamically Adaptable Systems. Moreover, he works on applying Formal Methods to several Modern Technologies like Smart Cities, Internet of Things (IoT), Smart Vehicles, Drones, Healthcare Systems, etc. Currently, he is also working on Formal Aspects related to Deep Learning, Data Mining, Block-chain, Smart Contracts and Optimization.

***

Peryala Abhinaya is a final year Bachelor of Engineering student, currently pursuing Computer Science and Technology at Stanley College of Engineering and Technology for Women. With a fervent passion for technology and a relentless drive for excellence,

Murtala Ismail Adakawa, CLN, has been a practicing librarian for a decade. He is currently working at the University Library, Bayero University Kano. He has participated in various committees at different capacities. He served as head, Mudi Sipikin Library, Mambayya House, Aminu Kano Centre for Democratic Research & Training, BUK. He has attended many international and local conferences and published several journal articles. He is currently a research fellow at the Department of Studies in Library and Information Science, University of Mysore, Manasagangothri, Mysuru, Karnataka, India. He authored a book (though in press) titled "Royal Library in Kano Palace: A Catalyst for Knowledge Society in Kano State.".

Aitelkadi Kenza is an associate Professor and head of Cartography-Photogrammetry-GIS-Remote sensing department from Geomatic Sciences and surveying engineering school, Agronomic and Veterinary Hassan II Institute; Rabat Morocco. She is graduated from IAV Hassan II in 2008. She obtained her PhD in 2016. The research, publishing and teaching she provides are photogrammetry, lasergrammetry, spatial data processing and deep learning approaches. Pr. Aitelkadi is member in several national and international project related to the technical topics such as Earth observation, Data mining, GIS, Spatial data infrastructure, etc. Since 2018, Pr. Aitelkadi has started on another complementary research axis complementary to their technical skills especially the Agrientrepreneurship and knowledge management. In this sense, she coordinated in the IAV the first projects around the entrepreneurship, innovation and knowledge management. Three projects were locally coordinated by Pr. Aitelkadi. The Agriengage ERASMUS + Capacity building project, Funded by European union, that aimed the strengthening Agri-Entrepreneurship and Community Engagement Training in East, West and North Africa. The SKIM project funded by IFAD with partners from Sudan, Moldova, Uzbekistan and Italy. The project facilitated and supported the growth of knowledge management and capacity development operations within the Near East and North Africa (NENA) and Central-Eastern Europe and Central Asia (CEN) regions. The third project was MunIE financed by DAAD fund with Koblenz University. MUnIE was a project within the framework of the Entrepreneurial Universities in Africa program, which involved higher education institutions working with African cooperation partners to design and implement reform measures that will increase the labor market orientation of African higher education institutions.

Ayoub Ouchlif is currently a Ph.D. student researcher in Geospatial Technologies for Intelligent Decision Making at the Hassan II Agronomic and Veterinary Institute (IAV HASSAN II) in Morocco. He is also a Senior Executive in Geomatics at the Urban Agency of Taroudannt-Tiznit-Tata. He obtained his Specialized Master's degree in Geospatial Sciences and Land Governance in 2020 from IAV HASSAN II, Rabat, Morocco. Previously, he obtained his Professional License in Geoinformation and Territorial Modeling in 2014 from the Faculty of Sciences (FSM), Moulay Ismail University, Meknes, Morocco. He also obtained his University Diploma of Technology in Computer Engineering from the Higher School of Technology (ESTM), Moulay Ismail University, Meknes. His research interests include geospatial technologies, GIS development, artificial intelligence, urban planning, land governance, and territorial modeling.

Hella Kaffel Ben Ayed received Engineering, PhD, and Habilitation Universitaire degrees in computer science from the Faculty of Sciences of Tunis, the University of Tunis El Manar (UTM). She's an Associate Professor in computer science at the Faculty of Science of Tunis, teaching graduate and undergraduate courses in computer networks, IT security, and Blockchain. She founded several teaching undergraduate and graduate programs in the fields of networking, cyber security, IoT, and Blockchain. She is the Director of the LIPAH research Laboratory and President of the Computer Science Doctorate Commission, UTM. She has supervised and co-sponsored several doctoral theses and Master of Research in Computer Science. She participated in the setting up of various national and international collaboration projects. She served as guest editor for many scientific journals. Her topics interests around Blockchain include access control and token management, Self-Sovereign identities, supply chain traceability, blockchain-based marketplaces and Blockchain for IoT systems. She is involved in two European projects on Blockchain-based traceability for agro-food supply chains.

Sarah El Himer is a Ph.D and Researcher Associate in Intelligent Systems, Georesources & Renewable Energies Laboratory. She is Ph.D, graduated in 2019 in renewable energies at the renewable energies and intelligent systems laboratory from the faculty of sciences and technology, university of Sidi Mohammed Ben Abdallah FEZ. She is a trainer and IT Consultant at IT Support and Consulting Center. She was working for faculty of science and technology FEZ as a Visiting Professor from 2015 to 2019. Dr. El Himer has made contributions in the fields of optical component of Concentrated photovoltaic and electrical vehicle. Her main research topics are Optical element for CPV, acceptance angle, optical efficiency micro CPV, Hybrid CPV. She has published over 25 papers (international journals, and conferences/workshops). She has served and continues to serve on executive and technical program committees and as a reviewer of numerous international conference and journals such as CPV conference, REEE2017

Hicham Hajji, PhD in Computer Sciences in 2005 and an MS in Computer Sciences from INSA LYON, France in 2001. In 1999, he received an engineer degree of Surveying from IAV Institute. He is now a full Professor in Geomatics and Surveying Engineering School, IAV Institute in Morocco, where he is conducting research in Spatial Big Data and in the integration of the state-of-the-art AI techniques into spatial applications. He has been involved in more than twenty projects ranging from financial data warehousing, geospatial Projects, Web Mapping to web 3.0 as Senior Technical Consultant and as Research Engineer.

Naser Abbas Hussein, University of Technology in Iraq, PhD student in Faculty of Science of Tunis, the University of Tunis El Manar, the research interests include, security, privacy, and identity management in IOT. and currently focus on the use of Blockchain for IOT Applications.

Hamid Khalifi, obtained a Ph.D. in Data Sciences and Artificial Intelligence from the Faculty of Sciences of Moulay Ismail University in Meknes, a State Engineer degree in Computer Science from the National Institute of Statistics and Applied Economics in Rabat, a Bachelor's degree in Computer Science, and a University Diploma in Software Engineering from Moulay Ismail University in Meknes. Currently, he is a Professor in the Department of Computer Science and Program Coordinator of the Applied Computer Science: Web and Mobile Development at the Faculty of Sciences, Mohammed V University in Rabat. He has supervised and co-supervised several End of Study Projects, Theses, and Dissertations for Bachelor's, Master's, and Ph.D. degrees. He is the author and co-author of numerous publications in internationally renowned journals and conferences, and has reviewed several research articles for various national and international journals and conferences. His research interests include Artificial Intelligence, Data Science, Machine & Deep Learning, Information Retrieval, Big Data, Optimization, Smart Cities, and Digitalization.

M. Vanitha is currently working as an Assistant Professor in the Department of Computer Science and Engineering, Srinivasa Ramanujan Centre, SASTRA Deemed University, Kumbakonam, Tamil Nadu, India. She has teaching experience for more than fifteen years. She has published several research works in peer-reviewed journals. Her research area includes Deep Learning, Machine learning and Time-series Analysis.

Geetha Manoharan is currently working in Telangana as an assistant professor at SR University. She is the university-level PhD program coordinator and has also been given the additional responsibility of In Charge Director of Publications and Patents under the Research Division at SR University. Under her tutelage, students are inspired to reach their full potential in all areas of their education and beyond through experiential learning. It creates an atmosphere conducive to the growth of students into independent thinkers and avid readers. She has more than ten years of experience across the board in the business world, academia,and the academy. She has a keen interest in the study of organizational behavior and management. More than forty articles and books have been published in scholarly venues such as UGC-refereed, SCOPUS, Web of Science, and Springer. Over the past six-plus years, she has participated in varied research and student exchange programs at both the national and international levels. A total of five of her collaborative innovations in this area have already been published and patented. Emotional intelligence, self-efficacy, and work-life balance are among her specialties. She organizes programs for academic organizations. She belongs to several professional organizations, including the CMA and the CPC. The TIPS GLOBAL Institute of Coimbatore has recognized her twice (in 2017 and 2018) for her outstanding academic performance.

Özen Özer earned her Bachelor's and Master's degrees in Mathematics from Trakya University in Edirne, Turkey. She also obtained her Ph.D. in Mathematics from Süleyman Demirel University in Isparta, Turkey. Currently, she holds the position of full Professor Doctor at the Department of Mathematics within the Faculty of Science and Arts at Kırklareli University. Her specialized research areas encompass a wide range of subjects, including the Theory of Real Quadratic Number Fields with practical applications, Diophantine and Pell Equations, Diophantine Sets, Arithmetic Functions, Fixed Point Theory, p-adic Analysis, q-Analysis, Special Integer Sequences, Nonlinear Analysis, C* Algebra, Matrix Theory, Optimization, Approximation Theory, Cryptography, Machine Learning, Artificial Intelligence, Differential Equations, Mathematical Models, Mathematical Education, Optimization, Fuzzy Set and Fuzzy Spaces, Statistics and more. With an extensive academic portfolio, she has authored or completed over 90 research papers published in prestigious international journals, alongside her participation in numerous national and international scientific projects. Özen ÖZER has also authored books and contributed chapters to international publishing houses on various subjects such as Number Theory, Algebra, Applied Mathematics, Analysis and Artificial Intelligence. She has further demonstrated her expertise as a keynote speaker at various national and international conferences held across different countries, covering a diverse array of topics. Her involvement extends to being a reviewer for more than 80 distinct publishing houses and journals, and she serves as a member of editorial boards for books, papers, and conference proceedings. Özen ÖZER is enthusiastic about engaging in academic and scientific collaborations, demonstrating a willingness to participate in cooperative endeavors across various domains.

Abhishek Ranjan has a PhD, M. Tech, and B.E in Electrical and Electronics Engineering, Computer Science and Engineering and Information technology respectively. In his current position as Deputy Pro-Vice-Chancellor, he oversees two international campuses of Botho University i.e., Ghana and Online Learning Campus operations, including Academic Services, Quality Management, Accreditation, Student Welfare, Admissions, Corporate Training and Marketing, and Stakeholder Engagement among others and Blended and Distance Learning Campus. Prof. Ranjan has about two-decade years of experience in educational policy development and reform, project management, education quality assurance, academic accreditation, strategic planning and institutional effectiveness, teaching, training, developing training materials, and administering, monitoring, and evaluating training programs in the fields of education and management. Prof Ranjan has served in various national and international level committees for quality assurance in higher education. Prof. Ranjan's experience spans Teaching, Research, Training, Quality Assurance, Faculty Management, Education Management, Financial management, and Campus Management roles

C Kishor Kumar Reddy, currently working as Associate Professor, Dept. of Computer Science and Engineering, Stanley College of Engineering and Technology for Women, Hyderabad, India. He has research and teaching experience of more than 10 years. He has published more than 50 research papers in National and International Conferences, Book Chapters, and Journals indexed by Scopus and others. He is an author for 2 text books and 3 co-edited books. He acted as the special session chair for Springer FICTA 2020, 2022, SCI 2021, INDIA 2022 and IEEE ICCSEA 2020 conferences. He is the corresponding editor of AMSE 2021 conferences, published by IoP Science JPCS. He is the member of ISTE, CSI, IAENG, UACEE, IACSIT

R. Renuga Devi received her Ph.D Degree in the Department of Information and Communication Engineering, Anna University. Her research interest includes wireless sensor network, mobile computing, Machine learning and Internet of Things. She published more than 20 papers in International journals and Conferences. Currently, She is working in SRM Institute of Technology. She has around 16 years of teaching experience in Engineering colleges. She is a life member of IAENG and ISTE.

Wasswa Shafik, a remarkable computer scientist, information technologist, and educator, hails from the vibrant capital city of Uganda, Kampala. His passion extends beyond technology, encompassing people's health, justice, and the responsible use of technological advancements, regardless of race, religion, sex, or any other. With a strong educational background, he holds a bachelor's degree in information technology engineering with a minor in mathematics from Ndejje University, Uganda. He further pursued a Master of Engineering degree in information technology engineering (MIT) with a computer and communication network option from Yazd University, Iran, and a PhD in Computer Science at Universiti Brunei Darussalam, Brunei Darussalam. As a member of the IEEE, Shafik has made significant contributions in various domains such as smart agriculture, health informatics, computer vision, network science, and security and privacy. His passion for sustainable solutions and his deep understanding of the health and environmental challenges we face today have propelled him to the forefront of sustainable measures. Shafik's interest extends to Smart Cities, where he investigates the intersection of Big Data, IoT, and Artificial Intelligence. His work delves into energy management, privacy, and performance evaluation in the context of Internet of Things (IoT) applications within smart urban environments. Additionally, Shafik explores fog computing architectures, privacy, and security solutions. His work also encompasses the evaluation and analysis of fog-mobile edge performance in IoT scenarios. Shafik's expertise extends to artificial intelligence and its application in cybersecurity and cyberdefense. He reviews the theoretical and practical understanding of deep learning in the context of analysis of UAV biomedical engineering technologies and health informatics. He has published more than 80 publications, including edited books, research articles, book sections, and conferences in many reputed journals, including Scopus Q1, Q2, and SCI in different publishers. He served as departmental support for Mathematics for Data Science, Advanced Topics in Computing, and Advanced Algorithms at the School of Digital Science, Universiti Brunei Darussalam. He served in different capacities, including a community data officer at Pace-Uganda, research associate at TechnoServe, research assistant at PSI-Uganda, research lead at Socio-economic Data Center (SEDC-Uganda), research lab head at the Dig Connectivity Research Laboratory (DCRLab), Kampala, Uganda, and ag. managing director at Asmaah Charity Organisation. Shafik's journey is one of curiosity, innovation, and a deep-rooted desire to positively impact the world through technology, education, and sustainability.

Col Ravinder Singh was commissioned in the Indian Army in June 1989 into Artillery Regiment and he is a serving officer with 35 years of service. An alumnus of National Defence Academy and Indian Military Academy, he has a vast military experience in various terrains. He holds a Master of Science in (Technology) Weapon Systems from Pune University and also qualified in Disaster Management from Gujarat Institute of Disaster Management, Ahmedabad. He also holds a Post Graduate Diploma in Mass Communication and Journalism from National Institute of Mass Communication and Journalism, Ahmedabad He conducts lectures on Disaster Management at Gujarat Institute of Disaster Management, Colleges, schools and for Associate NCC officers and NCC cadets. He is also holding a degree in MBA from ICFAI University. He has also completed Post Graduate Diploma in Disaster Management from IGNOU and currently he is research scholar with SR University.

Samrath Singh is a student currently pursuing a Bachelor's degree in Computer Science and Engineering. With a fervent passion for innovation and technology, Samrath has actively participated in several competitions to showcase his talents and contribute to societal betterment. One of his notable achievements is his participation in the esteemed IEEE Yesist Junior track, an international event that celebrates innovation and creativity. Here, he demonstrated his compassion and technical acumen by developing Ultrasonic Obstacle Detecting Goggles tailored specifically for the visually impaired. His groundbreaking project not only showcased his technical skills but also highlighted his commitment to making a positive impact on the lives of others. For this innovative solution, Samrath received honorable recognition, earning acclaim for his dedication and ingenuity. In addition to his international endeavors, Samrath has also made significant contributions at the local level. He participated in the Inter APS Science Competition, where his innovative spirit shone once again.

His project, a wireless firecracker igniter, captured the attention of the judges and earned him the prestigious first prize. His ability to combine technical expertise with creative problem-solving underscores his potential to effect meaningful change in the world. Through his participation in various competitions and projects, Samrath Singh exemplifies the qualities of a forward-thinking innovator, driven by a desire to make a positive difference in society. As he continues his academic journey in Computer Science and Engineering, Samrath remains committed to leveraging technology for the betterment of humanity.

Muhammad Usman Tariq has more than 16+ year's experience in industry and academia. He has authored more than 200+ research articles, 100+ case studies, 50+ book chapters and several books other than 4 patents. He has been working as a consultant and trainer for industries representing six sigma, quality, health and safety, environmental systems, project management, and information security standards. His work has encompassed sectors in aviation, manufacturing, food, hospitality, education, finance, research, software and transportation. He has diverse and significant experience working with accreditation agencies of ABET, ACBSP, AACSB, WASC, CAA, EFQM and NCEAC. Additionally, Dr. Tariq has operational experience in incubators, research labs, government research projects, private sector startups, program creation and management at various industrial and academic levels. He is Certified Higher Education Teacher from Harvard University, USA, Certified Online Educator from HMBSU, Certified Six Sigma Master Black Belt, Lead Auditor ISO 9001 Certified, ISO 14001, IOSH MS, OSHA 30, and OSHA 48. He has been awarded Principal Fellowship from Advance HE UK & Chartered Fellowship of CIPD.

Index

Ensure Quality Research is Introduced to the Academic Community

Become a Reviewer for IGI Global Authored Book Projects

 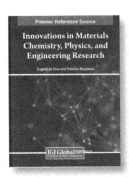 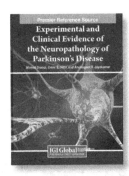 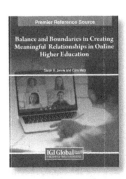

The overall success of an authored book project is dependent on quality and timely manuscript evaluations.

Applications and Inquiries may be sent to:
development@igi-global.com

Applicants must have a doctorate (or equivalent degree) as well as publishing, research, and reviewing experience. Authored Book Evaluators are appointed for one-year terms and are expected to complete at least three evaluations per term. Upon successful completion of this term, evaluators can be considered for an additional term.

If you have a colleague that may be interested in this opportunity, we encourage you to share this information with them.

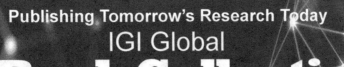

Individual Article & Chapter Downloads

US$ 37.50/each

Printed in the United States
by Baker & Taylor Publisher Services